LONDON MATHEMATICAL SOCIETY LECTURE NOTE SERIES

Managing Editor: Professor N.J. Hitchin, Mathematics Institute,
University of Oxford, 24–29 St Giles, Oxford OX1 3LB, United Kingdom

The titles below are available from booksellers, or, in case of difficulty, from Cambridge University Press.

- 46 *p*-adic Analysis: a short course on recent work, N. KOBLITZ
- 59 Applicable differential geometry, M. CRAMPIN & F.A.E. PIRANI
- 66 Several complex variables and complex manifolds II, M.J. FIELD
- 86 Topological topics, I.M. JAMES (ed)
- 87 Surveys in set theory, A.R.D. MATHIAS (ed)
- 88 FPF ring theory, C. FAITH & S. PAGE
- 89 An F-space sampler, N.J. KALTON, N.T. PECK & J.W. ROBERTS
- 90 Polytopes and symmetry, S.A. ROBERTSON
- 92 Representations of rings over skew fields, A.H. SCHOFIELD
- 93 Aspects of topology, I.M. JAMES & E.H. KRONHEIMER (eds)
- 96 Diophantine equations over function fields, R.C. MASON
- 97 Varieties of constructive mathematics, D.S. BRIDGES & F. RICHMAN
- 98 Localization in Noetherian rings, A.V. JATEGAONKAR
- 99 Methods of differential geometry in algebraic topology, M. KAROUBI & C. LERUSTE
- 100 Stopping time techniques for analysts and probabilists, L. EGGHE
- 104 Elliptic structures on 3-manifolds, C.B. THOMAS
- 105 A local spectral theory for closed operators, I. ERDELYI & WANG SHENGWANG
- 107 Compactification of Siegel moduli schemes, C.-L. CHAI
- 109 Diophantine analysis, J. LOXTON & A. VAN DER POORTEN (eds)
- 113 Lectures on the asymptotic theory of ideals, D. REES
- 114 Lectures on Bochner-Riesz means, K.M. DAVIS & Y.-C. CHANG
- 116 Representations of algebras, P.J. WEBB (ed)
- 119 Triangulated categories in the representation-theory of finite-dimensional algebras, D. HAPPEL
- 121 Proceedings of *Groups - St Andrews 1985*, E. ROBERTSON & C. CAMPBELL (eds)
- 128 Descriptive set theory and the structure of sets of uniqueness, A.S. KECHRIS & A. LOUVEAU
- 130 Model theory and modules, M. PREST
- 131 Algebraic, extremal & metric combinatorics, M.-M. DEZA, P. FRANKL & I.G. ROSENBERG (eds)
- 132 Whitehead groups of finite groups, ROBERT OLIVER
- 133 Linear algebraic monoids, MOHAN S. PUTCHA
- 134 Number theory and dynamical systems, M. DODSON & J. VICKERS (eds)
- 137 Analysis at Urbana, I, E. BERKSON, T. PECK, & J. UHL (eds)
- 138 Analysis at Urbana, II, E. BERKSON, T. PECK, & J. UHL (eds)
- 139 Advances in homotopy theory, S. SALAMON, B. STEER & W. SUTHERLAND (eds)
- 140 Geometric aspects of Banach spaces, E.M. PEINADOR & A. RODES (eds)
- 141 Surveys in combinatorics 1989, J. SIEMONS (ed)
- 144 Introduction to uniform spaces, I.M. JAMES
- 146 Cohen-Macaulay modules over Cohen-Macaulay rings, Y. YOSHINO
- 148 Helices and vector bundles, A.N. RUDAKOV *et al*
- 149 Solitons, nonlinear evolution equations and inverse scattering, M. ABLOWITZ & P. CLARKSON
- 150 Geometry of low-dimensional manifolds 1, S. DONALDSON & C.B. THOMAS (eds)
- 151 Geometry of low-dimensional manifolds 2, S. DONALDSON & C.B. THOMAS (eds)
- 152 Oligomorphic permutation groups, P. CAMERON
- 153 L-functions and arithmetic, J. COATES & M.J. TAYLOR (eds)
- 155 Classification theories of polarized varieties, TAKAO FUJITA
- 156 Twistors in mathematics and physics, T.N. BAILEY & R.J. BASTON (eds)
- 158 Geometry of Banach spaces, P.F.X. MÜLLER & W. SCHACHERMAYER (eds)
- 159 Groups St Andrews 1989 volume 1, C.M. CAMPBELL & E.F. ROBERTSON (eds)
- 160 Groups St Andrews 1989 volume 2, C.M. CAMPBELL & E.F. ROBERTSON (eds)
- 161 Lectures on block theory, BURKHARD KÜLSHAMMER
- 162 Harmonic analysis and representation theory, A. FIGA-TALAMANCA & C. NEBBIA
- 163 Topics in varieties of group representations, S.M. VOVSI
- 164 Quasi-symmetric designs, M.S. SHRIKANDE & S.S. SANE
- 166 Surveys in combinatorics, 1991, A.D. KEEDWELL (ed)
- 168 Representations of algebras, H. TACHIKAWA & S. BRENNER (eds)
- 169 Boolean function complexity, M.S. PATERSON (ed)
- 170 Manifolds with singularities and the Adams-Novikov spectral sequence, B. BOTVINNIK
- 171 Squares, A.R. RAJWADE
- 172 Algebraic varieties, GEORGE R. KEMPF
- 173 Discrete groups and geometry, W.J. HARVEY & C. MACLACHLAN (eds)
- 174 Lectures on mechanics, J.E. MARSDEN
- 175 Adams memorial symposium on algebraic topology 1, N. RAY & G. WALKER (eds)
- 176 Adams memorial symposium on algebraic topology 2, N. RAY & G. WALKER (eds)
- 177 Applications of categories in computer science, M. FOURMAN, P. JOHNSTONE & A. PITTS (eds)
- 178 Lower K- and L-theory, A. RANICKI
- 179 Complex projective geometry, G. ELLINGSRUD *et al*
- 180 Lectures on ergodic theory and Pesin theory on compact manifolds, M. POLLICOTT
- 181 Geometric group theory I, G.A. NIBLO & M.A. ROLLER (eds)
- 182 Geometric group theory II, G.A. NIBLO & M.A. ROLLER (eds)
- 183 Shintani zeta functions, A. YUKIE

184	Arithmetical functions, W. SCHWARZ & J. SPILKER
185	Representations of solvable groups, O. MANZ & T.R. WOLF
186	Complexity: knots, colourings and counting, D.J.A. WELSH
187	Surveys in combinatorics, 1993, K. WALKER (ed)
188	Local analysis for the odd order theorem, H. BENDER & G. GLAUBERMAN
189	Locally presentable and accessible categories, J. ADAMEK & J. ROSICKY
190	Polynomial invariants of finite groups, D.J. BENSON
191	Finite geometry and combinatorics, F. DE CLERCK et al
192	Symplectic geometry, D. SALAMON (ed)
194	Independent random variables and rearrangement invariant spaces, M. BRAVERMAN
195	Arithmetic of blowup algebras, WOLMER VASCONCELOS
196	Microlocal analysis for differential operators, A. GRIGIS & J. SJÖSTRAND
197	Two-dimensional homotopy and combinatorial group theory, C. HOG-ANGELONI et al
198	The algebraic characterization of geometric 4-manifolds, J.A. HILLMAN
199	Invariant potential theory in the unit ball of C^n, MANFRED STOLL
200	The Grothendieck theory of dessins d'enfant, L. SCHNEPS (ed)
201	Singularities, JEAN-PAUL BRASSELET (ed)
202	The technique of pseudodifferential operators, H.O. CORDES
203	Hochschild cohomology of von Neumann algebras, A. SINCLAIR & R. SMITH
204	Combinatorial and geometric group theory, A.J. DUNCAN, N.D. GILBERT & J. HOWIE (eds)
205	Ergodic theory and its connections with harmonic analysis, K. PETERSEN & I. SALAMA (eds)
207	Groups of Lie type and their geometries, W.M. KANTOR & L. DI MARTINO (eds)
208	Vector bundles in algebraic geometry, N.J. HITCHIN, P. NEWSTEAD & W.M. OXBURY (eds)
209	Arithmetic of diagonal hypersurfaces over finite fields, F.Q. GOUVÊA & N. YUI
210	Hilbert C*-modules, E.C. LANCE
211	Groups 93 Galway / St Andrews I, C.M. CAMPBELL et al (eds)
212	Groups 93 Galway / St Andrews II, C.M. CAMPBELL et al (eds)
214	Generalised Euler-Jacobi inversion formula and asymptotics beyond all orders, V. KOWALENKO et al
215	Number theory 1992–93, S. DAVID (ed)
216	Stochastic partial differential equations, A. ETHERIDGE (ed)
217	Quadratic forms with applications to algebraic geometry and topology, A. PFISTER
218	Surveys in combinatorics, 1995, PETER ROWLINSON (ed)
220	Algebraic set theory, A. JOYAL & I. MOERDIJK
221	Harmonic approximation, S.J. GARDINER
222	Advances in linear logic, J.-Y. GIRARD, Y. LAFONT & L. REGNIER (eds)
223	Analytic semigroups and semilinear initial boundary value problems, KAZUAKI TAIRA
224	Computability, enumerability, unsolvability, S.B. COOPER, T.A. SLAMAN & S.S. WAINER (eds)
225	A mathematical introduction to string theory, S. ALBEVERIO, J. JOST, S. PAYCHA, S. SCARLATTI
226	Novikov conjectures, index theorems and rigidity I, S. FERRY, A. RANICKI & J. ROSENBERG (eds)
227	Novikov conjectures, index theorems and rigidity II, S. FERRY, A. RANICKI & J. ROSENBERG (eds)
228	Ergodic theory of Z^d actions, M. POLLICOTT & K. SCHMIDT (eds)
229	Ergodicity for infinite dimensional systems, G. DA PRATO & J. ZABCZYK
230	Prolegomena to a middlebrow arithmetic of curves of genus 2, J.W.S. CASSELS & E.V. FLYNN
231	Semigroup theory and its applications, K.H. HOFMANN & M.W. MISLOVE (eds)
232	The descriptive set theory of Polish group actions, H. BECKER & A.S. KECHRIS
233	Finite fields and applications, S. COHEN & H. NIEDERREITER (eds)
234	Introduction to subfactors, V. JONES & V.S. SUNDER
235	Number theory 1993–94, S. DAVID (ed)
236	The James forest, H. FETTER & B. GAMBOA DE BUEN
237	Sieve methods, exponential sums, and their applications in number theory, G.R.H. GREAVES et al
238	Representation theory and algebraic geometry, A. MARTSINKOVSKY & G. TODOROV (eds)
239	Clifford algebras and spinors, P. LOUNESTO
240	Stable groups, FRANK O. WAGNER
241	Surveys in combinatorics, 1997, R.A. BAILEY (ed)
242	Geometric Galois actions I, L. SCHNEPS & P. LOCHAK (eds)
243	Geometric Galois actions II, L. SCHNEPS & P. LOCHAK (eds)
244	Model theory of groups and automorphism groups, D. EVANS (ed)
245	Geometry, combinatorial designs and related structures, J.W.P. HIRSCHFELD et al
246	p-Automorphisms of finite p-groups, E.I. KHUKHRO
247	Analytic number theory, Y. MOTOHASHI (ed)
248	Tame topology and o-minimal structures, LOU VAN DEN DRIES
249	The atlas of finite groups: ten years on, ROBERT CURTIS & ROBERT WILSON (eds)
250	Characters and blocks of finite groups, G. NAVARRO
251	Gröbner bases and applications, B. BUCHBERGER & F. WINKLER (eds)
252	Geometry and cohomology in group theory, P. KROPHOLLER, G. NIBLO, R. STÖHR (eds)
253	The q-Schur algebra, S. DONKIN
254	Galois representations in arithmetic algebraic geometry, A.J. SCHOLL & R.L. TAYLOR (eds)
255	Symmetries and integrability of difference equations, P.A. CLARKSON & F.W. NIJHOFF (eds)
256	Aspects of Galois theory, HELMUT VÖLKLEIN et al
257	An introduction to noncommutative differential geometry and its physical applications 2ed, J. MADORE
258	Sets and proofs, S.B. COOPER & J. TRUSS (eds)
259	Models and computability, S.B. COOPER & J. TRUSS (eds)
260	Groups St Andrews 1997 in Bath, I, C.M. CAMPBELL et al
261	Groups St Andrews 1997 in Bath, II, C.M. CAMPBELL et al
263	Singularity theory, BILL BRUCE & DAVID MOND (eds)
264	New trends in algebraic geometry, K. HULEK, F. CATANESE, C. PETERS & M. REID (eds)

London Mathematical Society Lecture Note Series. 267

Surveys in Combinatorics, 1999

Edited by

J. D. Lamb
University of Kent at Canterbury

D. A. Preece
University of Kent at Canterbury

PUBLISHED BY THE PRESS SYNDICATE OF THE UNIVERSITY OF CAMBRIDGE
The Pitt Building, Trumpington Street, Cambridge CB2 1RP, United Kingdom

CAMBRIDGE UNIVERSITY PRESS
The Edinburgh Building, Cambridge, CB2 2RU, UK   http://www.cup.cam.ac.uk
40 West 20th Street, New York, NY 10011-4211, USA   http://www.cup.org
10 Stamford Road, Oakleigh, Melbourne 3166, Australia

© Cambridge University Press 1999

This book is in copyright. Subject to statutory exception
and to the provisions of relevant collective licensing agreements,
no reproduction of any part may take place without
the written permission of Cambridge University Press.

First published 1999

Printed in the United Kingdom at the University Press, Cambridge

A catalogue record for this book is available from the British Library

ISBN   0 521 65376 2   paperback

Contents

The Rado lecture   1

The Coming of the Matroids by W. T. Tutte   3
    Appendix I: geometrical terminology 11
    Appendix II: binary and regular matroids 13

The Invited Lectures   15

Polynomials in Finite Geometries by S. Ball   17
    1   Introduction 17
    2   Definitions and useful polynomials 18
    3   Nuclei 21
    4   Affine blocking sets 23
    5   Non-Desarguesian planes 25
    6   Maximal arcs 27
    7   Unitals 30

Applications of Combinatorial Designs to Communications, Cryptography, and Networking by C. J. Colbourn, J. H. Dinitz, D. R. Stinson   37
    0   Background 38
    1   Optical orthogonal codes 39
    2   Synchronous multiple access to channels 43
    3   Group testing and superimposed codes 45
    4   Erasure codes and information dispersal 48
    5   Threshold and ramp schemes 54
    6   Authentication codes 59
    7   Resilient and correlation-immune functions 62
    8   Multidrop networks 65
    9   Channel graphs and interconnection networks 68
    10   Partial match queries on files 72
    11   Software testing 76
    12   Disk layout and striping 78
    13   (t, m, s)-nets and numerical integration 80
    14   About things not said 87

Random Walks on Combinatorial Objects by Martin Dyer and Catherine Greenhill   101
    1   Introduction 101
    2   Notation and preliminaries 102
    3   A computational framework 102

4	Review	109
5	Coupling	114
6	Path coupling	116
7	Perfect sampling	120
8	Negative results	128

Bose–Burton Type Theorems for Finite Projective, Affine and Polar Spaces by Klaus Metsch   137

1	Introduction	137
2	Blocking configurations for projective spaces	139
3	Variations of the Bose–Burton result in projective spaces	142
4	Spreads and partial spreads in $PG(d,q)$	144
5	A result in affine spaces	146
6	Ovoids and Spreads of finite classical polar spaces	146
7	Blocking lines by points in the polar spaces $Q^+(2n+1,q)$, $U(2n+1,q)$ and $Q(2n,q)$	150
8	The unitary polar spaces $U(2n,q)$	155
9	Unsolved problems	163

Geometric Graph Theory by János Pach   167

1	Introduction, basic definitions	167
2	Crossing-free geometric graphs	168
3	Unavoidable crossings	170
4	Forbidden geometric subgraphs—Multiple crossings	173
5	Forbidden geometric subgraphs—Non-crossing configurations	176
6	Ramsey-type results	181
7	Applications	185
8	Geometric hypergraphs	190

Recent Excluded Minor Theorems for Graphs by Robin Thomas   201

1	Introduction	201
2	Seymour's splitter theorem	202
3	A splitter theorem for internally 4-connected graphs	204
4	A splitter theorem for cyclically 5-connected cubic graphs	207
5	Excluding a general graph	208
6	The graph minor theorem	209
7	Linklessly embeddable graphs	210
8	The four colour theorem	212
9	Hadwiger's conjecture	213
10	Tutte's edge 3-colouring conjecture	214
11	Pfaffian orientations	215

Parity, Cycle Space, and K_4-Subdivisions in Graphs by C. Thomassen — 223

    1   Introduction 223
    2   The cycle space of a graph and generating sets of cycles 224
    3   The cycle space and collections of cycles determining uniquely a graph up to isomorphism 226
    4   The cycle space generated by the cycles through two fixed edges 228
    5   The cycle space of a graph and K_4-subdivisions 229
    6   Towards a characterization of the graphs containing no totally odd K_4-subdivisions 231
    7   Open problems 234

Models of Random Regular Graphs by N. C. Wormald — 239

    1   Introduction 239
    2   Uniform model for random regular graphs 241
    3   Other uniform models 265
    4   The small subgraph conditioning method, contiguity, and superposition models 268
    5   The generation problem 283
    6   Algorithmically defined models 286
    7   A wider perspective 287

Preface

The British Combinatorial Conference, a biennial event, took place at the University of Kent at Canterbury in 1999, with ourselves as the Local Organisers. This volume contains the texts of the Invited Talks from this seventeenth Conference in the series. As at the previous Conferences, the Invited Speakers were distinguished research workers chosen by the British Combinatorial Committee to provide seminal surveys of topics representative of the main areas of present-day combinatorial mathematics. We are delighted with the excellence of the papers that have been produced.

In preparing this volume, we have been greatly assisted by being able to use the LaTeX style-file and other documentation prepared by Professor Rosemary A. Bailey for the Invited Talks for the 1997 Conference [London Mathematical Society Lecture Note Series 241].

The Conference was very grateful for financial support from the London Mathematical Society and from the Institute of Mathematics and Its Applications.

John D. Lamb
Donald A. Preece

University of Kent at Canterbury

J.D.Lamb@ukc.ac.uk
D.A.Preece@ukc.ac.uk

Richard Rado FRS (1906–1989)

The Rado Lecture

Professor W. T. Tutte FRS

The Coming of the Matroids

W. T. Tutte

Summary The author rehearses his role in the development of the theory of matroids. The story starts in 1935 when he became an undergraduate at Trinity College, Cambridge, and started to collaborate with Leonard Brooks, Cedric Smith and Arthur Stone. It continues through his war-time work with codes and ciphers, followed by his return to Trinity in 1945, where his PhD thesis entitled "An Algebraic Theory of Graphs" foreshadowed his matroid papers published in 1958 and 1959. He describes the context in which he obtained the now well-known excluded minor conditions for a binary matroid to be regular and for a regular matroid to be graphic. He subsequently invented the whirl, and lectured on matroids at the 1964 Conference where the theory of matroids was first proclaimed to the world. This paper has two appendices: "Geometrical Terminology" and "Binary and Regular Matroids".

As we all know, matroids made their appearance in the mathematical literature in 1935, in a paper of Hassler Whitney entitled "On the abstract properties of linear dependence" [17].

Whitney looked at a matrix and saw that some sets of columns were independent and some were not. There were even simple rules about this distinction. For example, any subset of an independent set of columns is independent—provided of course that you count the null set as independent. Also, if you had an independent set A you could make it into a bigger one by adding the right member of any independent set that was bigger than A. In a flash of genius, Whitney said "Let us make these statements the axioms of a theoretical structure that is like a matrix and yet more than a matrix!"

I imagine him reflecting: "A spheroid is something like a sphere. A cycloid is something like a cycle. So something like a matrix could be a 'matroid'. So be it!"

Now Whitney, only a few years before, had published some important papers on graph theory—papers making up the nearest thing we had to a textbook on the subject [12–16]. Everyone knows nowadays that a graph has associated matroids, and it seems reasonable that matroid theory should develop out of graph theory. Whitney remarks on the connection between the two subjects in his introduction and carries over some terms from one theory to the other. That is why he calls a circuit a "circuit", though it does not always look like one.

But Whitney left graph theory, and perhaps matroid theory was the path by which he left—though we should note that having sojourned forty years in the wilderness he emerged to part-write a paper on the Four Colour Problem [18].

But let all that be introductory. I suppose you are not anxious for me to make this lecture a recitation from the literature. Better that I should tell of experiences of my own, of how I myself encountered matroids and other abstractions from graph theory.

1935 was the year when I became an undergraduate at Trinity College, Cambridge. There I joined the Trinity Mathematical Society and formed an informal association with three other members, Leonard Brooks, Cedric Smith and Arthur Stone. The object of this association was the study of out-of-the-way mathematical problems, notably that of dissecting a square into unequal squares.

Since this problem depended on a knowledge of Kirchhoff's Laws for electric currents, it gave us an excellent grounding in graph theory. We therefore began to look at other graph-theoretical problems too.

Our association received different names from time to time. The Important Members, The Four Horsemen, The Gang of Four. Take your pick.

A set of currents in a graph obeying Kirchhoff's Laws we called a "Kirchhoff Chain". Arthur Stone, the topologist, pointed out that this chain was "a cycle modulo the poles" and "an absolute cocycle". Some years later, in a course on combinatorial topology, I found out what he meant.

Smith began to abstract from Four Colour Theory. He started with Tait's variation on that theory, which considers 3-colourings of the edges of a cubic graph G so that all three colours meet at each vertex. In Smith's first abstraction the edges of G became geometrical points associated in threes. Each triad of course corresponded to a vertex of G. He called his abstraction a "3-net".

Now if the edges of G are properly coloured in three colours a, b and c, then those of any two colours a and b form a "Tait cycle", that is, a union of one or more disjoint even circuits that takes in all the vertices. Smith now constructed a second 3-net that he called the "derivative" of the first. Its points were the Tait cycles—and they came in triads, one triad to each Tait colouring. Soon he had a derivative of the derivative, and so on [4].

I have told elsewhere [10] of how others of the Four watched apprehensively this process of abstraction and generalization. Soon Smith's 3-nets had become mere sets of points in a finite vector space, and a Tait colouring was an assignment to those points of non-zero members of the 4-group, the coefficients conforming to the linear relations of the points. It seemed to us other three that our beloved graph theory had vanished into a mist of algebra. "Graph theory" we would explain to our friends "is like the Cheshire cat: the cat has vanished but the grin remains".

Smith's 3-nets as I see them are not matroids but are based on the same principle. Generalize some aspect of graph theory; it shall undergo a change "Into something rich and strange".

I left Cambridge in 1941 with the idea that graph theory could be reduced to abstract algebra, but that it might not be the conventional kind of algebra.

There had been developments since 1935. The Four solved their problem about dissected squares. They wrote a paper about such dissections, eventually published in the Duke Mathematical Journal [1]. They researched on

Hamiltonian circuits. Then a war broke out. I found myself at Bletchley Park, in Buckinghamshire, studying some Continental codes and ciphers. (The work at Bletchley Park was part of the activities of GC & CS, the Government Code and Cipher School.)

I mention this because some of those ciphers posed problems that I thought involved a kind of linear algebra. We would receive an intercepted cipher message that was a long string of letters or teleprinter symbols. That could be called a vector. Call it C for "cipher". In the relevant cases C was formed from two other vectors, P for "plain language" and K for "key". We would have the simple equation
$$C = P + K$$
in some chosen finite arithmetic. The key K would be constructed on some secret machine.

There one had an equation in linear algebra and to start with we, in the Research Section at Bletchley, would know only C. Sometimes mistakes at the European end, such as sending two messages on the same key, would enable us to solve for K and the two P's [2].

In one case of importance, K was a sum of subkeys. Some of these were periodic, advancing one step for each letter. The others were basically periodic but for each new letter they sometimes advanced and sometimes stayed still. These subkeys each involved only two symbols, known to us as "dot" and "cross". The patterns of dot and cross were changed from time to time but the periods of the subkeys were fixed. We called the subkeys "wheels".

Sometimes, knowing C and assuming some statistical properties of P, we were able to disentangle the subkeys of K and determine them all, using a curious mixture of statistics and linear algebra. The problem would be simplified when we knew the cyclic patterns of the wheels and had thereafter only to determine their settings.

The point I want to make is that at Bletchley I was learning an odd new kind of linear algebra; I was still being prepared for the Coming of the Matroids.

I have been warned that this Conference is oriented towards Computer Science. That gives me another reason to mention Bletchley. For there an electronic computer was invented more than half a century ago.

I remember a particular problem of the time and place. We would have a sequence of dots and crosses, at least 2000 long, derived from a cipher message. We would have also a periodic sequence of dots and crosses, of period $31 \times 41 = 1271$, derived from our knowledge of the ciphering machine and its current wheel-patterns. We would want to compare the two in all their 1271 relative settings and pick the setting that gave the best agreement. If there was then a statistically significant agreement we would infer the setting of two wheels and go on to the next stage. Now 1271 comparisons per message were rather too many for the biological computers initially available, so electrical ones were invented and constructed, first with relays and then with thermionic valves (also called vacuum tubes) [2].

A replica of one of the later models can be seen now at Bletchley Park. Over and over again it finds the two sequences of dots and crosses, of periods 31 and 41, whose combination gives the best agreement with a long, long sequence derived from a genuine wartime cipher message.

Late in 1945 I found myself back at Trinity as a Fellow of the College. I now had to work for my PhD degree. What should be the subject of my thesis? Why not that abstractifying of graph theory in a reduction to linear algebra?

Following in the ways of Arthur Stone, I contemplated the additive group of cycles of a graph G, with coefficients in a ring R. Perhaps the coefficients would be integers as with our Kirchhoff chains. Perhaps they would be residues mod 2, as used for Tait cycles. Or perhaps even they would be elements of the four-group of 2-vectors mod 2. Such cycles in cubic graphs, with no zero coefficients, defined the Tait colourings. It seemed that many graph-theoretical entities could be described in terms of these additive cycle-groups.

Each cycle had its "support", the set of graph-edges with non-zero coefficients in it. It seemed good to define an elementary cycle as a non-zero cycle whose support contained that of no other non-zero cycle. That could be abbreviated as a "cycle with minimal (non-null) support". Those elementary cycles corresponded to the circuits of the graph. What a pleasing theorem!

Now I ventured to abstract after the manner of Smith. Forget the graph-structure. Replace it by a finite set S of objects called "cells". Giving a coefficient in R to each cell one got a "chain on S over R". A set of such chains, closed under addition and multiplication by elements of R, was a "chain-group". A cycle-group of a graph was merely a special case of a chain-group.

I went on happily developing a theory of chain-groups and their elementary chains, these latter of course being defined by minimal supports. The method was to select theorems about graphs and try to generalize them to chain-groups. These was not too difficult for theorems expressible in terms of circuits. But theorems about 1-factors imposed problems.

As I look back on this episode I am grieved to recall that I still did not appreciate the work of Whitney. Yet these chain-groups were half-way to matroids and their minimal supports were Whitney's matroid-circuits, his "minimal dependent sets". Perhaps if I had read, marked and learned that paper of Whitney's [17] I would have said "Look, Whitney has done this stuff better already; I will abandon chain-groups and write about other things". That, I think now, would have been a pity.

I understand that Richard Rado was once in a somewhat similar position writing about abstract linear dependence but unaware of the earlier work of Whitney. I get an urge of fellow-feeling with that great man to whom this lecture is dedicated. But my linear dependence was not yet abstract. It was still thoroughly and unadventurously orthodox.

Returning to my own story, I went on to a second stage in chain-group theory. I discussed "binary" chain-groups, those over the ring of residues mod 2. Binary chain-groups have the interesting property that each chain is uniquely determined by its support. Then I wrote of "regular" chain-groups. These are over the ring of integers. But each elementary chain is restricted to the coefficients 0, 1 and −1, or is an integral multiple of such a chain of the group. An equivalent definition derives a regular chain-group from a totally unimodular matrix, that is, a matrix in which each square submatrix has determinant 1, −1 or 0. The chains correspond to the rows of the matrix and their linear combinations (with integral coefficients). Regular chain-groups were interesting because the cycle-group of a graph, over the integers, was always regular. Last came the "graphic" chain-groups, those that could be represented as integral cycle-groups of graphs.

The later part of my thesis established what would now be called "excluded minor" conditions first for a binary chain-group to be regularizable, that is, to correspond cell to cell and circuit to circuit with a regular chain-group, and then for a regular chain-group to be graphic [5]. All that foreshadowed my own contribution to matroid theory, published some ten years later [6–8].

In the interval I had my thesis accepted and received my PhD degree in 1948. I then became a lecturer at the University of Toronto. By 1958 I was an "Assistant Professor". In the interval I had learned to appreciate matroids. I put the work in my thesis into matroid terminology and generalized from chain-groups to matroids. I found conditions for a given matroid to be "binary", that is, the matroid of a binary chain-group. Then from the thesis-theorems I got the now well-known excluded minor conditions for a binary matroid to be regular and for a regular matroid to be graphic.

I published this work in a 2-part paper entitled "A homotopy theorem for matroids" [6, 7]. I do not find that homotopy theorem in the later literature. Perhaps it is mentioned with a warning that it is terribly long and then the author tells of some shorter, slicker proof of the excluded minor conditions. That is the way of Mathematics.

Yet I feel some sadness at the disappearance of the process of homotopy. It began with a geometrical representation of a matroid. The points were the circuits of a matroid, that is, its minimal dependent sets. To any set U of cells could be assigned a "rank", the least number of cells whose removal destroyed all the circuits in U. A union of circuits of rank 2 was a "line" and one of rank 3 was a "plane". And so on. With this terminology you could study matroid theory in a geometrical context provided you bore in mind that two points did not necessarily determine a line, nor three non-collinear points a plane. However two lines in a plane were conventional enough to intersect in a point.

There is a distinction between connected and disconnected lines. A disconnected line has two points only. And these, considered as circuits of the matroid, are disjoint. A connected line has three points. Since the theorem is about binary matroids there cannot be more than three points on a line.

Such combinatorial geometries are still met with. The homotopy paper went on to define a "linear subclass" as a set K of points which with any two points on a connected line contained also the third. Then attention was directed to those re-entrant paths along the connected lines of the geometry that were "off K", that is, passed through no point of K. To me the most interesting part of the work described in the paper was showing that any such path could be reduced to a null path by a sequence of elementary operations of four kinds.

The first operation replaced

$$XYZYT \quad \text{by} \quad XYT \quad \text{or conversely}.$$

The second replaced

$$XYZTYU \quad \text{by} \quad XYU \quad \text{or conversely},$$

provided that Y, Z and T were coplanar. The third replaced

$$XYZTUYV \quad \text{by} \quad XYV \quad \text{or conversely},$$

provided that Y, Z, T and U were coplanar. There were two other points in the plane and these belonged to K.

The fourth operation uses a configuration that can be described briefly as projectively equivalent to a cube with its edges and faces extended to three points at infinity. Any two of these three make up a disconnected line. Four vertices of the cube, no two on the same edge, belong to C. An elementary path of the fourth kind is of the form $AXBYA$ where A and B are "points at infinity" and X and Y are distinct vertices of the cube not in C.

That was the homotopy theorem and I was able to use it to characterize regular matroids. The later result saying when a regular matroid was graphic was guided, in the usual vague graph-to-matroid way, by Kuratowski's Theorem and my favourite proof thereof [8].

One aspect of this work rather upset me. I had valued matroids as generalizations of graphs. All graph theory, I had supposed, would be derivable from matroid theory and so there would be no need to do independent graph theory any more. Yet what was this homotopy theorem, with its plucking of bits of circuit across elementary configurations, but a result in pure graph theory? Was I reducing matroid theory to graph theory in an attempt to do the opposite? Perhaps it was this jolt that diverted me from matroids back to graphs.

Yet I did do some more work on matroids. I can claim to have invented the whirl if not the wheel. And I lectured on matroids at the first formal conference devoted to them [4]. That conference was organised by Jack Edmonds and his colleagues at the National Bureau of Standards in Washington in 1964. To me that was the year of the Coming of the Matroids. Then and there the theory of matroids was proclaimed to the mathematical world. And outside the halls of lecture there arose the repeated cry: "What the hell is a matroid?" In their text-books Dominic Welsh and James Oxley have attempted to answer that question [3, 11].

Richard Rado took a keen interest in abstract linear dependence, and his name is attached to an important theorem in the theory of transversals and transversal matroids [3, 11]. By the time I met him I was back with graphs and maps, trying to enumerate rooted planar maps of various kinds. My wife and I met Richard and Louise Rado quite often at Waterloo, at Reading and at conferences elsewhere. I enjoyed many stimulating conversations with him. When I spoke of my enumerative work he advised me earnestly to use exponential generating functions. Alas, I still have not found a way of doing that.

References

[1] R. L. Brooks, C. A. B. Smith, A. H. Stone & W. T. Tutte, The dissection of rectangles into squares, *Duke Mathematical Journal*, **7** (1940), 312–340.

[2] F. H. Hinsley and Alan Stripp, editors, *Codebreakers*, Oxford University Press (1993).

[3] J. G. Oxley, *Matroid Theory*, Oxford University Press (1992).

[4] C. A. B. Smith, Map colourings and linear mappings, in *Combinatorial Mathematics and its Applications, Proceedings of a Conference held at the Mathematical Institute, Oxford, from 7–10 July 1969*, (ed. D. J. A. Welsh), Academic Press, London (1969), pp. 259–283.

[5] W. T. Tutte, An Algebraic Theory of Graphs, PhD Thesis, Cambridge, 1948.

[6] W. T. Tutte, A homotopy theorem for matroids, I, *Transactions of the American Mathematical Society*, **88** (1958), 144–160.

[7] W. T. Tutte, A homotopy theorem for matroids, II, *Transactions of the American Mathematical Society*, **88** (1958), 161–174.

[8] W. T. Tutte, Matroids and Graphs, *Transactions of the American Mathematical Society*, **90** (1959), 527–552.

[9] W. T. Tutte, Lectures on Matroids, *Journal of Research of The National Bureau of Standards, Series B, Mathematics and Mathematical Physics*, **69B** (1965), 1–47.

[10] W. T. Tutte, *Graph Theory As I have Known It*, Oxford University Press (1998).

[11] D. J. A. Welsh, *Matroid Theory*, Academic Press, London (1976).

[12] H. Whitney, A theorem on graphs, *Annals of Mathematics 2*, **32** (1931), 378–390.

[13] H. Whitney, The colouring of graphs, *Annals of Mathematics 2*, **33** (1932), 688–718.

[14] H. Whitney, A logical expansion in mathematics, *Bulletin of the American Mathematical Society*, **38** (1932), 572–579.

[15] H. Whitney, 2-isomorphic graphs, *American Journal of Mathematics*, **55** (1933), 245–254.

[16] H. Whitney, Non-separable and planar graphs, *Transactions of the American Mathematical Society*, **34** (1932), 339–362.

[17] H. Whitney, On the abstract properties of linear dependence, *American Journal of Mathematics*, **57** (1935), 509–533.

[18] H. Whitney & W. T. Tutte, Kempe chains and the four colour problem, *Utilitas Mathematica*, **2** (1972), 241–281.

151 Manderston Road
Newmarket
Suffolk CB8 0NS

Appendix I: geometrical terminology

Given a matroid M on a set E, a rank $r(S)$ can be assigned to each subset S of E. In terms of circuits, the rank is the least number of cells of S whose deletion destroys all the circuits of M contained in S. (The "cells" of M are the elements of E.) The rank of E is called also the rank $r(M)$ of M, and similarly $r(S)$ is the rank of $M \times S$, the matroid on S whose circuits are those of M contained in S.

There is another matroid $M \cdot S$ on S. Its circuits are those non-null intersections with S of circuits of M that contain no other such intersections.

The rank-function has the following important properties, where S and T are subsets of E:

$$r(M \times S) + r(M \cdot (E - S)) = r(M), \tag{1}$$

$$r(S \cup T) + r(S \cap T) \geq r(S) + r(T). \tag{2}$$

In [9] it is found convenient to define a "flat" as a union of circuits, the null subset of E being counted as a flat of rank zero. Each $S \subseteq E$ defines a flat $\langle S \rangle$, the union of the circuits contained in S. Noticing some geometrical analogies the author of [9] experimented with a geometrical terminology in which circuits were "points", flats of rank 2 were "lines", those of rank 3 were "planes" and those of rank 4 were "3-spaces". Even the rank $r(S)$ was replaced by the more geometrical "dimension" $d(S) = r(S) - 1$.

The geometrical analogy is not perfect. Two distinct points determine a unique flat, their union as subsets of E, but this flat is not necessarily a line. However, if flat S is properly contained in flat T (as a subset of E) then

$$d(S) < d(T), \tag{3}$$

by (1). Two lines in the same plane intersect in a unique point, and two planes in the same 3-space intersect in a unique line, by (2).

A feature of this geometry was that a distinction had to be made between connected and disconnected flats. A "separator" U of a flat S is a subset of S such that each circuit of M in S is contained either in U or in $S - U$. Thus S and the null subset of E are always separators of S. If S has any other separator it is disconnected; otherwise it is connected.

Elementary matroid theory established some properties of connected and disconnected flats. For example a disconnected line was on exactly two points and a connected one on at least three. Also, a connected d-flat S, that is, a flat of dimension d, on a connected $(d+2)$-flat T was on two distinct connected $(d+1)$-flats contained in T, the union of the two being T.

At this stage the writer's attention was drawn to the graph $G(M)$ of M. The vertices of this graph are the points of M. Two points are adjacent in $G(M)$ if and only if they are on the same connected line of M. Graph-theoretical concepts could now be introduced, such as paths. There were simple paths (that use no vertex twice), re-entrant paths (which return to their starting points) and the conventional degenerate paths (each confined to a single vertex and counted as re-entrant). There was the pleasing theorem that a flat S is connected if and only if any two distinct points on S can be joined by a simple path in S. Attempts could be made to construct homotopy theorems saying that any re-entrant path could be reduced by suitably defined "elementary operations" to a degenerate path.

Alas, an important possible application made necessary a further complication. In an inductive argument a matroid M_1 on a set E_1 was to be reduced to
$$M = M_1 \cdot (E_1 - \{a\})$$
where $a \in E_1$ and $E_1 - \{a\}$ can be identified with E. The object was to show that if M had a certain property P then so did M_1. But it was found necessary to impose on M a sort of shadow of M_1. This took the form of a subset Q of the set of points of M, the set of all circuits of M that were also circuits of M_1 (without requiring the adjunction of a).

In order to cope with Q without mentioning M_1 a new definition had to be made. A "linear subclass" of M was defined as a set C of points of M such that if two points on a line L belonged to C then so did all the other points of L. Then Q was just one of the linear subclasses of M.

A new homotopy theorem was needed. The matroid M was taken with an arbitrarily chosen linear subclass C and the theorem concerned only those re-entrant paths that were "off C", that is passed through no point of C. Elementary operations were defined within structures of three or fewer dimensions that might include points of C. In the final theorem there were four of these basic structures. They are briefly described in the main text of the Lecture. In [9] the theorem is proved in the geometrical terminology and each main step in the proof is illustrated by a geometrical diagram.

Appendix II: binary and regular matroids

In [9] we considered a binary matroid M on a set E. The distinguishing feature of a binary matroid is that it has exactly three points on each connected line. Any two of these three points must have a non-null intersection as subsets of E, by the connection of the line. A binary matroid becomes "regular" if it can be "co-ordinatized" in the following way. With each circuit S of M we associate a chain $f(S)$ on E with support S and with coefficients restricted to the integers 1, −1 and 0. Now on any connected line L of M there are just three points, S, T and U say. As part of the co-ordinatization we require that their chains $f(S)$, $f(T)$ and $f(U)$ shall be linearly dependent. This means that each of them is a sum or difference of the other two. Note the following implication: the product of its coefficients in $f(S)$ and $f(T)$ has the same value +1 or −1 for each cell of $S \cap T$.

If such a co-ordinatization exists its chains $f(S)$ generate a chain-group N on E, and Theorem 5.11 of [9] assures us that its matroid $M(N)$ is identical with M. We can then say that M is regular as well as binary, and that N is a regular chain-group with matroid M.

With regard to the line L of the preceding paragraph and its points S, T and U we write $\mu(S,T)$ for the product of the coefficients in S and T of a cell of the non-null set $S \cap T$.

The main application of the homotopy theorem in [9] is in a proof that a binary matroid is regular if it has no Fano matroid or dual thereof as a minor. In that proof we assume a binary matroid M_1 (on a set E_1) that has neither of these forbidden minors and yet is not regular, and we take M_1 to have the least number of cells consistent with this description. We choose a cell $a \in E_1$ and write

$$M = M_1 \cdot E, \quad \text{where} \quad E = E_1 - \{a\}.$$

We then note that the binary matroid M is regular since it has no forbidden minor and has fewer cells than M_1.

We may assume that the set $\{a\}$ is not a circuit of M_1; if it were, the regularity of M_1 would follow at once from that of M. Some circuits of M will be circuits also of M_1. These constitute a linear subclass C of M. The other circuits of M are made into the remaining circuits of M_1 by the adjunction of a. A line L of M either has all its points of the first kind, that is, in C, or it has one point of the first kind and two of the second.

In [9] the homotopy theorem is used to prove the following result: either the product of the numbers $\mu(S,T)$ around any re-entrant path off C is 1 or one of the four kinds of basic structures occurs in M in such a way as to impose a forbidden minor on M_1. But the latter alternative is ruled out by the definition of M_1.

A co-ordinatization of M can now be extended to M_1. For consider any connected line L in M with one point U in C and two points S and T not in C. By adjunction of a this becomes a line L_a of M_1 with points U, S_a and T_a, the two last being extensions of S and T. In the co-ordinate-extension each of S_a and T_a receives a number 1 or -1 as the co-ordinate of a. And if we are to prove M_1 regular the product of the two numbers must be $\mu(S,T)$. The theorem noted in the preceding paragraph shows that the coefficients of a can be assigned consistently with this requirement. So, by another application of Theorem 5.11 of [9], M_1 is regular, contrary to assumption.

The Invited Lectures

Polynomials in Finite Geometries

S. Ball

Summary A method of using polynomials to describe objects in finite geometries is outlined and the problems where this method has led to a solution are surveyed. These problems concern nuclei, affine blocking sets, maximal arcs and unitals. In the case of nuclei these methods give lower bounds on the number of nuclei to a set of points in $PG(n,q)$, usually dependent on some binomial coefficient not vanishing modulo the characteristic of the field. These lower bounds on nuclei lead directly to lower bounds on affine blocking sets with respect to lines. A short description of how linear polynomials can be used to construct maximal arcs in certain translation planes is included. A proof of the non-existence of maximal arcs in $PG(2,q)$ when q is odd is outlined and some bounds are given as to when a (k,n)-arc can be extended to a maximal arc in $PG(2,q)$. These methods can also be applied to unitals embedded in $PG(2,q)$. One implication of this is that when q is the square of a prime a non-classical unital has a limited number of Baer sublines amongst its secants.

1 Introduction

The effectiveness of polynomials as a means of studying problems in finite geometries has become increasingly evident in the 1990's, although the first examples seem to date back to R. Jamison [38] in 1977 and A. E. Brouwer and A. Schrijver [19] in 1978. Indeed in [22] A. A. Bruen and J. C. Fisher described the "Jamison method" as the following: reformulate the problem in terms of points of an affine space and associate suitable polynomials defined over the corresponding finite field; calculate. This is the approach employed in [19] too; in fact the main difference between [38] and [19] is that Jamison viewed the points of an affine space as elements of a finite field. In effect, this has the advantage of reducing the number of variables in the polynomials and allowing one to use simple arguments concerning the degree or the coefficients of a polynomial. Earlier survey papers covering polynomial applications to finite geometries include [11], [12] and [53] and in some ways the present paper is an update of those, although there is much material in those articles that is not covered here.

In general, we are interested in solving problems of the form: Given a set of subspaces (usually points) in a Desarguesian space with restricted intersections with larger subspaces (usually lines), what can we say about the size of the set and can we characterise the extremal cases? Historically this stems from the famous proof of B. Segre [49] that any set of $q+1$ points in the Desarguesian plane of odd order q having at most two points on a line is a conic.

Section 2 considers polynomials whose zeros correspond to subspaces of Desarguesian affine and projective spaces. This leads us to define polynomials, given an arbitrary set of points $\mathcal{S}$, whose properties reflect the properties of $\mathcal{S}$. These polynomials are fundamental to many of the proofs of the results covered in this paper.

Section 3 updates results concerning nuclei. It is not a complete survey; indeed emphasis is given to those results for which the polynomials in Section 2 have been the most useful. The intriguing conjecture from [16] is included. Following on directly from the bounds in Section 3, lower bounds on the size of affine blocking sets are detailed. I include a general definition for blocking sets in affine and projective spaces in the hope that this will be adopted. Since the early 1990's there have appeared conflicting definitions by various authors, which has led to some confusion. I have not surveyed recent developments in projective blocking sets, there being too much material for the scope of this paper. However a survey from 1997 can be found in [37, Chapter 13]. The recent constructions by G. Lunardon [40] and by P. Polito and O. Polverino [47] concerning linear blocking sets are the most notable developments since then.

Section 5 leaves surveying aside and gives details of how one can view translation planes with polynomials using the construction of André [1] and Bruck and Bose [20], in the hope of proving algebraic results previously only possible in Desarguesian planes. Returning to the surveying, Section 6 contains recent results and constructions concerning maximal arcs, including a sketch of the non-existence proof for Desarguesian planes. A construction of some maximal arcs in translation planes using polynomials is also included.

Finally Section 7 considers unitals embedded in a Desarguesian plane. The classification of such objects appears to be a very hard problem; some characterisations can be obtained from polynomial arguments.

Where possible I have put definitions in their relevant sections in such a way that each section is self-standing. However, the construction in Section 6 is dependent on Section 5 and Section 4 is closely related to Section 3.

2 Definitions and useful polynomials

Let π_n denote a projective space of dimension n and $\mathrm{PG}(n,q)$ the Desarguesian space of order q. Let $\mathcal{A}_n$ denote an affine space of dimension n and $\mathrm{AG}(n,q)$ the Desarguesian space of order q. Throughout, $\theta_n = (q^{n+1}-1)/(q-1)$, the number of points of π_n, and $q = p^h$ for some prime p.

2.1 Affine spaces

The elements of $\mathrm{GF}(q^n)$, where $q = p^h$ for some prime p, can be viewed as the points of $\mathrm{AG}(n,q)$. The points lying on a hyperplane are given by the zeros of equations

$$\mathrm{Tr}_{q^n \to q}(ax) + b = 0,$$

Polynomials in Finite Geometries

where b is an element of $\mathrm{GF}(q)$ and $\mathrm{Tr}_{q^n \to q}(x) = x^{q^{n-1}} + x^{q^{n-2}} + \ldots + x^q + x$ is the trace function from $\mathrm{GF}(q^n)$ to $\mathrm{GF}(q)$. To see this, note that the polynomial should have degree q^{n-1}. Every hyperplane in $\mathrm{AG}(n,q)$ is a translate of a hyperplane through the origin; this translate can be seen as an $(n-1)$-dimensional subspace over $\mathrm{GF}(q)$, and the corresponding polynomial is therefore $\mathrm{GF}(q)$-linear and so of the form

$$H(x) := \sum_{j=0}^{n-1} a_j x^{q^j} + b.$$

R. Jamison provided a proof of this [38, Lemma A, p. 259] which he credited to O. Ore, who wrote two expositions on polynomials of the form (1) [44, 45]. These polynomials are called linearized polynomials, see [39, Chapter 3, Section 4]. The polynomial

$$a_{n-1} H^q - a_{n-1}^{q+1}(x^{q^n} - x) - a_{n-2}^q H$$

has degree at most q^{n-2} and since all the points of the hyperplane are zeros it is identically zero. Equating coefficients of x^{q^i} for $0 \leq i \leq n-2$ implies the trace function form above.

A suitable linear combination of k hyperplane polynomials will give an equation of the form

$$\sum_{j=0}^{n-k-1} \alpha_j x^{q^j} + \beta = 0, \tag{1}$$

whose zeros correspond to a subspace of dimension $n-k-1$ that is, the intersection of the corresponding k hyperplanes. In particular, lines are given by the sets of zeros of equations of the form

$$x^q - \alpha x + \beta = 0,$$

and for a line joining a point x and a point y (viewed as elements of $\mathrm{GF}(q^n)$) we have $\alpha = (x-y)^{q-1}$. The non-zero $(q-1)$-th powers are θ_{n-1}-th roots of unity in $\mathrm{GF}(q^n)$, so there is a one-to-one correspondence between the θ_{n-1}-th roots of unity in $\mathrm{GF}(q^n)$ and the θ_{n-1} directions of lines in $\mathrm{AG}(n,q)$.

Given a set of points $\mathcal{S}$, a subset of $\mathrm{AG}(n,q)$, viewed as elements of $\mathrm{GF}(q^n)$ and not containing the zero element, define the *locator polynomial* (Jamison would call this the *root polynomial* and were $\mathcal{S}$ to be a subspace the *Ore polynomial*) of $\mathcal{S}$ to be

$$S(x) := \prod_{s \in \mathcal{S}}(1 - sx) = \sum_{j=0}^{|\mathcal{S}|}(-1)^j \sigma_j x^j,$$

where σ_j is the j-th symmetric function of the set $\mathcal{S}$. Strictly speaking this is the locator polynomial for the set $\{1/s \mid s \in \mathcal{S}\}$ since these are the zeros of $S(x)$, but we choose to define it this way simply so that the coefficient of $(-1)^j x^j$ in $S(x)$ is the j-th symmetric function.

Define the *direction polynomial* of a set S to be

$$F(u,x) := \prod_{s \in S}(1-(1-sx)^{q-1}u) = \sum_{j=0}^{|S|}(-1)^j \chi_j(x) u^j,$$

where $\chi_j(x)$ is the j-th symmetric function of the set $\{(1-sx)^{q-1} \mid s \in S\}$, a polynomial in x of degree at most $k(q-1)$. If $F(u,x_0)$ is viewed as a polynomial in u, its zeros are θ_{n-1}-th roots of unity and moreover

$$(1-s_1 x_0)^{q-1} = (1-s_2 x_0)^{q-1}$$

if and only if $(1/x_0 - s_1)^{q-1} = (1/x_0 - s_2)^{q-1}$ if and only if $1/x_0$, s_1 and s_2 are collinear.

2.2 Projective spaces

The $(q-1)$-th powers of the elements of $GF(q^{n+1})$ can be viewed as the directions of the lines through the origin in $AG(n+1,q)$ and hence the points of $PG(n,q)$. The hyperplanes through the origin are given by zeros of equations of the form

$$\operatorname{Tr}_{q^{n+1} \to q}(AX) = 0 = AX \sum_{i=0}^{n} A^{q^i-1} X^{q^i-1},$$

and by writing $x = X^{q-1}$ and $a = A^{q-1}$ the hyperplanes of $PG(n,q)$ are given by the zeros of equations of the form

$$\sum_{i=0}^{n} a^{(q^i-1)/(q-1)} x^{(q^i-1)/(q-1)} = 0.$$

As in the affine case, taking a suitable linear combination of k hyperplane polynomials, one can obtain an equation of the form

$$\sum_{j=0}^{n-k-1} \alpha_j x^{\theta_j} + \beta = 0$$

whose zeros correspond to the points of a subspace of dimension $n-k-1$, that is the intersection of the corresponding k hyperplanes. In particular, lines are given by the sets of zeros of equations of the form

$$x^{q+1} - \alpha x + \beta = 0,$$

where there exist relations between α and β depending on the dimension, and for a line joining a point x and a point y (viewed as $(q-1)$-th power of $GF(q^{n+1})$) we have

$$\alpha = (x^{q+1} - y^{q+1})/(x-y).$$

Polynomials in Finite Geometries

Given a set S we define the locator polynomial $S(x)$ as for the affine spaces but define the direction polynomial slightly differently as

$$F(u,x) := \prod_{s \in S}(1 - sx(1-sx)^{q-1}u) = \sum_{j=0}^{|S|}(-1)^j \Delta_j(x) u^j,$$

where $\Delta_j(x)$ is the j-th symmetric function of the set $\{sx(1-sx)^{q-1} \mid s \in S\}$, a polynomial in x of degree at most jq. The linear factors of $F(u, x_0)$ have zeros of the form $u_0^{-1} = sx_0(1-sx_0)^{q-1}$ which satisfies

$$1 + u + u^{q+1} + u^{q^2+q+1} + \ldots + u^{\theta_{n-1}} = 0 \qquad (2)$$

for $u = u_0^{-1}$. Moreover $s_1 x_0 (1 - s_1 x_0)^{q-1} = s_2 x_0 (1 - s_2 x_0)^{q-1}$ if and only if $1 + s_1 x_0 (1 - s_1 x_0)^{q-1} = 1 + s_2 x_0 (1 - s_2 x_0)^{q-1}$ if and only if

$$\frac{1/x_0^{q+1} - s_1^{q+1}}{1/x_0 - s_1} = \frac{1/x_0^{q+1} - s_2^{q+1}}{1/x_0 - s_2}$$

if and only if $1/x_0$, s_1 and s_2 are collinear. Therefore there is a one-to-one correspondence between the θ_{n-1} zeros of (2) and the directions of the lines through $1/x_0$.

3 Nuclei

A point P in π_n is a *t-fold nucleus* of a set of points S if $P \notin S$ and on every line through P there are at least t points of S. A 1-fold nucleus is called a *nucleus*. The original definition of nucleus stems from a $(q+1)$-arc in $PG(2, 2^h)$ which has one such nucleus.

3.1 Nuclei in affine planes

The following theorem is a consequence of results proved for q even by Bruen and Thas [23] and for q odd by Segre and Korchmáros [50].

Result 3.1 *Let S be a set of $q+1$ points of $AG(2, q)$. The set of nuclei of S cannot contain a conic.*

Motivated by this the question of how many nuclei a $(q+1)$-set can have was posed. In the Desarguesian case this was answered by Blokhuis and Wilbrink [18] who gave a simple proof using polynomials which contains many of the basic ideas used to prove much of what is contained in this article. For that reason the proof is included, by way of an example of how the polynomials work.

Theorem 3.2 *Let S be a set of $q+1$ points in $AG(2, q)$. The set of nuclei of S has size at most $q-1$.*

Proof Consider the direction polynomial of such a set $\mathcal{S}$

$$F(u,x) := \prod_{s \in \mathcal{S}}(1-(1-sx)^{q-1}u) = \sum_{j=0}^{q+1}(-1)^j \chi_j(x) u^j,$$

and an x_0 such that $1/x_0$ is a nucleus of $\mathcal{S}$. By definition, there is exactly one point of $\mathcal{S}$ on each line through $1/x_0$ and so

$$F(u,x_0) = 1 - u^{q+1}$$

for all such x_0. Comparing the coefficient of u in the two above equations implies $\chi_1(x_0) = 0$ and since $\chi_1(x)$ has degree at most $q-1$ it is identically zero if there exist more than $q-1$ nuclei. However

$$F(u,0) = (1-u)^{q+1} = 1 - u - u^q + u^{q+1}$$

and in particular $\chi_1(0) = 1$, and hence $\chi_1(x)$ cannot be identically zero.

In $\mathrm{AG}(2,q)$ the only known examples of $(q+1)$-sets having exactly $q-1$ nuclei are a set consisting of a line together with a point, and a sporadic example in $\mathrm{AG}(2,5)$ where the 10 points of a Desargues configuration can be partitioned into a set of size 6 and 4 nuclei. This leads to the following conjecture [16].

Conjecture 3.3 *In* $\mathrm{AG}(2,q)$ *these are the only* $(q+1)$*-sets having exactly* $q-1$ *nuclei.*

Partial results towards this conjecture appear in [16]. The following lemma of Segre and Korchmáros may well prove to be a powerful tool in settling the above conjecture.

Lemma 3.4 *Let* $\mathcal{S}$ *be a set of* $q+1$ *points in* $\mathrm{PG}(2,q)$*. For any three non-collinear nuclei* N_1*,* N_2 *and* N_3*, the points of* $\mathcal{S}$ *on the three lines* $N_i N_j$ *are collinear.*

3.2 Generalised and non-planar nuclei

Following on from Theorem 3.2, various generalisations were proven. All of these are contained in Result 3.5, which appears in [4].

Result 3.5 *Suppose there exists a hyperplane* $\mathcal{H}$ *containing exactly* i *points of* $\mathcal{S}$*, a set of* $t\theta_{n-1} + k - 1$ *points in* $\mathrm{PG}(n,q)$*. The number of* t*-fold nuclei in* $\mathrm{AG}(n,q) = \mathrm{PG}(n,q) \setminus \mathcal{H}$ *is at most* $(k+r)(q-1)$ *provided that the binomial coefficient*

$$\binom{t\theta_{n-1}+k-i-1}{k+r} \not\equiv 0 \pmod{p},$$

for some $r \geq 0$.

Polynomials in Finite Geometries

The case $t = 1$ and $i = r = 0$ and $n = 2$ comes from [13], and general t from [14], and for general n and i is due to Sziklai [52]. In all cases the essence of the proof follows Theorem 3.2.

Let us consider the case $t = k = 1$. Result 3.5 implies that a set S of size θ_{n-1} in $AG(n,q)$ has at most $q - 1$ nuclei. However, in the case where S intersects every hyperplane of $PG(n,q)$, Blokhuis and Mazzocca [17] proved the following.

Result 3.6 *If S intersects every hyperplane then the number of nuclei is at most $q^{n-1} - q^{n-2}$; moreover sets exist that attain the bound.*

Consider such a set S of size θ_{n-1} and with $q^{n-1} - q^{n-2}$ nuclei. Result 3.5 implies that S has at most $(r+1)(q-1)$ nuclei provided that the binomial coefficient
$$\binom{\theta_{n-1} - i}{r+1} \neq 0 \pmod{p}.$$
In the extremal case when $r = q^{n-2} - 1$ we have that $q^{n-2} | \theta_{n-1} - i$ or we would have been able to find a smaller r for which the binomial coefficient was non-zero. It follows then that $i \equiv \theta_{n-3} \pmod{q^{n-2}}$ or in words that every hyperplane meets S in $\theta_{n-3} \pmod{q^{n-2}}$ points. As a consequence of [17, Proposition 14] we have that the classification of θ_{n-1}-sets in $PG(n,q)$ ($n > 2$) with $q^{n-1} - q^{n-2}$ nuclei is equivalent to the case of $(q+1)$-sets in $PG(2,q)$ having $q - 1$ nuclei, that is Conjecture 3.3.

3.3 Extension of the nucleus definition

In [52], and for $n = 2$ earlier in [30], a point P in π_n is called a $(\leq t)$-fold nucleus of a set S of size *less* than $t\theta_{n-1}$ if $P \notin S$ and on every line through P there are at *most* t points of S. The following result appears in [30] for $r = 1$ and $n = 2$, and in [52] for general n.

Result 3.7 *Suppose there exists a hyperplane $\mathcal{H}$ containing exactly i points of S, a set of $t\theta_{n-1} - k$ points in $PG(n,q)$. The number of $(\leq t)$-fold nuclei in $AG(n,q) = PG(n,q) \setminus \mathcal{H}$ is at most $(k+r)(q-1)$ provided that the binomial coefficient*
$$\binom{-t\theta_{n-1} + k + i}{k+r} \neq 0 \pmod{p},$$
for some $r > 0$.

4 Affine blocking sets

A set S is a *t-fold blocking set with respect to s-dimensional subspaces* if every s-dimensional subspace contains at least t points of S. Some confusion in the definition and notation surrounding blocking sets has arisen in the 1990's and some clarification is required. Historically the term blocking set arises from a "blocking coalition" in game theory. Originally it signified a 1-fold blocking set of lines in the planar (projective) case; hence the term blocking set held no ambiguity.

However, conflicting definitions have appeared in for example [2, 21, 36, 40, 47] which is far from ideal. I propose to use the above definition along with calling a 1-fold blocking set with respect to s-dimensional subspaces a *blocking set with respect to s-dimensional subspaces*, and a t-fold blocking set with respect to 1-dimensional subspaces a *t-fold blocking set*. This is consistent with earlier definitions from [14] and [52].

A (k,r)-arc in π_2 or $\mathcal{A}_2$ is a set of k points having at most r points on a line. A $(k,2)$-arc is called a k-arc. In $\mathcal{A}_2$ the complement of a (k,r)-arc is a $(q-r)$-fold blocking set with $q^2 - k$ points and in π_2 the complement is a $(q+1-r)$-fold blocking set with $q^2 + q + 1 - k$ points.

4.1 Line blocking sets

Consider a set $\mathcal{S}$ of $t\theta_{n-1} + k - 1$ points of $\mathrm{AG}(n,q)$. For $\mathcal{S}$ to be a t-fold blocking set is equivalent to saying that every point of $\mathrm{AG}(n,q) \setminus \mathcal{S}$ is a t-fold nucleus. Since $\mathcal{S}$ is contained in $\mathrm{AG}(n,q)$ we can set $i = 0$ and apply Result 3.5. The following appears in [4].

Theorem 4.1 *Let $\mathcal{S}$ be a t-fold blocking set of $\mathrm{AG}(n,q)$ and let $e(t)$ be maximal such that $p^{e(t)}$ divides t. Then the set $\mathcal{S}$ has at least $(t+1)q^{n-1} - p^{e(t)}$ points.*

Proof Put $k = q^{n-1} - t\theta_{n-2} - p^{e(t)}$ in Result 3.5 and write $t = \gamma p^{e(t)}$ such that p does not divide γ. Consider the binomial coefficient

$$\binom{t\theta_{n-1} + k - 1}{k} = \binom{t\theta_{n-1} + k - 1}{t\theta_{n-1} - 1} = \binom{tq^{n-1} + q(q^{n-2} - 1) + q - p^{e(t)} - 1}{tq^{n-1} + tq^{n-2} + \ldots + tq + t - 1}.$$

A simple application of Lucas' Theorem implies that this binomial coefficient is non-zero (modulo p) precisely when

$$\binom{q - p^{e(t)} - 1}{\gamma p^{e(t)} - 1} = \binom{q - 2p^{e(t)} + p^{e(t)} - 1}{(\gamma - 1)p^{e(t)} + p^{e(t)} - 1} = \binom{q/p^{e(t)} - 2}{\gamma - 1} \pmod{p}$$

is non-zero (modulo p), and it is non-zero (modulo p) since $\gamma \neq 0 \pmod{p}$. Hence $\mathcal{S}$ cannot be a t-fold blocking set when $k = q^{n-1} - t\theta_{n-2} - p^{e(t)}$ since $\mathcal{S}$ has at most $k(q-1)$ t-fold nuclei and

$$t\theta_{n-1} + k - 1 + k(q-1)$$
$$= t\theta_{n-1} + kq - 1$$
$$= (q^{n-1} - t\theta_{n-2} - p^{e(t)})q + t\theta_{n-1} - 1$$
$$= q^n + t - p^{e(t)}q - 1 < q^n.$$

Polynomials in Finite Geometries

This theorem brings together several bounds: for general t and q and $n = 2$, Bruen [21]; for $(t,q) = 1$ and $n = 2$, Blokhuis [14]; and for $(t,q) = 1$ and $n > 2$, Sziklai [52]. The bound is sharp in the planar case $n = 2$ in some cases. The following infinite families are good examples.

1. (Mason [41]) The affine complements of Mason's $((q-p^m)(q-1), q-p^m)$-arcs are p^m-fold blocking sets in $AG(2,q)$, $q = p^h$ for some h, of size $p^m q - p^m + q = (t+1)q - p^m$ where $t = p^m$ and hence $e(t) = m$.

2. (Denniston [28]) The affine complements of the maximal arcs constructed by Denniston are $(q - 2^m)$-fold blocking sets in $AG(2,q)$, $q = 2^h$ for some h, of size $(q - 2^m + 1)q - 2^m = (t+1)q - 2^{e(t)}$ where $t = q - 2^m$ and hence $e(t) = m$.

3. (Barlotti [10]) The external points to a conic together with all but one points of the conic form a $(q+1)/2$-fold blocking set in $PG(2,q)$ whenever q is odd. Moreover this set contains a line and by deletion we can form a $(q-1)/2$-fold blocking set of size $q(q+1)/2 + q - (q+1) = (t+1)q - 1$ in $AG(2,q)$ where $t = (q-1)/2$ and hence $e(t) = 0$.

4.2 Hyperplane blocking sets

J. Doyen [29] conjectured that an affine blocking set in $AG(2,q)$ has at least $2q - 1$ points. This was proven by Jamison [38] and independently by Brouwer and Schrijver [19]. Generalising these methods, Bruen [21] proved the following.

Result 4.2 *Let S be a t-fold blocking set with respect to hyperplanes of $AG(n,q)$. Then S has at least $(n+t-1)(q-1) + 1$ points.*

In the case $n = 2$ this is a weak version of Result 4.1 and in most cases when $n > 2$ this bound can also be improved. For small t there is the following which is from [5].

Result 4.3 *For $t < q$ a t-fold blocking set with respect to hyperplanes in $AG(n,q)$ has at least $(t+n-1)(q-1) + k$ points provided there exists a j such that $k - 1 \leq j < t$ and the binomial coefficient*

$$\binom{k-n-t}{j} \not\equiv 0 \pmod{p}.$$

5 Non-Desarguesian planes

Owing to their lack of regularity, polynomials seem hard to use for tackling problems in non-Desarguesian planes. To my knowledge there are no examples where they have been used. This section outlines a model for translation planes of order q^2 by way of an example of how polynomials might be used. Section 6 considers how such a model can be used to construct maximal arcs.

5.1 Spreads of $\mathrm{PG}(3,q)$

A *1-spread* of $\mathrm{PG}(3,q)$ is a collection of q^2+1 lines that partitions the space. Consider the model for $\mathrm{PG}(3,q)$ described in Section 2. As points we have the (q^3+q^2+q+1)-th roots of unity (or alternatively the non-zero $(q-1)$-th powers) in $\mathrm{GF}(q^4)$. Hyperplanes (planes) are given by the zeros of equations of the form

$$a^{q^2+q+1}x^{q^2+q+1} + \overline{a^{q+1}x^{q+1}} + ax + 1 = 0,$$

where a is also a (q^3+q^2+q+1)-th root of unity. Lines are given by the zeros of polynomials of the form

$$L_{\alpha\beta}(x) := x^{q+1} - \alpha x + \beta,$$

where α and β satisfy certain condition which we shall calculate. All the points of $\mathrm{PG}(3,q)$ satisfy $x^{q^3+q^2+q+1} - 1 = 0$ and so

$$L_{\alpha\beta}^{q^2} x^{q+1} - (x^{q^3+q^2+q+1} - 1) + \alpha^{q^2} x L_{\alpha\beta}^q - (\beta^{q^2} - \alpha^{q^2+q}) L_{\alpha\beta}$$
$$= (\alpha^{q^2}\beta^q + \alpha\beta^{q^2} - \alpha^{q^2+q+1})x - (\beta^{q^2+1} - \alpha^{q^2+q}\beta - 1)$$

is identically zero since it is a polynomial of degree at most 1 and has $q+1$ distinct zeros corresponding to the points on the line $L_{\alpha\beta}$. By manipulating the coefficients of the right-hand side we have that $L_{\alpha\beta}$ is a line of $\mathrm{PG}(3,q)$ precisely when

$$\beta^{q^3+q^2+q+1} = 1 \quad \text{and} \quad \alpha^{q+1} = \beta^q - \beta^{q^2+q+1}. \tag{3}$$

If $\beta^{q^2+1} = 1$ then $\alpha = 0$ and if $\beta^{q^2+1} \neq 1$ then there are $q+1$ possibilities for α. This gives $(q^3+q)(q+1) + q^2 + 1 = (q^2+1)(q^2+q+1)$ lines, so these restrictions are sufficient as well as necessary.

5.2 Translation planes

Given a 1-spread of $\mathrm{PG}(3,q)$ one can construct a translation plane of order q^2 via the construction of André [1] and Bruck and Bose [20]. Consider a hyperplane Σ of $\mathrm{PG}(4,q)$ and let $\mathcal{S}$ be a 1-spread of Σ. The following incidence structure is a translation plane $\pi(\mathcal{S})$ of order q^2.

Let points of $\mathrm{PG}(4,q) \setminus \Sigma$ be points of $\pi(\mathcal{S})$ together with the q^2+1 spread elements of $\mathcal{S}$. The lines of $\pi(\mathcal{S})$ are the planes of $\mathrm{PG}(4,q)$ meeting Σ in an (spread) element of $\mathcal{S}$, together with a line l_∞ consisting of points that are elements of $\mathcal{S}$. The incidence relation of $\pi(\mathcal{S})$ is induced by incidence in $\mathrm{PG}(4,q)$.

Given a spread $\mathcal{S}$ let (α_i, β_i) for $i = 1, \ldots, q^2+1$ be such that the zeros of the polynomials $x^{q+1} - \alpha_i x + \beta_i$ correspond to the lines of $\mathcal{S}$, those zeros coming from the non-zero $(q-1)$-th powers in $\mathrm{GF}(q^4)$. We put $x = X^{q-1}$ and multiply through by X to recover the polynomial whose zeros correspond to a plane of $\mathrm{AG}(4,q)$. This plane together with its translations give the lines of $\pi(\mathcal{S})$; that is, the plane and its translations are given by the zeros in $\mathrm{GF}(q^4)$ of equations of the form

$$X^{q^2} - \alpha_i X^q + \beta_i X + \gamma = 0$$

together with a point P_i on l_∞ corresponding to the spread element given by the pair (α_i, β_i). One can verify that γ satisfies the equation $\gamma^{q^2} - \alpha_i^{q^2}\gamma^q - \beta_i^{-1}\gamma = 0$, and the q^2 solutions for γ together with l_∞ give the q^2+1 lines through the point P_i.

5.3 Symplectic spreads

For every point λ of $\mathrm{PG}(3,q)$, viewed as a (q^3+q^2+q+1)-th root of unity in $\mathrm{GF}(q^4)$, let

$$\omega_\lambda(x) := (-\lambda^{q^2})^{q^2+q+1}x^{q^2+q+1} + (-\lambda^{q^2})^{q+1}x^{q+1} + (-\lambda^{q^2})x + 1$$

be a polynomial over $\mathrm{GF}(q^4)$ whose zeros correspond to the points of a hyperplane. It is easy to verify that

$$\omega_\lambda(\epsilon)\lambda^{q+1} = \omega_\epsilon(\lambda)\epsilon^{q+1} \quad \text{and} \quad \omega_\lambda(\lambda) = 0$$

and it follows from this that ω defines a symplectic polarity on $\mathrm{PG}(3,q)$, see Dembowski [27, p. 43]. The planes $\omega_\lambda(x)$ and $\omega_\epsilon(x)$ intersect in the line given by the zeros of the equation

$$x^{q+1} + \epsilon\lambda\frac{(\lambda^{q+1} - \epsilon^{q+1})^q}{\lambda - \epsilon}x + \frac{\epsilon\lambda^{q+1} - \epsilon^{q+1}\lambda}{\epsilon - \lambda} = 0.$$

The line joining the points λ and ϵ is given by the zeros of the equation $x^{q+1} - \alpha x + \beta = 0$ where

$$\alpha = \frac{\lambda^{q+1} - \epsilon^{q+1}}{\lambda - \epsilon} \quad \text{and} \quad \beta = \frac{\lambda^{q+1}\epsilon - \epsilon^{q+1}\lambda}{\lambda - \epsilon}.$$

The two lines coincide whenever $\alpha^q\beta = -\alpha$ and the lines for which this condition hold are the totally isotropic lines, again see Dembowski [27, p. 41].

A *symplectic 1-spread* is a spread whose elements are totally isotropic and a translation plane arising from a symplectic 1-spread is called a *symplectic translation plane*.

6 Maximal arcs

Recall that a (k,r)-arc in π_2 is a set $\mathcal{K}$ of k points with at most r points on a line. A line through a point $P \in \mathcal{K}$ has at most $r-1$ other points of $\mathcal{K}$

and therefore $k \leq (r-1)(q+1) + 1 = rq - q + r$. If equality occurs then $\mathcal{K}$ is called a *maximal arc*. It follows that each line has either 0 or r points of a maximal arc. The *degree* of a maximal arc is r and the dual of the external lines to a maximal arc in π_2 is a maximal arc of degree q/r in the plane dual to π_2.

6.1 Constructions for q even

In $\mathrm{PG}(2,q)$, q even, R. H. F. Denniston constructed maximal arcs for all r dividing q [28]. Following this, J. A. Thas [55] constructed maximal arcs in certain symplectic translation planes from ovoids in $\mathrm{PG}(3,q)$, see [43] for a survey on ovoids in $\mathrm{PG}(3,q)$. In [31], [32] and [33] N. Hamilton proved that the construction of Thas works in derived dual translation planes. N. Hamilton and C. Quinn [35] constructed maximal arcs from m-systems of polar spaces incorporating both the constructions by J. A. Thas [55, 56]. For details on m-systems and partial m-systems see [51].

6.2 A construction of maximal arcs using polynomials

Theorem 6.1 *For q even, the distinct zeros of the polynomial*

$$M(x) := \mathrm{Tr}_{q^2 \to 2}(x^{q^2+1}) = x^{q^2+1} + x^{2(q^2+1)} + x^{4(q^2+1)} + \ldots + x^{(q^2/2)(q^2+1)}$$

in $\mathrm{GF}(q^4)$ form a maximal arc of degree $q^2/2$ in the translation plane of order q^2 arising from a symplectic 1-spread via the construction of André [1] and Bruck and Bose [20].

Proof Firstly let us prove that $M(x)$ has the correct number of distinct zeros in $\mathrm{GF}(q^4)$.

$$M(x)^2 + M(x) = x^{q^4+q^2} + x^{q^2+1} = x^{q^2}(x + x^{q^4})$$

and hence $M(x)/x^{q^2}$ divides $x + x^{q^4}$ and is therefore fully reducible into distinct linear factors over $\mathrm{GF}(q^4)$. Thus it has exactly $(q^2/2)(q^2+1) - q^2$ distinct zeros.

Secondly we have to prove that the distinct zeros of $M(x)$ have at most $q^2/2$ points on any line. This is sufficient by a simple counting argument; we do not need to prove equality. Any line of a translation plane of order q^2 is given by the zeros of polynomials of the form

$$L_{\alpha\beta\gamma}(x) := x^{q^2} + \alpha x^q + \beta x + \gamma.$$

We continue by reducing the polynomial $M(x)$ modulo $L_{\alpha\beta\gamma}(x)$. The characteristic is 2.

$$M(x) = \mathrm{Tr}_{q^2 \to 2}(x(\alpha x^q + \beta x + \gamma)) \pmod{L_{\alpha\beta\gamma}}$$

$$= \sum_{i=0}^{h-1} \alpha^{2^i q} x^{2^i q} (\alpha^{2^i} x^{2^i q} + \beta^{2^i} x^{2^i} + \gamma^{2^i}) + \beta^{q^2/2}(\alpha x^q + \beta x + \gamma) + \alpha^{q/2} x^{q^2/2 + q/2}$$

Polynomials in Finite Geometries

$$+ \text{ terms of degree at most } q^2/2 \pmod{L_{\alpha\beta\gamma}}$$
$$= (\alpha^{q^2/2}\beta^{q/2} + \alpha^{q/2})x^{q^2/2+q/2} + \text{ terms of degree at most } q^2/2 \pmod{L_{\alpha\beta\gamma}}.$$

Hence we have that for translation planes at most $q^2/2 + q/2$ of the distinct zeros of $M(x)$ lie on a line. Moreover if the corresponding spread is symplectic, $(\alpha^{q^2/2}\beta^{q/2} + \alpha^{q/2}) = (\alpha^q\beta + \alpha)^{q/2} = 0$, at most $q^2/2$ of the distinct zeros of $M(x)$ lie on a line. We have to check that this polynomial is not identically zero; this would imply that the the zeros of $M(x)$ contain all of the points on the line $L_{\alpha\beta\gamma}(x)$. However, the constant term in this reduction is

$$\gamma(\alpha^{q+1} + \beta^q)^{q/2} = \gamma\beta^{q/2(q^2+q+1)}$$

by (3) and the coefficient of x is

$$\gamma + \beta^{q^2/2+1} + \beta\alpha^{q^2/2+q/2} = \gamma + \beta^{q/2(q^2+q+1)},$$

again by (3). If both these are zero then $\alpha = \beta = \gamma = 0$ which is ridiculous.

This construction can be extended insofar as for each subfield $\mathrm{GF}(2^r)$ of $\mathrm{GF}(q^2)$ the distinct zeros of $\mathrm{Tr}_{q^2 \to 2^r}(x^{q^2+1})$ form a maximal arc of degree $q^2/2^r$ in the translation plane of order q^2 arising from a symplectic 1-spread. The proof is similar but involves slightly more calculations. N. Hamilton [34] pointed out that this construction yields the maximal arcs constructed by J. A. Thas [56].

6.3 Non-existence for q odd

The non-existence of maximal arcs in Desarguesian planes was a conjecture dating back to the 1960's. A. Cossu [26] proved the initial case $(r, q) = (3, 9)$ and J. A. Thas [57] proved non-existence for $(r, q) = (3, 3^h)$. The conjecture was proven initially in [6], but a shorter proof is given in [7].

Let us consider a sketch of the proof. For S a maximal $(rq - q + r, r)$-arc in $\mathrm{AG}(2, q)$, we consider S as elements of $\mathrm{GF}(q^2)$ as in Section 2. The polynomial $F(u, x)$ defined there will then have the property

$$F(u, x_0) = (1 - u^{q+1})^{r-1}$$

whenever $1/x_0$ is an element of S. This follows since every line through a point of the maximal arc has exactly $r - 1$ other points of S on it. The coefficient in $F(u, x)$ of u^r is $\chi_r(x)$, a polynomial of degree at most $r(q - 1)$ by definition. This polynomial is divisible by the locator polynomial $S(x)$ since $\chi_r(x_0) = 0$ for all $1/x_0 \in S$. However, $\chi_r(0) = \binom{rq-q+r}{r} = 1$, so clearly χ_r cannot be identically zero. After some calculations one can show that not only S divides χ_r but S^{p-1} divides χ_r as well. This then implies that χ_r has more zeros than its degree whenever $p > 2$ and is therefore identically zero, a contradiction which implies that maximal arcs do not exist in Desarguesian planes of odd order.

The only non-existence results for non-Desarguesian planes come from the exhaustive computer searches of T. Penttila and G. Royle [46] who searched all planes of order 9, and A. Blokhuis, N. Hamilton and H. Wilbrink [15] who showed that Thas' constructions [55, 56] do not extend to odd order planes.

6.4 Incompleteness results

The method used in [6] and [7] can be extended to show that large (k,r)-arcs in $\mathrm{PG}(2,q)$ can be extended to maximal arcs whenever (necessarily) r divides q. In the case q is odd this simply extends the non-existence to smaller arcs. More precisely we have the following, which appears in [8].

Result 6.2 *Let S be a $(rq - q + r - \varepsilon, r)$-arc in $\mathrm{PG}(2,q)$ where r divides q.*

1. *(q even.) For $\varepsilon < r/2$ and $q/r > 2$ or $\varepsilon < 0.381r$ and $q/r = 2$, S can be extended uniquely to some maximal arc containing $rq - q + r$ points.*

2. *(q odd.) If $q/r > 3$ then $\varepsilon > r/2$ and if $q/r = 3$ then $\varepsilon > 0.476r$.*

In what is essentially also a polynomial proof, although involving some basic properties of algebraic curves, Szőnyi [54] proved the following.

Result 6.3 *Let S be a $(pq - q + p - \varepsilon, p)$-arc in $\mathrm{PG}(2,q)$. Then for q odd $\varepsilon > q^{1/4}/2$.*

7 Unitals

A *unital* in π_2 of square order q is a set $\mathcal{U}$ of $q\sqrt{q}+1$ points, such that each line meets it in either 1 or $\sqrt{q}+1$ points. A line is a *tangent* or a *secant* of $\mathcal{U}$ if it contains 1 or $\sqrt{q}+1$ points of $\mathcal{U}$ respectively. A point P of $\mathcal{U}$ lies on one tangent and q secants, while a point Q not on $\mathcal{U}$ lies on $\sqrt{q}+1$ tangents and $q - \sqrt{q}$ secants. It follows that $\mathcal{U}$ has $q\sqrt{q}+1$ tangents and $q^2 - q\sqrt{q} + q$ secants, and that the set of tangents of $\mathcal{U}$ form a unital in the dual plane.

7.1 Unitals in $\mathrm{PG}(2,q)$

An example of a unital in $\mathrm{PG}(2,q)$ is given by the set of absolute points of a unitary polarity (see Hirschfeld [37, p. 36–37]). This is called a *classical unital* (or *Hermitian curve*), and any classical unital is the image under an element of $\mathrm{P}\Gamma\mathrm{L}(3,q)$ of the set of points (x_0, x_1, x_2) satisfying the equation

$$x_0^{\sqrt{q}+1} + x_1^{\sqrt{q}+1} + x_2^{\sqrt{q}+1} = 0.$$

In 1976, F. Buekenhout [24] proved the existence of unitals in every translation plane π of square order q with kernel containing $\mathrm{GF}(\sqrt{q})$. In particular, he noted that his construction gave a family of non-classical unitals in $\mathrm{PG}(2,q)$ for $\sqrt{q} > 2$ even and not a square. R. Metz [42], in 1979, extended this observation to the case of $\sqrt{q}$ even and square, and $\sqrt{q}$ odd; hence, for any prime power $\sqrt{q} > 2$, there exist non-classical unitals in $\mathrm{PG}(2,q)$. A *Buekenhout–Metz unital* in π is a unital which arises by the construction due to Buekenhout [24, Section 4, Remark (4)]. Since the classical unital in $\mathrm{PG}(2,q)$ can be constructed in this way, it is included in the class of Buekenhout–Metz unitals.

The following characterisation of Buekenhout–Metz unitals comes from [25, 48].

Result 7.1 *Let $\mathcal{U}$ be a unital in $\mathrm{PG}(2,q)$, where q is square. Then $\mathcal{U}$ is a Buekenhout–Metz unital if and only if there exists a point T of $\mathcal{U}$ such that the points of $\mathcal{U}$ on each of the q secants to $\mathcal{U}$ through T form a Baer subline.*

There are many other characterisations of Buekenhout–Metz and classical unitals in $\mathrm{PG}(2,q)$. However, the classification appears to be a very hard problem and still somewhat out of reach. It may well be that all unitals embeddable in $\mathrm{PG}(2,q)$ are Buekenhout–Metz unitals. The following result again uses the application of the polynomials in Section 2 and appears in [9].

Result 7.2 *Let $\mathcal{U}$ be a unital in $\mathrm{PG}(2,q)$, where $q = p^2$ and p is a prime. Then $\mathcal{U}$ is a classical unital if and only if it admits at least $(q-2)\sqrt{q}$ Baer sublines among its secants.*

7.2 Partial unitals

A *partial unital* is a $(k, \sqrt{q}+1)$-arc $\mathcal{X}$ such that each point of $\mathcal{X}$ lies on a tangent. It would be useful to know (and we are a long way from knowing) how large a partial unital can be, such that it is not part of a unital, i.e. cannot be extended to a unital. The current bound which appears in [3] and applies only to $\mathrm{PG}(2,q)$ again uses the polynomials from Section 2.

Result 7.3 *A partial unital $\mathcal{X}$ in $\mathrm{PG}(2,q)$ with*

$$q\sqrt{q} + 1 - \sqrt{q} < |\mathcal{X}| < q\sqrt{q} + 1$$

can be extended to a unital.

There is no evidence that this is the right lower bound however. Indeed, the only construction that I know for a partial unital that cannot be extended to a unital is the following. Let $\mathcal{U}$ be a unital and P a point not in the unital. If we remove a point of $\mathcal{U}$ on each of the lines through P and add the point P, the remaining $q\sqrt{q} - q + 1$ points form a partial unital and one can verify that this cannot be extended to a larger partial unital.

Acknowledgements

I thank Aart Blokhuis, Nick Hamilton, Dieter Jungnickel, Michel Lavrauw and Tamas Szőnyi for their helpful comments during the preparation of this paper.

References

[1] J. André, Über nicht-Desarguessche Ebenen mit transitiver Translationgruppe, *Mathematische Zeitschrift*, **60** (1954), 156–186.

[2] S. Ball, Multiple blocking sets and arcs in finite planes, *Journal of the London Mathematical Society*, **54** (1996), 581–593.

[3] S. Ball, Partial unitals and related structures in Desarguesian planes, *Designs, Codes and Cryptography*, **15** (1998), 231–236.

[4] S. Ball, On nuclei and blocking sets in Desarguesian spaces, *Journal of Combinatorial Theory, Series A*, to appear.

[5] S. Ball, On intersection sets in Desarguesian affine spaces, preprint.

[6] S. Ball, A. Blokhuis & F. Mazzocca, Maximal arcs in Desarguesian planes of odd order do not exist, *Combinatorica*, **17** (1997), 31–41.

[7] S. Ball & A. Blokhuis, An easier proof of the maximal arcs conjecture, *Proceedings of the American Mathematical Society*, **126** (1998), 3377–3380.

[8] S. Ball & A. Blokhuis, On the incompleteness of (k,n)-arcs in Desarguesian planes of order q where n divides q, *Geometriae Dedicata*, to appear.

[9] S. Ball, A. Blokhuis & C. M. O'Keefe, Unitals with many Baer sublines, *Designs, Codes and Cryptography*, to appear.

[10] A. Barlotti, *Some topics in finite geometrical structures*, Institute of Statistics, University of California, mimeo series 439 (1965).

[11] A. Blokhuis, Extremal problems in finite geometries, *Bolyai Society Mathematical Studies*, **3** (1991), 111–135.

[12] A. Blokhuis, Polynomials in finite geometries and combinatorics, in *Surveys in Combinatorics* (eds. K. Walker), *London Mathematical Society Lecture Note Series*, 187, Cambridge University Press, Cambridge (1993), pp. 35–52.

[13] A. Blokhuis, On nuclei and affine blocking sets, *Journal of Combinatorial Theory, Series A*, **67** (1994), 273–275.

[14] A. Blokhuis, On multiple nuclei and a conjecture of Lunelli-Sce, *Bulletin of the Belgian Mathematical Society*, **3** (1994), 349–353.

[15] A. Blokhuis, N. Hamilton & H. Wilbrink, The non-existence of Thas maximal arcs in translation planes of odd order, *European Journal of Combinatorics*, **19** (1998), 413–417.

[16] A. Blokhuis & F. Mazzocca, On maximal sets of nuclei in PG(2, q) and quasi-odd sets in AG(2, q), in *Advances in Finite Geometries and Designs* (eds. J. W. P. Hirschfeld, D. R. Hughes & J. A. Thas), Oxford University Press, Oxford (1991), pp. 27–34.

[17] A. Blokhuis & F. Mazzocca, Special point sets in PG(n, q) and the structure of sets with the maximal number of nuclei, *Journal of Geometry*, **41** (1991), 33–41.

[18] A. Blokhuis & H. A. Wilbrink, A characterization of exterior lines of certain sets of points in PG(2, q), *Geometriae Dedicata*, **23** (1987), 253–254.

[19] A. E. Brouwer & A. Schrijver, The blocking number of an affine space, *Journal of Combinatorial Theory, Series A*, **24** (1978), 251–253.

[20] R. H. Bruck & R. C. Bose, Linear representations of projective planes in projective spaces, *Journal of Algebra*, **1** (1966), 117–172.

[21] A. A. Bruen, Polynomial multiplicities over finite fields and intersection sets, *Journal of Combinatorial Theory, Series A*, **60** (1992), 19–33.

[22] A. A. Bruen & J. C. Fisher, The Jamison method in Galois geometries, *Designs, Codes and Cryptography*, **1** (1991), 199–205.

[23] A. A. Bruen & J. A. Thas, Flocks, chains and configurations in finite geometries, *Atti della Accademia dei Lincei*, **59** (1975), 744–748.

[24] F. Buekenhout, Existence of unitals in finite translation planes of order q^2 with a kernel of order q, *Geometriae Dedicata*, **5** (1976), 189–194.

[25] L. R. A. Casse, C. M. O'Keefe & T. Penttila, Characterizations of Buekenhout–Metz unitals, *Geometriae Dedicata*, **59** (1996), 29–42.

[26] A. Cossu, Su alcune proprietà dei $\{k; n\}$-archi di un piano proiettivo sopra un corpo finito, *Rendiconti di Matematica e delle sue Applicazioni*, **20** (1961), 271–277.

[27] P. Dembowski, *Finite Geometries*, Springer, Berlin (1968).

[28] R. H. F. Denniston, Some maximal arcs in finite projective planes, *Journal of Combinatorial Theory*, **6** (1969), 317–319.

[29] J. Doyen, Lecture at Oberwolfach, 1976.

[30] A. Gács, P. Sziklai & T. Szőnyi, Two remarks on blocking sets and nuclei in planes of prime order, *Designs, Codes and Cryptography*, **10** (1997), 29–39.

[31] N. Hamilton, Some maximal arcs in derived dual Hall planes, *European Journal of Combinatorics*, **15** (1994), 525–532.

[32] N. Hamilton, Some inherited maximal arcs in derived dual translation planes, *Geometriae Dedicata*, **55** (1995), 165–173.

[33] N. Hamilton, Some maximal arcs in Hall planes, *Journal of Geometry*, **52** (1995), 101–107.

[34] N. Hamilton, Personal communication, 1998.

[35] N. Hamilton & C. Quinn, m-systems of polar spaces and maximal arcs in projective planes, *Bulletin of the Belgian Mathematical Society*, submitted.

[36] U. Heim, Proper blocking sets in projective spaces, *Discrete Mathematics*, **174** (1994), 167–176.

[37] J. W. P. Hirschfeld, *Projective Geometries over Finite Fields*, Second edition, Oxford University Press, New York (1998).

[38] R. Jamison, Covering finite fields with cosets of subspaces, *Journal of Combinatorial Theory, Series A*, **22** (1977), 253–266.

[39] R. Lidl & H. Niederreiter, *Introduction to Finite Fields and their Applications*, Cambridge University Press, Cambridge (1986).

[40] G. Lunardon, Linear k-blocking sets, *Combinatorica*, submitted.

[41] J. R. M. Mason, A class of $((p^n - p^m)(p^n - 1), p^n - p^m)$-arcs in PG$(2, p^n)$, *Geometriae Dedicata*, **15** (1984), 355–361.

[42] R. Metz, On a class of unitals, *Geometriae Dedicata*, **8** (1979), 125–126.

[43] C. M. O'Keefe, Ovoids of PG$(3, q)$: a survey, *Discrete Mathematics*, **151** (1996), 175–188.

[44] O. Ore, On a special class of polynomials, *Transactions of the American Mathematical Society*, **35** (1933), 559–584.

[45] O. Ore, Contributions to the theory of finite fields, *Transactions of the American Mathematical Society*, **36** (1934), 243–274.

[46] T. Penttila & G. Royle, Sets of type (m, n) in the affine and projective planes of order nine, *Designs, Codes and Cryptography*, **6** (1995), 229–245.

[47] P. Polito & O. Polverino, Small blocking sets, *Combinatorica*, **18** (1998), 133–137.

[48] C. Quinn & L. R. A. Casse, Concerning a characterisation of Buekenhout-Metz unitals, *Journal of Geometry*, **52** (1995), 159–167.

[49] B. Segre, Ovals in a finite projective plane, *Canadian Journal of Mathematics*, **7** (1955), 414–416.

[50] B. Segre & G. Korchmáros, Una proprietà degli insiemi di punti di un piano di Galois caratterizzante quelli formati dei punti delle singole rette esterne ad una conica, *Atti della Accademia dei Lincei*, **62** (1977), 613–619.

[51] E. E. Shult & J. A. Thas, m-systems and partial m-systems of polar spaces, *Designs, Codes and Cryptography*, **8** (1996), 229–238.

[52] P. Sziklai, Nuclei of point sets in PG(n, q), *Discrete Mathematics*, **174** (1997), 323–327.

[53] T. Szőnyi, Some applications of algebraic curves in finite geometry and combinatorics, in *Surveys in Combinatorics* (ed. R. A. Bailey), *London Mathematical Society Lecture Note Series*, 241, Cambridge University Press, Cambridge (1997), pp. 197–236.

[54] T. Szőnyi, On the embeddability of (k, p)-arcs, *Designs, Codes and Cryptography*, submitted.

[55] J. A. Thas, Construction of maximal arcs and partial geometries, *Geometriae Dedicata*, **3** (1974), 61–64.

[56] J. A. Thas, Construction of maximal arcs and dual ovals in translation planes, *European Journal of Combinatorics*, **1** (1980), 189–192.

[57] J. A. Thas, Some results concerning $\{(q+1)(n-1); n\}$-arcs and $\{(q+1)(n-1)+1; n\}$-arcs in finite projective planes of order q, *Journal of Combinatorial Theory, Series A*, **19** (1975), 228–232.

Discrete Wiskunde
Technical University of Eindhoven
Den Dolech 2
5600 MB Eindhoven
The Netherlands
simeon@win.tue.nl

Applications of Combinatorial Designs to Communications, Cryptography, and Networking

C. J. Colbourn, J. H. Dinitz, D. R. Stinson

Summary Combinatorial designs have long had substantial application in the statistical design of experiments and in the theory of error-correcting codes. Applications in experimental and theoretical computer science have emerged, along with connections with the theory of cryptographic communication. This paper focuses on applications in the general area of communications, including cryptography and networking. Applications have been chosen to represent those in which design theory plays a useful, and sometimes central, role. Moreover, applications have been chosen to reflect in addition the genesis of new and interesting problems in design theory in order to treat the practical concerns. Of many candidates, thirteen applications areas have been included:

1. Optical orthogonal codes
2. Synchronous multiple access to channels
3. Group testing and superimposed codes
4. Erasure codes and information dispersal
5. Threshold and ramp schemes
6. Authentication codes
7. Resilient and correlation-immune functions
8. Multidrop networks
9. Channel graphs and interconnection networks
10. Partial match queries on files
11. Software testing
12. Disk layout and striping
13. (t, m, s)-nets and numerical integration

The theory of combinatorial designs continues to grow, in part as a consequence of the variety of these applications and the increasing depth of the connections with challenging problems on designs.

0 Background

The theory of combinatorial designs has a long and rich history. Its origins lie in somewhat specialized problems that arose in algebra, geometry, topology, and number theory. However, applications are found in the design of experiments [8] and in the theory of error-correcting codes [7]. Both fields of application served as sources for a wide variety of research directions.

In the past few decades, combinatorial design theory has grown to encompass a wider variety of investigations, many of which are not apparently motivated by any practical application. Rather they are motivated by a desire to obtain a coherent and powerful theory of existence and properties of designs. Nevertheless, it comes as no surprise that applications in experimental design and in coding theory continue to arise, and also that designs have found applications in new areas. Cryptography in particular has provided a new source of applications of designs, and simultaneously a source of new and challenging problems in design theory [9]. Across the spectrum of theoretical and experimental computer science, there are similar connections [3].

Arguably, many of the connections that arise are somewhat superficial, and appear to require only the translation of elementary combinatorial properties to the application domain. Naturally, the limited application does not then provide evidence of an important role for combinatorial design theory. However, we believe that there is ample evidence not only of superficial connections of theoretical investigations on designs to applications, but of deeper and more substantial connections. The importance of these connections cannot be overstated. While we can never know which results will find a genuine application, we expect to see such applications arise. Moreover, the evolution of the field depends largely upon its ability to make contributions both to other theories and to applications.

This paper presents some applications in which the connection with designs appears to be substantial. We have selected applications primarily from the area of communications, including cryptography and networking but avoiding for the most part the well understood connections with error-correcting codes. Our objective is to present evidence that combinatorial designs continue to arise in applications areas, often in unexpected ways; that the connections involve difficult aspects of the theory of designs; and that the applications motivate new research in design theory.

Accordingly, we do not provide an overview of design theory, but assume that the reader is familiar with the major topics in the area; see [1, 2, 4–6] for comprehensive treatments. In general, we attempt to give a sufficient introduction to the applications problem and then outline the connection with designs. Our interpretation of what is a 'combinatorial design' is a liberal one. There are numerous objects that do not share the general structure of a balanced set system or an array of symbols, yet they are an integral part of combinatorial design theory. We adopt an inclusive view, including for

Combinatorial Designs in Communications

example all of the specialized types of 'designs' discussed in [2].

1 Optical orthogonal codes

A fibre-optic channel must have the ability for multiple users to simultaneously share the channel without interference. In order to facilitate this, optical orthogonal codes were developed by Salehi [19]. Viewing these codes as sets of integers modulo n leads to interesting design theoretic questions.

1.1 The application

The study of optical orthogonal codes was first motivated by an application in a fibre-optic code-division multiple access channel. Many users wish to transmit information over a common wide-band optical channel. The objective is to design a system that allows the users to share the common channel. Other approaches have included frequency division, time division, collision detection or some type of network synchronization. Each required frequent conversions between the optical domain and the electrical domain. However, employing a code-division multiple access system with optical orthogonal codes reduces the complexity of the system, enabling implementation with available technology and with potentially higher transmission efficiency [15].

An $(n, w, \lambda_a, \lambda_c)$ *optical orthogonal code* (OOC), C, is a family of $(0,1)$-sequences of length n and weight w satisfying the following two properties (all subscripts are reduced modulo n):

1. $\sum_{0 \leq t \leq n-1} x_t x_{t+i} \leq \lambda_a$ for any $\mathbf{x} = (x_0, x_1, \ldots, x_{n-1})$ and any integer $i \not\equiv 0 \bmod n$ (the *auto-correlation property*);

2. $\sum_{0 \leq t \leq n-1} x_t y_{t+i} \leq \lambda_c$ for any $\mathbf{x} = (x_0, \ldots, x_{n-1})$, $\mathbf{y} = (y_0, \ldots, y_{n-1})$, and any integer $i \not\equiv 0 \bmod n$ (the *cross-correlation property*).

When $\lambda_a = \lambda_c = \lambda$ the code is an (n, w, λ) OOC. Research has concentrated on this case. To simplify the discussion of the model, let C be an $(n, w, 1)$ OOC with m codewords. The communications system can handle up to m simultaneous transmitters. Each transmitter is assigned one codeword from C, so that transmitter T_i is assigned codeword $\{s_1, s_2, \ldots, s_w\} = c_i \in C$ (s_j indicates the position of the jth 1 in the (0,1) sequence). At the transmitter, every information bit of a signal is encoded into a frame of n optical chips: If the information bit is 1, then in the corresponding frame (consisting of n optical chips), photon pulses are sent at exactly the s_1th, s_2th, ..., s_wth chips. In the other $n - w$ chips, no photon pulses are sent. If, however, the information bit is 0, then no photon pulses are sent in the corresponding frame (still consisting of n chips). For example, if transmitter T_i wishes to send the message 101, this gets encoded as the sequence of 3 frames of length $3n$ where photon pulses are sent at times $s_1, s_2, \ldots, s_w, 2n + s_1, 2n + s_2, \ldots, 2n + s_w$.

All m users are allowed to transmit at any time; there is no network synchronization required. At the receiving end, decoders are used to separate the transmitted signals. The decoder consists of a bank of m tapped delay-lines, one for each codeword (so say D_i is the decoder for transmitter T_i). These delay taps on decoder D_i are a (possibly null) cyclic shift modulo n of those on T_i.

Each tapped delay-line can effectively calculate the correlation of the received waveform with its signature sequence. By the properties of OOCs, the correlation between different signature sequences is low. The delay-line output is high only when the intended transmitter's information bit is a 1. In particular, the output is w when decoder D_i receives the information bit 1 from transmitter T_i and they are synchronized correctly. The output is s for some $s \leq m$ when D_i is not synchronized with T_i and T_i is sending a 1 or when T_i is sending a 0. Thus the receiver can effectively determine when the corresponding transmitter is transmitting an information bit of 1. "Bit stuffing" inserts a '1' in a prescribed manner after a specified number of consecutive '0's are transmitted, to ensure that the receiver can determine when long strings of 0's are sent.

OOCs consist of truly (0,1) sequences and are intended for environments that have no negative components. Most other correlation sequences are $(+1, -1)$ sequences intended for systems having both positive and negative components, which can be an important distinction [15].

Research has also been done on using optical orthogonal codes for multimedia transmission in fibre-optic LANs [18] and in multirate fibre-optic CDMA systems [17]. The mathematical theory is quite similar to that described here.

1.2 The connection to designs

A convenient way of viewing OOCs is from a set-theoretic perspective. An $(n, w, \lambda_a, \lambda_c)$ optical orthogonal code C can be considered as a family of w-sets of integers modulo n, in which each w-set corresponds to a codeword and the numbers in each w-set specify the nonzero bits of the codeword. The correlation properties can be rephrased in this set-theoretic framework. As an example, $C = \{1100100000000, 1010000100000\}$ is a (13,3,1) code with two codewords. In set theoretic notation, $C = \{\{0,1,4\},\{0,2,7\}\}$ mod 13. The code is equivalent to a (13, 3, 1) difference family in $\mathbb{Z}_{13}$ (which gives a Steiner triple system of order 13). Considering the set-theoretic interpretation of OOCs, it is to be expected that many of the constructions for OOCs are design theoretic in nature. This section outlines some constructions for OOCs that involve designs. To be consistent with design theoretic notation, we write $v = n$ and $k = w$, to speak of (v, k, λ) OOCs.

The main connection is to difference packings. Let $\mathcal{B} = \{B_1, B_2, \ldots, B_t\}$, where $B_i = \{b_{i1}, b_{i2}, \ldots b_{ik}\}, b_{ij} \in \mathbb{Z}_v, 1 \leq i \leq t$ and $1 \leq j \leq k$. The differences in $\mathcal{B}$ are $D = \{b_{ij} - b_{is} : 1 \leq i \leq t,\ 1 \leq j,\ s \leq k,\ j \neq s\}$. The pair $(\mathbb{Z}_v, \mathcal{B})$

is called a *cyclic difference packing* or $CP(v, k, \lambda)$ if the cardinality of D is exactly $\lambda k(k-1)t$ and $0 \notin D$. It can be easily verified that a (v, k, λ) cyclic difference packing gives a (v, k, λ) optical orthogonal code. A $CP(v, k, 1)$ is termed *g-regular* if the *difference leave* $(\mathbb{Z}_v \setminus D)$ along with 0 forms an additive subgroup of $\mathbb{Z}_v$ having order g. When $\lambda = 1$ a cyclic difference packing satisfies the bound $t \leq \lfloor (v-1)/(k(k-1)) \rfloor$, and is *optimal* if $t = \lfloor (v-1)/(k(k-1)) \rfloor$.

A similar bound pertains to OOCs. Let $\Phi(v, k, \lambda)$ denote the maximum number of codewords in a (v, k, λ) OOC. Analogous to the Johnson bound from coding theory (see [7]), we have:

Theorem 1.1 ([15]) $\Phi(v, k, \lambda) \leq ((v-1)(v-2) \cdots (v-\lambda))/(k(k-1)(k-2) \cdots (k-\lambda))$.

When $\lambda = 1$ this reduces to $\Phi(v, k, 1) \leq (v-1)/(k(k-1))$. When $|C| = \lfloor (v-1)/(k(k-1)) \rfloor$ the code is an *optimal OOC*.

Theorem 1.2 ([21]) *The existence of an optimal $(v, k, 1)$ OOC is equivalent to the existence of an optimal $(v, k, 1)$ cyclic difference packing.*

Hence results on (optimal) cyclic difference packings directly relate to results on (optimal) OOCs. When $k = 3$ and $\lambda = 1$, optimal OOCs arise from cyclic Steiner triple systems. In fact, when $v \equiv 1, 3 \pmod{6}$ they coincide. Chung, Salehi and Wei [15] solved all cases when $v \not\equiv 2 \pmod{6}$:

Theorem 1.3 $\Phi(v, 3, 1) = \lfloor \frac{(v-1)}{6} \rfloor$ *if* $v \not\equiv 2 \pmod{6}$.

The proof is by a direct construction of a $(v, 3, 1)$ cyclic difference packing and is very reminiscent of the construction of cyclic triple systems from Skolem sequences (see [4]). In a similar vein, Yin [21] summarized results concerning optimal $(v, k, 1)$ cyclic difference packings, and hence gives many optimal OOCs. Most of these have $k = 4$, but there are some results for other $k \leq 11$.

Another class of optimal OOCs comes from projective geometry. Chung, Salehi and Wei [15] used $PG(d, q)$ to construct a $(v, k, 1)$ OOC where $v = (q^{d+1} - 1)/(q-1)$ and $k = q+1$. Let α be a primitive element of $GF(q^{d+1})$ and say that $\log \beta = e$ if $\beta = \alpha^e$. Now, in the vector space $V(d+1, q)$ the nonzero vectors on a line ℓ through the origin are $\ell = \{\alpha^i, \alpha^{i+v}, \alpha^{i+2v}, \ldots, \alpha^{i+(q-2)v}\}$ where again $v = (q^{d+1} - 1)/(q-1)$. For any point $p \in PG(d, q)$, let $\log p$ denote the log of any vector on the line corresponding to p in $V(d+1, q)$ modulo v. Hence each line in the projective geometry corresponds to a subset of the integers modulo v.

Let a *cyclic shift* of a line L in $PG(d, q)$ be the set of points

$$\{p \colon \log p = 1 + \log p' \pmod{v} \text{ for some point } p' \in L\}.$$

The cyclic shift of a line is also a line in PG(d, q) so this creates a number of orbits of lines. If the number of lines in an orbit is v, then the orbit is termed *full*; otherwise it is a *short* orbit.

To construct an $(v, k, 1)$ OOC from the projective geometry, take one representative line from each full orbit and map each of these lines to the set of integers modulo v under the action of the log. These sets satisfy the autocorrelation and cross-correlation restrictions. The OOCs formed in this manner are optimal:

Theorem 1.4 *Suppose that q is an odd prime power, $v = (q^{d+1} - 1)/(q - 1)$ and $k = q + 1$. Then $\Phi(v, k, 1) = \lfloor \frac{(v-1)}{k(k-1)} \rfloor$.*

This construction can be extended to $\lambda > 1$ by using s-dimensional subspaces instead of 1-dimensional subspaces (lines).

A construction by Chen, Ge and Zhu [14] gives $(6v, 4, 1)$ optimal OOCs for infinitely many odd values of v. They construct these OOCs directly from skew starters. A *starter* in the cyclic group $\mathbb{Z}_v$ (v odd) is a set of unordered pairs $S = \{\{x_i, y_i\} : 1 \leq i \leq (v-1)/2\}$ which satisfies the two properties (1)$\{x_i : 1 \leq i \leq (v-1)/2\} \cup \{y_i : 1 \leq i \leq (v-1)/2\} = \mathbb{Z}_v \setminus \{0\}$ and (2) $\{\pm(x_i - y_i) : 1 \leq i \leq (v-1)/2\} = \mathbb{Z}_v \setminus \{0\}$. The starter is a *skew starter* if in addition it satisfies the property (3) $\{\pm(x_i + y_i) : 1 \leq i \leq (v-1)/2\} = \mathbb{Z}_v \setminus \{0\}$. Skew starters have been useful in the construction of Room squares, Hamiltonian path balanced tournament designs and other combinatorial designs. See [16] for a survey.

Assume that $\gcd(v, 6) = 1$ and that $S = \{\{x_i, y_i\} : 1 \leq i \leq (v-1)/2\}$ is a skew starter in $\mathbb{Z}_v$. In $\mathbb{Z}_v \times \mathbb{Z}_6$ (which is isomorphic to $\mathbb{Z}_{6v}$), let $c_i = \{(x_i, 0), (y_i, 0), (x_i + y_i, 1), (0, 4)\}$ for $1 \leq i \leq (v-1)/2$. They show that the set $C = \{c_1, c_2, \ldots, c_{(v-1)/2}\}$ forms an $(6v, 4, 1)$ optimal OOC. Using results on the existence of skew starters they derive:

Theorem 1.5 *There exists an optimal $(6v, 4, 1)$ OOC in $\mathbb{Z}_{6v}$ for all v such that $\gcd(v, 6) = 1$.*

Many combinatorial constructions for optimal $(v, k, 1)$ OOCs are given by Yin [21]. The following is a sample. For the definition of difference matrices see [2].

Theorem 1.6 *Suppose that there exists a g-regular $CP(v, k, 1)$, a (k, m)-difference matrix and an optimal $CP(gm, k, 1)$. Then there exists an optimal $(mv, k, 1)$ OOC.*

Yin also gave a recursive construction for optimal OOCs using group divisible designs.

Combinatorial Designs in Communications   43

2 Synchronous multiple access to channels

The previous section examined an application involving sharing an optical channel. In that context, users acted asynchronously, and hence the channel decoding involved cyclic shifts of codewords. This section examines an analogous problem. In this variant, however, rather than multiplexing by partitioning the channel's capacity into discrete time slots, we employ multiplexing based on available frequencies. This avoids some of the issues that arise in synchronization, but introduces some additional complexities.

2.1 The application

Each user is to be able to send one of m different messages in a channel, or can remain silent. The channel is capable of carrying any subset of v different pulses or tones simultaneously, and can be equipped with intensity detection devices that determine not only the presence of a particular tone, but also the intensity with which this tone was employed. The latter is usually measured in multiples of some basic nonzero intensity, and accurately distinguishes large variations in intensity. However, small variations are not considered to be significant, in order to allow for noise.

Each message for each user is mapped to a *codeword*, which indicates a selection of k of the v available tones (i.e. the scheme is *multi-tone*). When the transmitter is silent, no signal is sent. When active, the transmitter sends the combination of k tones corresponding to the desired message. Typically, few transmitters are active. As with optical orthogonal codes, a receiver must be able to detect the presence of a message from a particular transmitter. For this reason, interference resulting from the simultaneous transmission of two (or more) codewords is to be kept to a minimum. When, for example, every transmitter is assigned only one message, we require that codewords for two different users share at most one tone. If all users are assigned one message only, and there is no intensity detection, then the task of the receiver is precisely that of solving a nonadaptive group testing problem (see Section 3). Intensity detection enables us to determine (with some degree of accuracy) the number of users who have transmitted a particular tone. Hence the problem is the variant of nonadaptive group testing in which tests report not just the presence of a defective, but also the number of defectives (see [38, Chapter 5]). Codes for this type of "spread spectrum" signalling system are described in [22, 23, 26, 27].

The m-ary problem is described in some detail in [28, 29]. Transmitters have a collection of m different messages and can choose any one to send or remain silent. The intended application here is to signalling systems in which, despite the large number of users, traffic from each user is bursty (i.e. high volume but short duration). The design of the system optimizes the handling of traffic when a single user is active, but permits multiple access by a small number of users. We retain the requirement that codewords assigned to different users share at most one tone. We enforce in addition a requirement that two codewords associated with different messages used by the same transmitter share no tone at all. This "orthogonality" requirement permits the most accurate decoding when a single user is active.

To accommodate multiple users, again it is necessary to be able to determine which combination of messages is present in the channel. For this reason, it is generally considered to be a poorer signalling design if one tone is used much more often than another. A secondary, but still important, criterion is therefore that all available tones appear in approximately the same number of assigned codewords.

2.2 The connection to designs

When every transmitter has a unique codeword, a simple design theoretic problem arises. Associate with each of the v tones an element, and with each codeword a block which is a subset of k elements. Blocks then have the property that they intersect in at most one element, and hence no pair occurs in more than one block. The result is a packing of index one, block size k, and order v. Maximizing the number of transmitters requires simply the choice of a maximum packing.

However, even this basic signalling problem poses some difficulties. When multiple transmitters are active concurrently, the received signal is the union of the transmitted signals. With intensity detection, the received signal is the multiset union. In the former situation, our task is to recover from the union the constituent sets; as noted, this is a nonadaptive group testing problem. It is not known whether, by assuming the availability of multiset unions rather than of set unions, better packings can be found. Codes can be found for which multiset unions permit proper reception while set unions do not; however, in the case $k = 3$, the largest number of codewords can be realized by a code for which set unions suffice [24].

The multi-tone systems lead to difficult problems in design theory as well. We concentrate on the case when blocks are triples. A code to be used in a signalling system is necessarily a packing by triples. Suppose that the packing to be used is on v elements and has b triples. If every user is to be assigned m of the triples, we require that the m triples assigned form a partial parallel class (i.e. any two triples in the class are disjoint). Then $m \leq \lfloor v/3 \rfloor$, and the maximum number of users that can be supported cannot exceed $\lfloor b/m \rfloor$. A suitable code for s users consists of a packing on v elements with ms triples, partitioned into s partial parallel classes of size m. Among such packings, those in which every two elements appear in approximately the same number of triples are preferred.

In the case $m = \lfloor v/3 \rfloor$, these packings have been extensively studied. For example, when $v \equiv 3 \pmod{6}$, the solutions are Kirkman triple systems (i.e. resolvable Steiner triple systems) [29]. Indeed, the "frame" obtained by deleting a single element in a Kirkman triple system provides a solution when $v \equiv 2 \pmod 6$. When $v \equiv 0 \pmod 6$, nearly Kirkman triple systems provide solutions, and when $v \equiv 1 \pmod 6$, Hanani triple systems provide the codes; see [4] for more details about these types of triple systems.

Colbourn and Zhao [25] completed the solution when $v \equiv 4, 5 \pmod 6$, so that the determination of codes when the number of messages is maximum is complete. In the intended application, although it is plausible that the number of messages coincides with the maximum permitted, this is unlikely. The primary application is in systems employing a digital to analog conversion, so that a chunk of ℓ bits in an incoming datastream is converted to a message in the form of the k tones selected. Typically, then, we find that $m = 2^\ell$, so that m is a power of two. Zhao [28] observed that, beginning with a packing partitioned into maximum partial parallel classes, simple heuristics usually suffice to partition the same packing into more and shorter partial parallel classes. Despite this evidence that partitioning is more difficult when m is large, and the use of such partitions in forming codes for smaller m, the current state of affairs in our knowledge of triple systems is quite incomplete. To begin with, the existence of solutions for large values of m does not always ensure the existence of solutions for smaller values of m. The Kirkman triple system of order 9, for instance, admits a partition with 4 classes of size 3, but does not admit a partition with 6 classes of size 2. More importantly, no method with a performance guarantee appears to be available at present which permits us to massage a packing with large partial parallel classes into one that has smaller but more partial parallel classes. Indeed, the existence question given m, s, and v asking for a packing by triples on v elements and ms blocks which has a partition into s partial parallel classes of v blocks has been solved only for certain restricted cases [25, 28, 29]. Among these are included the cases for all small values of v when m is a power of two, which form the principal cases employed in the application.

3 Group testing and superimposed codes

This section examines a collection of applications to the design of codes for simultaneous communication, and to experimental design of pooling strategies. Du and Hwang [38] provides a much more detailed treatment.

3.1 The application

A population $\mathcal{P}$ of b items contains a number d of *defective* items, and the remaining $b - d$ items are *good*. Items can be pooled together for testing: for a subset $X \subseteq \mathcal{P}$, the *group test* reports "yes" if X contains one or more defective elements, and reports "no" otherwise. The objective is to determine, using a number of group tests, precisely which items are defective. When group tests are all undertaken in parallel, the problem is *nonadaptive*; otherwise it is *adaptive*. Then results from one or more tests are available while constructing further pools to be tested. Among adaptive testing methods, some operate in a limited number of stages or rounds.

Group testing was first studied in screening large populations for disease [37], and with the advent of large-scale HIV screening, it has grown in importance. It has also arisen in satellite communications [33, 48]. In this application, a large number of ground stations which rarely communicate share a satellite link. Rather than polling the ground stations individually, pools of the ground stations are formed as part of the system design. When the satellite enters a phase of accepting requests for reservations of time slots, it polls each pool and from the positive results on the pools it determines which ground stations wish to transmit. The satellite may have many positive responses within one pool, but detects only that there is at least one response. Hence, while cosmetically similar to the optical communication situation, this problem encounters unions rather than sums of colliding signals.

Another primary application arises in mapping genomes. In the human genome project, for example, information about long strings of genetic material is obtained by first forming a library of subsegments (*clones*) of the material. To determine where a particular sequence is located within the genetic material, we conduct a test to determine which of the clones it appears in. Pooling of different clones can be used [30, 31, 34].

A further application arises in the construction of frameproof codes, which are designed to avoid coalitions of users forging the signature of a user not in the coalition; see [45, 46].

3.2 The connection to designs

Let $\mathcal{P}$ be a set of b items, and let $\mathcal{X}$ be a collection of subsets of $\mathcal{P}$ corresponding to the group tests performed. Then $(\mathcal{P}, \mathcal{X})$ is a solution to the nonadaptive group testing problem if and only if, for any possible sets D_1 and D_2 of defective items, $\{X : D_1 \cap X \neq \emptyset, X \in \mathcal{X}\} = \{X : D_2 \cap X \neq \emptyset, X \in \mathcal{X}\}$ only if $D_1 = D_2$.

The dual of a solution $(\mathcal{P}, \mathcal{X})$ is a pair $(V, \mathcal{B})$, where the v group tests of $\mathcal{X}$ are in one-to-one correspondence with the points of V, and the b items are in correspondence with the blocks of $\mathcal{B}$ (for each item, the corresponding block contains the elements corresponding to the group tests containing the item). Typically $(V, \mathcal{B})$ is referred to as a solution to the group testing problem; the goal is to maximize the number of blocks (items tested) as a function of the number of points (group tests performed).

Often it is known with high probability that the number of defectives d does not exceed some threshold value p. In the *hypergeometric* problem, the number of defectives is assumed never to exceed p, and hence it is necessary that $(V, \mathcal{B})$ has the union of any two distinct sets, each containing at most p blocks, themselves distinct. In the *strict* problem, it is necessary to identify the set of defective items correctly when $d \leq p$ and to report when $d > p$. In the latter case, the specific set of defective items need not be determined, however.

Now consider a solution $(V, \mathcal{B})$ to the nonadaptive group testing problem with d defectives. Form a $|V| \times |\mathcal{B}|$ incidence matrix. This matrix has the property that the unions of two sets of at most d columns are distinct. The matrix is then called $\overline{d}$-*separable* [38], and the corresponding set system is d-*union free* [40, 41]. The columns of a $\overline{d}$-separable matrix form a superimposed code [39,42] which permits up to d simultaneously transmitted codewords to be unambiguously decoded. The decoding technique appears somewhat involved, because we could in principle be required to examine all unions of up to d columns. Hence a related family of matrices (or codes, or set systems) arises. If the incidence matrix contains no collection of d columns whose union covers a column not in the collection, then M is a d-*disjunct* matrix. If a disjunct matrix is employed, there is a simple decoding mechanism, observing that all codewords covered by the received union are 'positive'. Equivalently, we can alter the condition on the set system to require that it is d-*cover free*, i.e. that no union of d or fewer blocks contains another. Evidently, a d-cover free family is also d-union free.

Probabilistic bounds on the maximum numbers of blocks in cover free and union free families are available [38]; see [39, 44] for upper bounds for cover free families, and [43] for lower bound. See [36] for progress in the union free case. Erdős, Frankl, and Füredi [40] established that among cover free families with constant block size, the maximum is realized by a Steiner t-design $S(\ell, 2\ell - 1, m)$; indeed Balding and Torney [31] recommend the use of an $S(3,5,65)$ in a genetic application. For union free families, Frankl and Füredi [41] noted that Steiner triple systems give the largest 2-union free families when the block size is three; by permitting block size *at most* three, Vakil and Parnes [47] established a somewhat larger exact bound using group divisible designs with block size three.

In the *error correction* version of group testing, some group tests are permitted to report "false positives"; an *a priori* bound q on the number of such false positives is assumed. Balding and Torney [30] observed that $(V, \mathcal{B})$ is a solution to the strict group testing problem with threshold p and error correction for q false positives if and only if, for every union of p or fewer blocks, every other block contains at least $q+1$ points not in this union. Any packing $(V, \mathcal{B})$ of t-sets into k-sets having $k \geq p(t-1) + q + 1$ is a solution to the strict group testing problem with threshold p and error correction for q false positives. A Steiner system $S(t, 2t - 1, v)$ is a solution to the strict group testing problem with $p = 2$ and $q = 0$ that has the maximum number of blocks of any solution [30].

Finally, we consider the use of combinatorial designs in two-stage group testing. Here the objective in a first stage of pools is not to identify all defectives precisely, but rather to identify a small subset of the items which is guaranteed to contain all defective items. Frankl and Füredi call a family of sets d-*weakly union free* if, whenever two *disjoint* sets of blocks are chosen, each containing d or fewer blocks, their unions are distinct. A 2-weakly union free family with block size three provides pools for a group testing method for $d = 2$, in which a set of at most three potential defectives are identified [35]. Moreover, while union free families have no more blocks than a Steiner triple system has, weakly union free families can have twice as many blocks [41]. Chee, Colbourn, and Ling [35] established that certain twofold triple systems realize the bound. Not any twofold triple system forms a weakly union free family; four forbidden configurations of four blocks each must be avoided. Again, while the bound of Frankl and Füredi [41] suggests that designs can realize the maximum, the particular designs needed require additional structural properties [35]. Applications of designs in general in two-stage group testing appear to be just being explored; see [32] for useful observations.

4 Erasure codes and information dispersal

Reliability is a major concern in the design of large disk arrays. Hellerstein et al. [57] examined erasure-resilient codes that allow us to reconstruct the original data even in the presence of disk failures. A set systems view of the problem of constructing erasure-resilient codes leads to interesting extremal problems. Solutions to some of these problems are characterized by well-known combinatorial designs. In other instances, combinatorial designs give asymptotically exact solutions.

4.1 The application

There has been a sustained exponential advance in the density and performance of semiconductor technology. With this progress came faster microprocessors as well as larger and faster primary memory devices. Improvements in secondary storage systems, on the other hand, have not kept pace. While the performance of RISC microprocessors has been increasing by more than 50% per year [60], disk transfer rates, which depend on the speed of mechanical movements and magnetic media densities, have only improved by about 20% each year [53]. This phenomenon has transformed many computationally-bound applications to being I/O-bound. Indeed, Amdahl [51] predicted about three decades ago that, unless accompanied by corresponding increases in secondary storage performance, big increases in microprocessor performance can only bring about marginal improvements in overall system performance. This disparity has led to the consideration of parallelism as a means to speed up secondary storage systems. Several ideas have been proposed as to how parallelism can be exploited. The most important and successful is the *disk array architecture*.

The disk array architecture organizes many independent small disks into one large logical disk. Small disks are preferable to large ones because they have a lower cost and consume less power. For improved performance, disk arrays employ the concept of *data striping* (see Section 12) which spreads data to multiple disks. This allows both single and multiple I/O requests to be processed in parallel by separate disks, thus improving effective transfer rates. A further advantage of disk striping is uniform load balance. The more disks we have in a disk array, the higher the performance we obtain. Unfortunately, large disk arrays have low reliability. For low disk failure rates, the failure rate of a disk array is directly proportional to the number of disks it contains. Many applications, notably database and transaction processing systems, require both high throughput and high data availability of their storage systems. The most demanding of these applications require continuous operation, which in terms of a storage system requires the ability to satisfy all requests for data even in the presence of disk failures, and the ability to reconstruct the content of a failed disk onto a replacement disk, thereby restoring itself to a fault-free state. These require redundancy to tolerate disk failures. Disk arrays which incorporate redundancy have come to be known as *Redundant Arrays of Independent Disks* (RAIDs).

There are three primary types of disk failures. Failures of the first type, *transient errors*, arise from noise corruption and are dealt with by repeating the requests. Those of the second, *media defects*, are caused by permanent defects in material, and are detected and masked by the manufacturer. Failures of the last type are *catastrophic failures*, such as head crashes and failures of the disk controller electronics. When a disk suffers a catastrophic failure, its data is rendered unreadable, and is effectively erased. We therefore call such a disk failure an *erasure*. For convenience, we also call a set of k disk failures a *k-erasure*. Error-correcting codes can be used to correct erasures. However, components in disk arrays allow us to determine exactly where erasures have occurred (this distinction between errors and erasures was apparently first drawn by Elias [54]). It is possible to take advantage of this additional information to derive codes that are better than those based on error-correcting codes.

Hellerstein et al. [57] pioneered the study of erasure-resilient codes for large disk arrays. Earlier, Rabin [61] investigated erasure-resilient codes for information dispersal, but his codes are not particularly suited for disk array applications. Alon et al. [50] also studied erasure-resilient codes to combat bursty losses in packet-switched networks. The parameters of interest there are also different from those for disk arrays.

4.2 The connection to designs

Let $\mathbf{x} = (x_1, \ldots, x_n) \in \{0,1\}^n$. The *weight* of $\mathbf{x}$, denoted wt($\mathbf{x}$), is the number $\sum_{i=1}^{n} x_i$. The *support* of $\mathbf{x}$, denoted supp($\mathbf{x}$), is the set $\{i : x_i = 1\}$.

A *data stripe*, or simply *stripe*, is the minimum amount of contiguous user data allocated to one disk before any data is allocated to any other disk. The size of a stripe must be an integral number of sectors, and is often the minimum unit of update used by system software. Because of this, we can view each disk as a collection of (disjoint) stripes.

An $[n, c, k]$-*erasure-resilient code*, or briefly an $[n, c, k]$-*ERC*, consists of an encoding algorithm $\mathcal{E}$ and a decoding algorithm $\mathcal{D}$ with the following properties. Given an n-tuple S of stripes, $\mathcal{E}$ produces an $(n + c)$-tuple or *codeword* $\mathcal{E}(S) = (\mathcal{E}_1(S), \ldots, \mathcal{E}_{n+c}(S))$ of stripes such that for any $I \subseteq \{1, \ldots, n\}$, where $|I| = n + c - k$, the decoding algorithm $\mathcal{D}$ is able to recover S from $(I, \{\mathcal{E}_i(S) \mid i \in I\})$. We often call an $[n, c, k]$-ERC a k-ERC when the parameters n and c are not important in the context.

To see the relevance of an $[n, c, k]$-ERC to the protection of data loss in a RAID, suppose that we have a piece of data which is partitioned into an n-tuple S of stripes. Given an $[n, c, k]$-ERC, we encode S into a codeword $(\mathcal{E}_1(S), \ldots, \mathcal{E}_{n+c}(S))$, and for $1 \leq i \leq n + c$, store $\mathcal{E}_i(S)$ on disk i of a disk array with $n + c$ disks. The definition of an $[n, c, k]$-ERC ensures that we can reconstruct the original data in the presence of up to k erasures.

For performance reasons, the erasure-resilient codes studied are assumed to satisfy two conditions:

1. We restrict ourselves to *systematic* codes. An $[n, c, k]$-ERC is *systematic* if $\mathcal{E}_i(S) = S_i$, for $1 \leq i \leq n$, where $S = (S_1, \ldots, S_n)$. The stripes $\mathcal{E}_i(S)$, for $n < i \leq n + c$, are called *checks*. This means that the encoding function leaves the data unmodified on some disks. This property is desirable to avoid read penalties associated with decoding when there are no disk failures.

2. We restrict ourselves to *linear* codes over the field GF_{2^L}, where L is the bit-size of a stripe. In this case, we interpret a stripe as an L-dimensional vector over GF_2, and $\mathcal{E}$ is a linear function. Hence, computations used to encode a stripe are restricted to component-wise modulo two arithmetic, that is, the parity operation $\oplus$. This restriction ensures that encodings and manipulations can be performed efficiently.

Restriction (1) above allows us to separate disks into *information disks*, which contain the original data, and *check disks*, which contain the checks. In fact, restrictions (1) and (2) imply that an $[n, c, k]$-ERC can be described in terms of a $c \times (n+c)$ matrix $H = [C \mid I]$ over GF_2, where I is the $c \times c$ identity matrix and C is a $c \times n$ matrix that determines the equations for the checks. This is a well-known result in the theory of error-correcting codes [7]. The matrix H is called the *parity-check matrix* of the code. Given the parity-check matrix $H = [C \mid I]$ of a k-ERC, we can think of the rows of C (as well as the rows and columns of I) as being indexed by the check disks of a disk array, and the columns of C as being indexed by the information disks. The content of check disk i is the modulo two sum of the content of those information disks, whose columns they index in C have a one in row i.

The following are some measures of an erasure-resilient code that are important for disk arrays.

Check disk overhead: This is the ratio of the number of check disks to information disks. An $[n, c, k]$-ERC has a check disk overhead of c/n.

Update penalty: This is the number of check disks whose content must be changed when an update is made in the content of a given information disk. We call these disks the *disks associated with the information disk*. If m check disks need to be involved in every write, then the parallelism of the disk array is reduced by a factor of $m + 1$. Since parallelism is the reason behind using disk arrays, update penalties should be kept as small as possible. The update penalties of an erasure-resilient code with parity-check matrix $H = [C \mid I]$ are the column sums of C.

Group size: This is the number of disks that must be accessed during the reconstruction of a single failed disk. The cost of reconstruction makes small group size desirable, while for load balancing reasons, uniform group size is desirable. The group sizes of an erasure-resilient code are the row sums of its parity-check matrix.

Since updates of data are usually much more frequent than the reconstruction of data lost in erasures, the update penalties are typically of more concern than the group sizes.

Suppose $H = [C \mid I]$ has a set of k or fewer linearly dependent columns (over GF_2). The failure of the corresponding disks makes reconstruction of data impossible. In fact, this is the only situation in which disk failures are irrecoverable [57]. It follows that H is the parity-check matrix of a k-ERC if and only if every set of k columns of H contains no nonempty set of linearly dependent columns. Precisely the same condition determines when H is the parity-check matrix of a k-error-detecting code [7].

This equivalence between k-ERCs and k-error-detecting codes means that results on error-detecting codes can be brought to bear. However, the study of codes for error detection has not focused on the same metrics. Indeed, as observed in [57], many of these codes are not suitable for disk array applications because they have large update penalties. If an erasure-resilient code is able to correct all k-erasures, then every update must affect the content of at least $k+1$ disks (one information disk and k check disks). Thus the update penalties of a k-ERC are at least k. In view of the importance of minimizing update penalties, we consider from here on only those k-ERCs for which the update penalties are all equal to k, the minimum possible. We speak, therefore, of the *update penalty*, instead of the update penalties of an erasure-resilient code. The corresponding parity-check matrix $H = [C \mid I]$ has column sums for C all equal to k.

A $(k+1)$-erasure is irrecoverable if it corresponds to the failure of an information disk and its k associated check disks. We call such $(k+1)$-erasures *bad*. With update penalty k, we can nonetheless hope to correct *all* $(k+1)$-erasures, except for bad ones [57]. In fact, it can happen that all t-erasures for some $t > k$ are recoverable except for those that contain bad $(k+1)$-erasures. A t-erasure, $t \geq k+1$, is *bad* if it includes the failure of an information disk and all of its k associated check disks. We extend the definition of ERCs to encompass this notion of higher resilience. An $[n, c, k, \ell]$-*ERC* is an $[n, c, k]$-ERC which can correct all t-erasures, for $k+1 \leq t \leq \ell$, except for bad t-erasures.

An alternative view of an $[n, c, k, \ell]$-ERC is that it is an erasure-resilient code with update penalty k that is able to correct all t-erasures, $t \leq \ell$, except bad ones. We often write (k, ℓ)-ERC for $[n, c, k, \ell]$-ERC when the parameters n and c are not important in the context. Requirements for higher reliability of disk arrays make (k, ℓ)-ERCs attractive. A (k, k)-ERC is simply a k-ERC.

Lemma 4.1 ([52]) *$H = [C \mid I]$ is the parity-check matrix of a (k, ℓ)-ERC if and only if for every t columns, $\mathbf{c}_1, \ldots, \mathbf{c}_t$ of C, where $2 \leq t \leq \ell$, the vector $\mathbf{x} = \bigoplus_{i=1}^{t} \mathbf{c}_i$ has weight at least $\ell + 1 - t$.*

Given c, k, and ℓ, define $F(c, k, \ell)$ to be the maximum n such that there exists an $[n, c, k, \ell]$-ERC. The maximum number of information disks that can be supported by c check disks is $F(c, k, \ell)$, if an update penalty of k is required, and we are to correct all t-erasures, $t \leq \ell$, except for bad ones. The important problem is: For given k and ℓ, determine the behaviour of $F(c, k, \ell)$ with respect to c; and construct $[n, c, k, \ell]$-ERCs having n as close to $F(c, k, \ell)$ as possible. An $[n, c, k, \ell]$-ERC with $n = F(c, k, \ell)$ is said to have *optimal check disk overhead*. We also abbreviate $F(c, k, k)$ to $F(c, k)$.

A set system $(X, \mathcal{A})$ *contains* a configuration $(Y, \mathcal{B})$ if there exists $Z \subseteq X$ and $\mathcal{C} \subseteq \mathcal{A}$ such that $(Z, \mathcal{C})$ is isomorphic to $(Y, \mathcal{B})$. If $(X, \mathcal{A})$ does not contain $(Y, \mathcal{B})$, then $(X, \mathcal{A})$ *avoids* $(Y, \mathcal{B})$. In this case, we also call $(Y, \mathcal{B})$ a *forbidden configuration* of $(X, \mathcal{A})$.

The *symmetric difference* of two sets A and B is denoted by $A \triangle B$. A *Túran-type problem* takes the form: Given a family $\mathcal{F}$ of configurations, determine the maximum number of blocks in a (k-uniform) set system of order n that avoids all the configurations in $\mathcal{F}$. We now explain the role of Túran-type problems in the design of erasure-resilient codes.

Given any matrix $M \in \{0, 1\}^{m \times n}$, define a set system $(X, \mathcal{A})$, where $X = \{1, \ldots, m\}$ and $\mathcal{A}$ contains precisely the supports of the columns of M. We call $(X, \mathcal{A})$ *the set system of M*.

Let $H = [C \mid I]$ be the parity-check matrix of an erasure-resilient code. We also call the set system of C *the set system of the erasure-resilient code*. If $(X, \mathcal{A})$ is the set system of an $[n, c, k, \ell]$-ERC, then $(X, \mathcal{A})$ is k-uniform, $|X| = c$, and $|\mathcal{A}| = n$. Therefore, the check disk overhead is $|X|/|\mathcal{A}|$, and the group sizes are one more than the replication numbers. This correspondence between set systems and parity-check matrices gives rise to Túran-type problems in erasure-resilient codes.

Lemma 4.2 ([52]) $(X, \mathcal{A})$ *is the set system of a* (k, ℓ)-*ERC if and only if for any* $2 \leq t \leq \ell$, *there do not exist* t *blocks* $A_1, \ldots, A_t \in \mathcal{A}$ *such that* $|\triangle_{i=1}^{t} A_i| \leq \ell - t$.

The construction of a (k, ℓ)-ERC with optimal check disk overhead is precisely the Túran-type problem of determining the maximum number of blocks in a set system satisfying the condition of Lemma 4.2.

When considering (k, ℓ)-ERCs, nontrivial cases arise only when $\ell \leq 2k - 1$. To see this, let $(X, \mathcal{A})$ be the set system of a (k, ℓ)-ERC. If $\mathcal{A}$ contains two blocks A and A' with nonempty intersection, then $|A \triangle A'| \leq 2k - 2$. By Lemma 4.2, $\ell - 2 < 2k - 2$, and so $\ell \leq 2k - 1$. Hence if $\ell \geq 2k$, then $\mathcal{A}$ must consist of pairwise disjoint blocks.

Chee, Colbourn and Ling [52] established that, for general k and ℓ with $1 \leq k \leq \ell$, there exist positive constants a_1 and a_2 such that

$$a_1 c^{(2k+1-\ell)/4} \leq F(c, k, \ell) \leq a_2 c^{k+1-\lfloor \ell/2 \rfloor},$$

for c a positive integer. The upper bound arises from the easy observation that the set system of a (k, ℓ)-ERC is a packing of strength $t = k + 1 - \lfloor \frac{\ell}{2} \rfloor$. Indeed, every packing with strength $t = 2$ and block size k underlies a k-ERC [57], but the same statement does not extend to strength $t = 3$, no matter how large the block size is [52].

The stronger connection with designs arises for particular small values of k. The set system of a 2-ERC is a graph, and indeed any graph without multiple edges suffices; hence $F(c, 2) = \binom{c}{2}$ using the complete graph. For such a 2-ERC to be a (2,3)-ERC, the graph cannot contain any triangles and hence the determination of $F(c, 2, 3)$ is the same as the maximum number of edges in a triangle-free graph on c vertices. This maximum is well known to be $\lfloor \frac{c}{2} \rfloor \cdot \lceil \frac{c}{2} \rceil$.

When $k = 3$, every set of triples forms a set system of a 3-ERC and hence $F(c, 3) = \binom{c}{3}$. However, the case of (3,4)- and (3,5)-ERCs are more interesting. The set systems associated with such ERCs are packings of strength two, and some classes of Steiner triple systems such as those arising from affine spaces do yield ERCs [57]. Chee, Colbourn, and Ling [52] noted that a 2-packing on c points underlies a (3,4)- or (3,5)-ERC if and only if it does not contain a *quadrilateral*, or *Pasch configuration*; this configuration consists of four blocks isomorphic to $\{\{a, c, e\}, \{a, d, f\}, \{b, c, f\}, \{b, d, e\}\}$. The existence of Steiner triple systems which are quadrilateral-free has been the subject of extensive study; see Ling et al. [59] for a statement of the current situation.

For $(4, \ell)$-ERCs with $4 \leq \ell \leq 7$, further classes of designs arise. When $\ell \in \{4, 5\}$, it appears that Steiner quadruple systems (SQSs) avoiding a large number of small configurations *might* exist, to reach the upper bound $F(c, 4, \ell) \leq \frac{c(c-1)(c-2)}{24}$, but the existence of such SQSs remains open (although the growth rate is known to be $O(c^3)$ when $\ell \in \{4,5\}$ [52]). When $\ell \in \{6,7\}$, block designs of block size four and index one arise. Numerous small examples of suitable designs establish certain situations in which $F(c,4,7) = \frac{c(c-1)}{12}$ [52]. Although one recursive construction is available, the existence question for such designs remains far from solved.

Related questions arise as a result of additional requirements in the application. Of particular note is the desire for scalability, such as bringing more disks online as they become available. In practice, this leads to imbalances in the disk load and hence degrades performance. Hence erasure codes whose underlying set systems are *resolvable* are of interest. One challenging question motivated by this is the existence of quadrilateral-free resolvable Steiner triple systems; currently, at least one third of the relevant cases are settled affirmatively [52].

The study of erasure codes suggests another major reason for the study of designs (and packings) with forbidden configurations.

5 Threshold and ramp schemes

The idea of distributed trust mechanisms is pervasive in information security and cryptography. This concept is realized using threshold schemes and various generalizations, such as secret sharing schemes for general access structures, and ramp schemes, to name two examples.

5.1 The application

Threshold schemes were invented independently by Blakley [62] and Shamir [69]. Let t and w be positive integers such that $t \leq w$. Informally, a (t,w)-*threshold scheme* is a method of splitting a *secret* into w shares, in such a way that the secret can be reconstructed from any t of the w shares, but no information about the secret is revealed by any $t-1$ shares. The value t is the *threshold* of the scheme. The secret is taken to be an element of a specified finite set, $\mathcal{K}$, and each share is an element of a finite set $\mathcal{S}$.

A (t,t)-threshold scheme can easily be constructed by taking $\mathcal{K} = \mathcal{S} = G$, where G is an additive abelian group. In this scheme, the secret is just the sum of the t shares.

Threshold schemes can be used in many common situations in which distributed trust is desirable. For example, the secret could be an access code for a secure area, a PIN (personal identification number) or password to enable a specified action, or a cryptographic key to be used for either signing or encrypting data. In many scenarios, it is not prudent to trust any one individual with sensitive information, and hence it is preferable that the secret information can be obtained only through the co-operation of a specified threshold of t out of w people. For example, access to a bank vault may require two out of three vice presidents; a (2,3)-threshold scheme could be used to set up an access mechanism of this type.

A threshold scheme specifies the method by which the secret is split into shares (a *share generation algorithm*) and the method used to determine the secret given t shares (a *share reconstruction algorithm*). The share reconstruction could be done either by the holders of the shares themselves, or, alternatively, by a trusted machine which is given the values of the shares (and possibly the identities of the individuals holding the shares). The second situation may be more desirable in many applications because it permits the secret to be "re-used" (provided that the value of the secret is not revealed by the machine that computes it, or by any subsequent action that is enabled as a result of the secret being reconstructed).

Threshold schemes are also commonly used as components in other cryptographic protocols. We mention a couple of illustrative examples. Our first concerns key escrow and key recovery. The idea of *key escrow* is to split a cryptographic key into shares, typically using a $(2,2)$-threshold scheme, which are stored securely in different locations. If at some later time, the individual possessing the key is suspected of criminal activity, then a law enforcement agency could obtain the shares, reconstruct the key, and decrypt any communications of the suspected individual. The use of the threshold scheme makes it difficult for someone to obtain the key who is not authorized to do so. (The entire idea of key escrow is very controversial. Law enforcement agencies and governments are strongly in favour of the idea, whereas most industry and civil liberties organizations are strongly opposed to it.) The idea of *key recovery* is similar and also involves splitting a secret key into shares using a threshold scheme. However, the motivation is different—a key recovery system is usually implemented by an individual or company in order to be able to recover (its own) lost keys. For example, it could be disastrous if keys that are used to store encrypted data are lost, and thus a key recovery scheme is essential.

Another application of $(2,2)$-threshold schemes is in *electronic cash*, in order to prevent *double spending*. Various schemes have been proposed in which the electronic cash is anonymous (much like regular cash). In contrast to a debit card, the cash is not withdrawn at the time of purchase, and the identity of the person spending the cash need not be known. Since electronic cash is nothing more than information, the problem of double spending arises: How do we prevent someone from spending the same electronic coin twice? One solution to this problem uses threshold schemes. When the coin is spent, a share of a $(2,2)$-threshold scheme is revealed. If the coin is spent twice, then the two shares revealed can be used to determine the identity of the person holding the coin. Thus anonymity is maintained as long as the coin is not spent twice.

Sometimes a "sharp" threshold, t, is not needed to secure a secret. This motivates the idea of a ramp scheme, introduced by Blakley and Meadows [63]. Let t_0, t_1 and w be non-negative integers such that $t_0 < t_1 \leq w$. A (t_0, t_1, w)-

ramp scheme is a method of splitting a secret into w shares, in such a way that the secret can be uniquely reconstructed from any t_1 of the w shares, but no information about the secret is revealed by any t_0 shares. Thus a (t, w) threshold scheme is nothing more than a $(t-1, t, w)$-ramp scheme. The reason for the term "ramp scheme" is as follows. Typically, as the number of shares is increased from t_0 to t_1, the amount of information about the secret gradually increases too.

In a $(0, t_1, w)$-ramp scheme, there is no security requirement. The resulting scheme is the same as an information dispersal algorithm (see Section 4). Thus ramp schemes interpolate between information dispersal schemes and threshold schemes.

So far, we have regarded the secret as an element of an arbitrary finite set, $\mathcal{S}$. Naor and Shamir [68] introduced the appealing idea of *visual cryptography* where the secret and each share consists of a collection of black and white (transparent) pixels on a transparency. The secret reconstruction algorithm in this scheme consists of stacking the transparencies, so the "computation" is performed by the human visual system!

5.2 The connection to designs

Threshold schemes are closely related to orthogonal arrays. Suppose that M is an $OA_1(t, w+1, v)$ on alphabet A. Thus M is a $(w+1) \times v^{t+1}$ array in which any t rows contain any t-tuple of symbols in exactly one column. The array M can be used to construct a (t, w)-threshold scheme in which $\mathcal{S} = \mathcal{K} = A$. Label the rows of M by the elements in $\{1, \ldots, w+1\}$. Each column r of M corresponds to a *distribution rule* as follows: The element $M(w+1, r)$ is the secret, and the w shares are the elements $M(i, r)$, $1 \leq i \leq w$.

The share generation algorithm works as follows. Given a secret $K \in \mathcal{K}$, a random distribution rule r is chosen such that $M(w+1, r) = K$ (there are v^{t-1} such rules to choose from, given K). Then the chosen rule r is used to determine the w shares.

The share reconstruction algorithm is straightforward. Suppose we are given the values of t shares, say $s_{i_j} = a_j$, $1 \leq j \leq t$, where $1 \leq i_1 < \ldots < i_t \leq w$. There is a unique column r of the orthogonal array such that a_j appears in row i_j for all j, $1 \leq j \leq t$. Hence, $M(w+1, r)$ is revealed as the secret.

It is also easy to see that a list of $t-1$ shares leaves the secret completely undetermined. In fact, given any possible guess K_0 as to the value of the secret, there is exactly one column r of the orthogonal array that is consistent with the $t-1$ given shares and such that $M(w+1, r) = K_0$.

It can be proved easily that $|\mathcal{S}| \geq |\mathcal{K}|$ in any (t, w)-threshold scheme. The construction described above provides schemes in which this bound is met with equality, i.e. in which the shares are as small as possible. A result of Martin [67] shows that the converse also holds:

Theorem 5.1 *There exists a (t, w)-threshold scheme in which $|\mathcal{S}| = |\mathcal{K}| = v$*

if and only if there exists an $OA_1(t, w+1, v)$.

We stated above that $|\mathcal{S}| \geq |\mathcal{K}|$ in any (t, w)-threshold scheme. This result can be easily generalized to ramp schemes, as follows:

Theorem 5.2 $|\mathcal{K}| \leq |\mathcal{S}|^{t_1-t_0}$ in any (t_0, t_1, w)-ramp scheme.

Ramp schemes meeting this bound can also be constructed using orthogonal arrays:

Theorem 5.3 If there exists an $OA_1(t, w + t_1 - t_0, v)$, then there exists a (t_0, t_1, w)-ramp scheme in which $|\mathcal{S}| = v$ and $|\mathcal{K}| = v^{t_1-t_0}$.

These and other results on ramp schemes are in [66].

The schemes based on orthogonal arrays require knowing which share is held by which participant in order to reconstruct the secret. This is because the share reconstruction algorithm needs to be given the t relevant rows of the orthogonal array (which correspond to the identities of the t participants reconstructing the secret). Stinson and Vanstone [71] considered *anonymous* schemes, in which the values of the shares alone are sufficient to reconstruct the secret. They gave a construction for anonymous schemes based on partitionable Steiner systems. A Steiner system $S(t, w, v)$ is *partitionable* if its block set can be partitioned into $N = (v - t + 1)/(w - t + 1)$ Steiner systems $S(t - 1, w, v)$. Examples of partitionable Steiner systems include resolvable 2-designs (the case $t = 2$), and large sets of Steiner triple systems (the case $t = w = 3$).

A partitionable Steiner system can be used to construct an anonymous (t, w)-threshold scheme, as follows. The systems $S(t - 1, w, v)$ are named $\mathcal{D}_i$, $1 \leq i \leq N$, the set of secrets is $\{1, \ldots, N\}$, and the set of shares is the set of points of the designs. If the secret to be shared is i_0, then a random block B of the system $\mathcal{D}_{i_0}$ is chosen, and the w shares are the w points in B. Any t shares determine B uniquely, and hence the secret is determined because i_0 is the unique value i such that B is a block in $\mathcal{D}_i$. On the other hand, if $t - 1$ shares are given, then, for every i, there is a unique block in $\mathcal{D}_i$ that contains the $t - 1$ shares. Thus the secret is completely undetermined.

Visual cryptography gives another example of anonymous schemes, since the reconstruction algorithm consists simply of stacking transparencies, and does not make use of the identities of the participants in the scheme. Realizing a visual cryptography scheme requires expanding each pixel into some number $m \geq 2$ of subpixels in each of the w shares. A reconstructed pixel (obtained by superimposing t transparencies) is not necessarily totally black or white. In general, there is a loss of contrast, which is measured by the difference in darkness between a reconstructed black and white pixel. Our goal is to maximize the contrast while minimizing the pixel expansion.

To illustrate, we describe the construction of the visual $(2,2)$-threshold scheme by Naor and Shamir [68]. Figure 5.2 illustrates the scheme, by specifying the algorithm for encoding one pixel. (This algorithm is to be applied for every pixel P in the image I in order to construct the two shares.) A pixel P is split into $m = 2$ subpixels in each of the two shares. If the given pixel P is white, then a random choice of one of the first two rows of Figure 5.2 is made. If the given pixel P is black, then a random choice of one of the last two rows of Figure 5.2 is made. Then the pixel P is encrypted as two subpixels in each of the two shares, as determined by the chosen row in Figure 5.2.

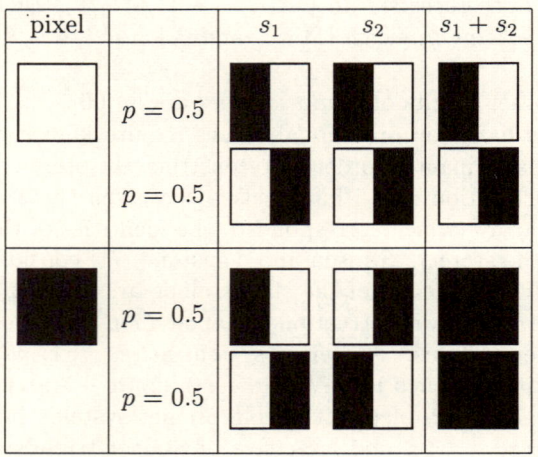

Figure 1: A 2-out-of-2 visual threshold scheme

To convince ourselves that the scheme works as desired, we first consider the security condition. Suppose we turn our attention to a pixel P in the share s_1. One of the two subpixels in P is black and the other is white. Moreover, each of the two possibilities black-white and white-black is equally likely to occur, independent of whether the corresponding pixel in the secret image I is black or white. Thus the share s_1 gives no clue as to whether the pixel is black or white. The same argument applies to the share s_2. Since all the pixels in I were encrypted using independent random coin flips, there is no information to be gained by looking at any group of pixels on a share, either. This demonstrates the security of the scheme.

Now consider what happens when we superimpose the two shares (here we refer to the last column of Figure 5.2). Consider one pixel P in the image I. If P is black, then we get two black subpixels when we superimpose the two shares; if P is white, then we get one black subpixel and one white subpixel when we superimpose the two shares. Thus we could say that the reconstructed pixel (consisting of two subpixels) has a grey level of 1 if P is black, and a grey

Combinatorial Designs in Communications

level of $1/2$ if P is white. There is a 50% loss of contrast in the reconstructed image, but it should still be visible.

In the above scheme, we say that the contrast is $\gamma = 0.5$. It was shown in [64] that, in any visual $(2,w)$-threshold scheme, it holds that $\gamma \leq \gamma^*(w)$, where
$$\gamma^*(w) = \frac{\left\lceil \frac{w}{2} \right\rceil \left\lfloor \frac{w}{2} \right\rfloor}{w(w-1)}.$$

If we examine the behaviour of the function $\gamma^*(w)$, we see that $\gamma^*(w) > 1/4$ for all $w \geq 2$, and $\lim_{w \to \infty} \gamma^*(w) = 1/4$. This raises the question if schemes can be constructed for all $w \geq 2$ which achieve relative contrast $\gamma^*(w)$. This is in fact possible [64]. Since the contrast of these schemes is always at least $1/4$, this means that the loss of contrast is at most 75%.

Several results on this problem are proved in [64] and [65] (also, see [70] for an elementary treatment). Here is an example of one of the results shown that involves designs.

Theorem 5.4 *Suppose that $w \equiv 3 \pmod 4$ and that there exists a visual $(2,w)$-threshold scheme with pixel expansion m and (optimal) relative difference $\gamma = \gamma^*(w)$. Then $m \geq w$, and $m = w$ if and only if there exists a $\left(w, \frac{w-1}{2}, \frac{w-3}{4}\right)$-BIBD (or, equivalently, a Hadamard matrix of order $w+1$).*

Here is an outline of how the design is used to obtain the desired scheme. Suppose first that the pixel P to be encrypted is black. Let I_1 denote the incidence matrix of the BIBD. A random permutation of the columns is first applied to I_1. Then the w rows of the resulting matrix are used to construct the w shares of P (each share consists of w subpixels).

If P is white, then we instead begin with a matrix I_0 in which each row consists of $(w-1)/2$ occurrences of 1 followed by $(w+1)/2$ occurrences of 0. As in the previous case, a random column permutation is performed, and then the rows of the resulting matrix determine the w shares.

6 Authentication codes

Unconditionally secure authentication codes were introduced in 1974 by Gilbert, MacWilliams and Sloane [72], and Simmons developed a comprehensive theory of authentication codes in the 1980's (see, for example, [76]).

6.1 The application

Two fundamental goals of cryptography are to ensure secrecy and integrity of sensitive data. Secrecy is achieved through encryption, while data integrity is accomplished by means such as signature schemes and message authentication codes. *Data integrity* provides a way to verify that data has not been changed by an adversary. The data in question could be communicated between two parties over an insecure network (email, for example), or it could be stored data (in a database, for example). Transmitted data is typically authenticated using a *signature scheme*, which allows anyone to verify an electronic signature using a public verification algorithm. The security of signature schemes depends on the assumed computational intractability of problems such as the discrete logarithm problem.

An alternative is to use a *message authentication code* (or MAC), which requires the used of a secret key to authenticate the data. As is the case when using a private-key cryptosystem, such as the Data Encryption Standard (DES), this key must be communicated to the intended receiver ahead of time, using a secure channel. A typical example of a MAC is constructed by using DES in *cipher block chaining* (CBC) mode. MACs can also be used to authenticate stored data, in which case the key should be stored in a separate, secure place from the data being authenticated.

The DES-based MAC is fast, but there is no known proof of security, not even one based on a plausible computational assumption. However, it is possible to construct MACs that can be proved secure, without any computational assumptions. This is called *unconditional security* and was introduced by Shannon to prove the security of the well-known one-time pad (which is used for encryption).

The simplest scenario to consider is authentication without secrecy. Here, a plaintext message s (called a *source state*) is authenticated by computing an *authentication tag*, a, which is appended to s. The tag a is computed as $a = h_K(s)$, where h_K is a hash function chosen from a specified set of hash functions using a random secret key K. An *authenticated message*, (s, a), is verified by computing $a_0 = h_K(s)$ and checking that $a_0 = a$.

Formally, we can describe the authentication code as a 4-tuple $(\mathcal{S}, \mathcal{A}, \mathcal{K}, \mathcal{H})$, where the following hold:

1. $\mathcal{S}$ is a finite set of possible source states;

2. $\mathcal{A}$ is a finite set of possible authentication tags;

3. $\mathcal{K}$ is a finite set of possible keys;

4. $\mathcal{H} = \{h_K : K \in \mathcal{K}\}$, where, for each $K \in \mathcal{K}$, $h_K : \mathcal{S} \to \mathcal{A}$.

As an example, $\mathcal{S}$ and $\mathcal{A}$ could consist of all binary strings of a certain length, i.e. $\mathcal{S} = (\mathbb{Z}_2)^n$ and $\mathcal{A} = (\mathbb{Z}_2)^m$. There is no requirement on the relative sizes of n and m, but the cases of greatest practical interest are when $n \gg m$, i.e. when we authenticate a "long" source with a "short" authenticator.

The security of an authentication code is measured by computing an opponent's *deception probabilities*. Suppose an opponent, Oscar, sees a sequence of $i \geq 0$ authenticated messages, all of which are authenticated using the same (unknown) key. Then Oscar creates a new forged message, (s', a'), which he hopes is an authentic message. That is, Oscar wins this game if $s' = h_K(a')$, where K is the secret key. This situation is called a *substitution* of order i. In a one-time scheme (i.e. where a key is used to authenticate only one message), the case $i = 0$ is called *impersonation* and the case $i = 1$ is called *substitution*. Oscar's probability of performing a successful deception of order i is denoted Pd_i.

6.2 The connection to designs

The main objectives of an authentication code are to minimize the deception probabilities Pd_i as a function of $|\mathcal{A}|$, and to minimize $|\mathcal{H}|$. Authentication codes in which these quantities are simultaneously minimized turn out to be closely related to designs. The following lemma is found in [77, Theorem 5.1].

Lemma 6.1 *For any authentication code without secrecy* $(\mathcal{S}, \mathcal{A}, \mathcal{K}, \mathcal{H})$, *and for any* $i \geq 0$, $\text{Pd}_i \geq 1/|\mathcal{A}|$.

If a key is chosen equiprobably from $\mathcal{K}$, then we have the following characterization shown in [77, Theorems 5.3 and 5.4].

Theorem 6.2 *An authentication code* $(\mathcal{S}, \mathcal{A}, \mathcal{K}, \mathcal{H})$ *in which keys are chosen equiprobably and in which* $\text{Pd}_i = 1/|\mathcal{A}|$ *for* $0 \leq i \leq t - 1$ *is equivalent to an orthogonal array* $\text{OA}_\lambda(t, |\mathcal{S}|, |\mathcal{A}|)$, *where* $\lambda = |\mathcal{H}|/|\mathcal{A}|^t$.

The construction of an authentication code from an orthogonal array is easy: Let M be an $\text{OA}_\lambda(t, k, v)$ on symbol set A, say. Suppose that the columns of M are labelled $1, \ldots, k$ and the rows are labelled $1, \ldots, \lambda v^t$. Define $\mathcal{S} = \{1, \ldots, k\}$, $\mathcal{A} = A$, $\mathcal{K} = \{1, \ldots, \lambda v^t\}$, and for any $K \in \mathcal{K}$, define $h_K(s) = M(K, s)$ for all $s \in \mathcal{S}$. Then $(\mathcal{S}, \mathcal{A}, \mathcal{K}, \mathcal{H})$ is the desired authentication code.

In view of Theorem 6.2, minimizing $|\mathcal{H}|$ is equivalent to finding the minimum value λ in the corresponding $\text{OA}_\lambda(t, k, v)$. Various classical bounds can be employed, e.g. the Rao and Bush bounds (see [7]). As stated previously, the situations of greatest practical interest occur when $k \gg v$, and an orthogonal array with $\lambda = 1$ does not exist in these cases. Hence it is often necessary to use OAs with $\lambda > 1$.

The results above all assume that keys are chosen equiprobably. Various theorems have been proven which show that there is no advantage in choosing keys using a different probability distribution; the optimal codes are still derived from orthogonal arrays. Here are two results of this type in the case $t = 2$.

Theorem 6.3 ([77]) *An authentication code,* $(\mathcal{S}, \mathcal{A}, \mathcal{K}, \mathcal{H})$, *in which* $\text{Pd}_i = 1/|\mathcal{A}|$ *for* $i = 0, 1$, *must have* $|\mathcal{H}| \geq |\mathcal{A}|^2$. *Further, if* $|\mathcal{H}| = |\mathcal{A}|^2$, *then there exists an orthogonal array* $\text{OA}_1(2, |\mathcal{S}|, |\mathcal{A}|)$.

Theorem 6.4 ([78]) *An authentication code,* $(\mathcal{S}, \mathcal{A}, \mathcal{K}, \mathcal{H})$, *in which* $\mathrm{Pd}_i = 1/|\mathcal{A}|$ *for* $i = 0, 1$, *must have* $|\mathcal{H}| \geq |\mathcal{S}|(|\mathcal{A}| - 1) + 1$. *Further, if* $|\mathcal{H}| = |\mathcal{S}|(|\mathcal{A}| - 1) + 1]$, *then there exists an orthogonal array* $\mathrm{OA}_\lambda(2, |\mathcal{S}|, |\mathcal{A}|)$ *in which* $\lambda = (|\mathcal{S}|(|\mathcal{A}| - 1) + 1)/|\mathcal{A}|^2$.

The orthogonal arrays in Theorem 6.3 correspond to mutually orthogonal latin squares; the OAs in Theorem 6.4 are those that meet the Rao bound with equality, e.g. simplex codes. Generalizations of these two theorems for higher values of t were proved in [77, Theorem 5.4] and [73, Theorem 1] respectively.

So far, we have considered authentication codes without secrecy, i.e. in which the source is authenticated with an appended authentication tag. A more general setting is to study deception probabilities when the source is "encrypted". This general setting allows the study of codes which provide secrecy and authentication simultaneously, as well as codes that provide authentication with no requirement with respect to secrecy or lack thereof. There are numerous results on codes of these types, analogous to results stated above for authentication codes without secrecy. Codes which achieve the minimum deception probabilities and which have a minimum possible number of keys are closely related to t-designs. We refer the reader to [73–75, 77, 78] for further information.

7 Resilient and correlation-immune functions

A boolean function exhibits correlation-immunity when, despite having specified the values of a certain number of bits of its input, the output bits retain their *a priori* distribution.

7.1 The application

Suppose $f: (\mathbb{Z}_2)^n \to (\mathbb{Z}_2)^m$. We refer to f as a *boolean function* with n inputs and m outputs. For an m-tuple $(y_1, \ldots, y_m) \in (\mathbb{Z}_2)^m$, let $\mathsf{p}(y_1, \ldots, y_m)$ denote the probability that $(y_1, \ldots, y_m)$ is the output of f when $(x_1, \ldots, x_n) \in (\mathbb{Z}_2)^n$ is chosen randomly. The function f is *balanced* if $\mathsf{p}(y_1, \ldots, y_m) = 2^{-m}$ for every $(y_1, \ldots, y_m) \in (\mathbb{Z}_2)^m$. Equivalently, for every output m-tuple $(y_1, \ldots, y_m) \in (\mathbb{Z}_2)^m$, there are precisely 2^{n-m} input n-tuples $(x_1, \ldots, x_n) \in (\mathbb{Z}_2)^n$ such that $f(x_1, \ldots, x_n) = (y_1, \ldots, y_m)$.

Suppose that $t \leq n - 1$, and the values of t of the inputs are fixed, say $x_{i_j} = c_j$ for $1 \leq j \leq t$, where $c_1, \ldots, c_t \in \mathbb{Z}_2$. The remaining $n - t$ inputs are chosen independently at random, as before. Let $\mathsf{p}(y_1, \ldots, y_m \mid x_{i_1} = c_1, \ldots, x_{i_t} = c_t)$ denote the probability that $(y_1, \ldots, y_m)$ is the output of f in this situation.

The function f is *correlation-immune* of order t if

$$\mathsf{p}(y_1, \ldots, y_m \mid x_{i_1} = c_1, \ldots, x_{i_t} = c_t) = \mathsf{p}(y_1, \ldots, y_m)$$

for all choices of $y_1, \ldots, y_m$, $x_{i_1}, \ldots, x_{i_t}$, and $c_1, \ldots, c_t$. The function f is *resilient* of order t if it is balanced and correlation-immune of order t; this is equivalent to saying that

$$\mathsf{p}(y_1, \ldots, y_m \mid x_{i_1} = c_1, \ldots, x_{i_t} = c_t) = 2^{-m}$$

for all choices of the relevant variables. In most applications, it is important that t be as large as possible (given n and m), or, equivalently, that m is as large as possible (given n and t).

Two examples of resilient functions are as follows. For any n, the function

$$f(x_1, \ldots, x_n) = x_1 + \ldots + x_n$$

is resilient of order $n - 1$. Also, for any h, the function

$$f(x_1, \ldots, x_{3h}) = (x_1 + \ldots + x_{2h}, x_{h+1} + \ldots + x_{3h})$$

is resilient of order $2h - 1$. (Addition is modulo 2 in both these examples.)

We now describe some cryptographic applications of these concepts. In a stream cipher, a binary sequence of plaintext is operated on by exclusive-or with a pseudorandom keystream in order to obtain ciphertext. A keystream is generated from a small random seed using a deterministic algorithm. One common way to generate a keystream is to combine the outputs of several linear feedback shift registers, which produce keystreams having maximum-length periods but which are not cryptographically secure. The combining function is a boolean function $f: (\mathbb{Z}_2)^n \to (\mathbb{Z}_2)^m$ where $m \leq n$ (usually $m = 1$ in this application). Such a combining function should satisfy several properties in order to produce a secure keystream. For example, the function should be balanced and should have high nonlinearity. Siegenthaler [85] suggested that correlation-immunity is another desirable property.

Block ciphers also use boolean functions in their construction. A typical example is a substitution box (or S-box) that is used in Feistel-type ciphers such as DES. An S-box can be described using a boolean function as defined above. In DES, for example, there are eight S-boxes used, each of which is a boolean function from $(\mathbb{Z}_2)^6$ to $(\mathbb{Z}_2)^4$. Among the design criteria satisfied by the DES S-boxes are certain balance and correlation-immune properties.

Another context in which resilient functions are useful is in renewing a partially leaked key. Cryptographic scenarios in which partially leaked keys are studied include quantum key exchange (Bennett, Brassard and Robert [79]) and key distribution patterns (Stinson [87]). In both these situations, resilient functions have been proposed as a useful method of obtaining a smaller secure key from a larger partially leaked key.

Suppose Alice and Bob share a random n-bit binary key, $x_1, \ldots, x_n$, and an opponent, Oscar, has learned the values of t of the n bits. Suppose further that Alice and Bob do not know which t bits have been leaked. If $f: (\mathbb{Z}_2)^n \to (\mathbb{Z}_2)^m$ is a t-resilient function, then $f(x_1, \ldots, x_n)$ is a random m-bit key about which Oscar has no information. (This is true since every output m-tuple is equally likely, given the value of t input variables.) The description of the function f does not need to be kept secret from Oscar.

7.2 The connection to designs

Let $f: (\mathbb{Z}_2)^n \to (\mathbb{Z}_2)^m$. For any $(y_1, \ldots, y_m) \in (\mathbb{Z}_2)^m$, $f^{-1}(y_1, \ldots, y_m)$ is a set (possibly empty) of binary n-tuples. We depict $f^{-1}(y_1, \ldots, y_m)$ as a binary array with n columns, in which the rows are the n-tuples in the set $f^{-1}(y_1, \ldots, y_m)$.

The following theorems give the connection between correlation-immune and resilient functions and orthogonal arrays (see [81, 84, 86]).

Theorem 7.1 *A function $f: (\mathbb{Z}_2)^n \to (\mathbb{Z}_2)^m$ is correlation-immune of order t if and only if each array $f^{-1}(y_1, \ldots, y_m)$ is an orthogonal array of strength t.*

Theorem 7.2 *A function $f: (\mathbb{Z}_2)^n \to (\mathbb{Z}_2)^m$ is resilient of order t if and only if each array $f^{-1}(y_1, \ldots, y_m)$ is an $\mathrm{OA}_{2^{n-m-t}}(t, n, 2)$.*

The difference between these two theorems is that the 2^m orthogonal arrays in Theorem 7.2 all have $\lambda = 2^{n-m-t}$, whereas in Theorem 7.1 the orthogonal arrays have (possibly) different values of λ. In both theorems, the orthogonal arrays are simple and disjoint, and the 2^m arrays form a partition of $(\mathbb{Z}_2)^n$. When the OAs in question all have the same λ value, the collection is *large set of orthogonal arrays* and is denoted LOA. The following corollary is immediate.

Corollary 7.3 ([86]) *There exists a function $f: (\mathbb{Z}_2)^n \to (\mathbb{Z}_2)^m$ that is resilient of order t if and only if there exists an $\mathrm{LOA}_{2^{n-m-t}}(t, n, 2)$.*

In the correlation-immune case, where the λ values need not identical, the corresponding collection of OAs has been termed a "large class".

Constructions of resilient functions are easily obtained from codes, using the following well-known result.

Proposition 7.4 *Suppose that $\mathcal{C}$ is a binary linear code of length n, dimension m and distance d. Then the dual code, $\mathcal{C}^\perp$, is an $\mathrm{OA}_{2^{n-m-d+1}}(d-1, n, 2)$.*

Since $\mathcal{C}^\perp$ is a subspace of $(\mathbb{Z}_2)^n$, every coset of $\mathcal{C}^\perp$ is also an $\mathrm{OA}_{2^{n-m-d+1}}(d-1, n, 2)$, and these 2^m OAs form a large set. Applying Corollary 7.3, we have the following result, shown in [79, 82].

Theorem 7.5 *If there exists a binary linear code of length n, dimension m and distance d, then there exists a function $f: (\mathbb{Z}_2)^n \to (\mathbb{Z}_2)^m$ that is resilient of order $d-1$.*

Simplex codes provide a good illustration of Theorem 7.5. For any integer $m \geq 2$, the simplex code $\mathcal{S}_m$ is a linear code of length $2^m - 1$, dimension m and distance 2^{m-1} (it is in fact the dual of a Hamming code). The following corollary of Theorem 7.5 is obtained.

Corollary 7.6 ([83]) *Let $m \geq 2$ be an integer. There exists a function $f: (\mathbb{Z}_2)^{2^m-1} \to (\mathbb{Z}_2)^m$ that is resilient of order $2^{m-1} - 1$.*

In order to know how good our constructions are, we need bounds (necessary conditions). The only known bounds for resilient functions are applications of bounds for orthogonal arrays. The two most important bounds follow:

Theorem 7.7 ([83], [80]) *If there exists a function $f: (\mathbb{Z}_2)^n \to (\mathbb{Z}_2)^m$ that is resilient of order t, then*

$$t \leq \left\lfloor \frac{2^{m-1}n}{2^m - 1} \right\rfloor - 1$$

and

$$t \leq 2 \left\lfloor \frac{2^{m-2}(n+1)}{2^{m-1}} \right\rfloor - 1.$$

Applying the first of these two bounds, we see that the resilient functions constructed in Corollary 7.6 are optimal, since the bound on t is met with equality. There are various other infinite classes of optimal resilient functions known. For example, complete results are known for $m = 1, 2$ and 3, as follows.

Theorem 7.8 ([80], [82], [83])

1. *There exists a function $f: (\mathbb{Z}_2)^n \to (\mathbb{Z}_2)^1$ that is resilient of order t if and only if $t \leq n - 1$.*

2. *There exists a function $f: (\mathbb{Z}_2)^n \to (\mathbb{Z}_2)^2$ that is resilient of order t if and only if $t \leq \lfloor 2n/3 \rfloor - 1$.*

3. *There exists a function $f: (\mathbb{Z}_2)^n \to (\mathbb{Z}_2)^3$ that is resilient of order t if and only if $t \leq \lfloor 4n/7 \rfloor - 1$ and $n \not\equiv 2 \bmod 7$, or $t \leq \lfloor 4n/7 \rfloor - 2$ and $n \equiv 2 \bmod 7$.*

We have discussed correlation-immune and resilient functions defined over $\mathbb{Z}_2$. These concepts can be generalized in an obvious way to non-binary alphabets, and many of the results discussed in this section hold true in a suitably generalized form; see [84].

8 Multidrop networks

The design of networks using broadcast media so that every two sites lie on a common link, subject to constraints on the number of links at each site (degree), and the number of sites on each link (link size), is examined. This leads to the examination of pairwise balanced designs and to the use of (k, n)-arcs in projective planes.

8.1 The application

The network design problem of interest is as follows. There are n network sites, to be connected using multidrop communication links such as Ethernets, token rings, or any broadcast medium. A *link* or *bus* is a subset of the n sites. In order to avoid congestion due to switching overhead from one link to another, it is required that every two sites appear together on at least one link. Typically, each site is equipped with a limited number of communication ports and hence can appear on at most some fixed number r of the links. Similarly, each link has a limit on the number of sites that it can connect. Reasons for such a limitation include capacity limits, and limits on acceptable routing delay within the link. With these constraints in mind, the problem can be informally stated as follows: Connect n sites so that every two sites appear together on at least one link, subject to the constraint that no link has more than k sites on it, and no site appears on more than r links.

Problems of this type have been studied extensively. Mickunas [102] considered the case when k and r are close to equal. Subsequently, Bermond and his colleagues [93–95] considered general network design problems of this type under the name "bus interconnection networks". They are primarily responsible for observing that numerous well studied combinatorial configurations lead to useful solutions to such network design problems; see also [3, 88, 98, 104, 106, 107].

8.2 The connection to designs

Block designs and pairwise balanced designs lead to optimal solutions for the network design problem when $k < r$ [93, 94]. When $k = r$, projective planes yield bus networks [102]. Practical concerns dictate that the replication number r be a fixed small number, while the block size k can be potentially much larger than r. Since block designs and PBDs always have $k \leq r$ by Fisher's inequality [1], a technique is needed to treat cases when $k > r$. This is one of the problems treated in [105].

Bermond, Bond, Paoli, and Peyrat [93] propose the following. Suppose that we are to construct a covering with replication number at most r and block size at most k, and our objective is to maximize the number of elements. Choose q so that q is a power of a prime, $q+1 \leq r$, and q is as large as possible subject to these constraints. Then form $\mathrm{PG}(2,q)$ on element set V of size $q^2 + q + 1$, and with block set $\mathcal{B}$. A *weight function* $\omega \colon V \to \mathbb{Z}^+$ from elements to positive integers is to be chosen, and a set of elements $W = \{(x,i) \colon x \in V \text{ and } 1 \leq i \leq \omega(x)\}$ defined. The weight of an element indicates the number of times that it is replicated in W. We choose ω so that the *weight*, $\omega(B)$, of block B satisfies $\omega(B) = \sum_{x \in B} \omega(x) \leq k$, for every $B \in \mathcal{B}$, and defines a new set of blocks

$$\mathcal{D} = \{\{(x,i) \colon x \in B \text{ and } 1 \leq i \leq \omega(x)\} \colon B \in \mathcal{B}\}.$$

Then $(W, \mathcal{D})$ is a covering with block sizes at most k, replication number $q + 1 \leq r$, and $\sum_{x \in V} \omega(x)$ elements. Naturally the problem is to determine ω so as to constrain the weight of each block to k while maximizing the number of elements. Bermond et al. [93,94] conjectured that the covering with replication number at most r, block sizes at most k, and the largest number of elements, arises in this manner when $r - 1$ is a prime power. It follows from a theorem of Füredi [99] that such a covering can have at most $rk - (r-1)\lceil \frac{k}{r} \rceil$ elements. Now choosing ω so that all element weights are as equal as possible subject to the constraint on block weight leads to coverings for which Füredi's bound is achieved infinitely often, and approaches this bound as $k \to \infty$ for fixed r when $r - 1$ is a prime power [94, 95]. Hence, although there are many potential methods for producing coverings, Füredi's result establishes that the asymptotically optimal coverings arise from replicating elements in projective planes.

Bermond et al. [94, 95] did not address the question of finding the largest number of elements in a covering with block size at most k and replication number at most r precisely. Yener, Ofek, and Yung [105], however, employed a similar underlying strategy but developed techniques for specifying the weight of each element in the projective plane so as to maximize the number of elements. Their method is a simple greedy strategy, which does not in general lead to the minimum block size [96].

When $k \leq r$, the existence of bus networks with restricted bus sizes amounts simply to the existence of pairwise balanced designs with specified block sizes. When $k > r$, two more specific problems arise.

The first arises when r is of the form $q + 1$ for q a prime power, so that a projective plane of order q exists. Then an apparently hard question is to determine the minimum increase in the maximum block size. A (k, n)-*arc* in a projective plane of order q is a nonempty set K of k elements such that n is the maximum number of elements in K that appear together on a block. A $(k, 2)$-arc is a k-*arc*. The existence of (k, n)-arcs was extensively studied, but their importance here is that the maximum number $m_n(q)$ of elements in a (k, n)-arc in PG$(2, q)$ is precisely the same as the maximum number of elements that can be replicated without increasing the maximum block size by more than n.

Barlotti [90] established that $m_n(q) \leq (n-1)(q+1) + 1$. A simple computation establishes that when PG$(2, q)$ contains a (k, n)-arc whose size meets this bound, replicating the elements of the arc yields equality in the asymptotic bound of Füredi [99] discussed earlier. Unfortunately, the determination of $m_n(q)$ is a very difficult problem in finite geometry that remains far from settled [100, 101]. Except when $n = q + 1$, equality in Barlotti's bound can be achieved only when n is a divisor of q [90], and is achieved in two *trivial cases*: when $n = 1$ (by a single element), and when $n = q$ by all elements not lying on a fixed block. When q is a second or higher power of a prime, nontrivial arcs meeting Barlotti's bound *always* exist when q is a power of 2 and n is a divisor of q [97]. However, in a 1997 breakthrough it has been shown that they *never* exist when q is a power of an odd prime [89].

When $m_n(q)$ does not realize Barlotti's bound, extensive research has attempted to obtain lower and upper bounds, and specific exact values; as an introduction to the literature, we suggest [100, 101]. Each lower bound can lead to a replication scheme for producing a covering, and each upper bound establishes a limit on how well such a replication scheme can do.

The second main problem arises when r is not one more than a prime power. Then the advisability of beginning with a plane of order less than r is open to question. Colbourn [96] observed that certain related combinatorial configurations can lead to better results. Consider, for example, the case when $r = 7$ and $v = 39$. There is no plane of order 6, and hence the plane $PG(2,5)$ can be used. Then block size 9 is obtained. However, there is a covering on 39 elements with replication number 7 and block size 7 [103], and hence in this case replication of elements in planes does not appear to lead to the best solution. For this reason, we mention a less well studied generalization of difference sets that can lead to (slightly) smaller block sizes for certain degree constraints.

A *difference cover* modulo v of *order* q, $D = \{d_0, \ldots, d_q\}$, has the property that $\{d_i - d_j : 0 \leq i, j \leq q \text{ and } i \neq j\}$, arithmetic modulo v, contains every nonzero integer in $\mathbb{Z}_v$. Hence every nonzero difference arises at least once as the difference modulo v of two elements in D. When $v = q^2 + q + 1$, a difference cover is a difference set. However, while difference sets only exist for certain values of q, difference covers exist for every value of q. Of course, the price is that the number of elements v is less than $q^2 + q + 1$ in general. Now adding each integer i to the elements of D in turn, we produce v blocks forming a covering with block size $q+1$ and replication number $q+1$. Wiedemann [103] gave a difference cover modulo 39 of order 6, which provides the illustration given above. He also presented a table of the smallest order difference cover modulo v for each value of $v \leq 133$. Unfortunately, these computational results are not at present accompanied by a useful theory. For certain small replication numbers such as 7 (a difference cover modulo 39) and 11 (a difference cover modulo 95 [103]), it appears that difference covers can improve upon the use of planes.

We have focused on the design of bus networks with diameter one. However, transversal designs arise also in the design of bus networks of larger diameter; see [92], for example.

9 Channel graphs and interconnection networks

This section examines interconnection strategies for computer networks and the input-output channel graphs that they contain.

Combinatorial Designs in Communications   69

9.1 The application

The generic *interconnection problem* is to establish communication paths between *input nodes* V_I and *output nodes* V_O; we allow the case that the input nodes and output nodes are the same. *Switch nodes* V_S may be used; these simply relay a message from one communications channel to another. Our task is to connect the nodes in $V = V_I \cup V_O \cup V_S$ using point-to-point *links*. The network must be *connecting*, in that there must be a communication path from each input node to each output node. The *distance* from an input node to an output node is the number of links in the shortest communication path connecting them. The *diameter* or *depth* is the maximum distance from an input to an output. To ensure small delay in communication, we require that the diameter be small.

When all input-output paths have length equal to the diameter, the switch nodes (if any) can be partitioned into *stages*, by placing all nodes at distance i from an input in the ith stage. Networks admitting such a partitioning are *multistage interconnection networks*. A multistage interconnection network with a single input node and a single output node is a *channel graph*. In the analysis of interconnection networks, the network designer is concerned with the overall network design. However, each input-output pair is concerned with the portion of the network containing the paths from the specified input to the specified output, i.e. the input-output *channel graph*.

In some applications, we require that the n input nodes can simultaneously communicate with the n output nodes, given a specified mapping of inputs to outputs. This requires *disjoint* communication paths, which share no common link or intermediate node. A good example of this situation arises in the design of shifting networks.

A *barrel shifter* is a network whose nodes are $\{0, 1, \ldots, n-1\}$, the integers modulo n. Given a shift distance s, $1 \leq s < n$, every node must transfer a value to the node whose label is s larger; more precisely, for each $0 \leq i < n$, node i must establish a connection to node $i+s$ (reduced modulo n as needed), and all n communication paths are to be disjoint. Kilian, Kipnis and Leiserson [113] developed a barrel shifter which has diameter one; when implemented in VLSI, the shift is accomplished in a single clock cycle.

We now consider an even stronger connection property of networks. An *n-superconcentrator* is a network with n inputs and n outputs in which disjoint communication paths can be established from the inputs to the outputs in *any* of the $n!$ possible orderings. We restrict superconcentrators to have only links, and no larger buses. A superconcentrator of depth one requires all n^2 connections (i.e. each input connected to each output); hence superconcentrators of depth greater than one are of interest. Nevertheless, superconcentrators are typically constructed using special types of depth-one networks in which every set of inputs is directly connected to a relatively large set of outputs (see, for example, [110]). More formally, a network $(V_I \cup V_O, E)$ with $V_I \cap V_O = \emptyset$ is a (n, α, β)-*expander* if every set of α inputs is directly connected to at least β output nodes.

The motivation for strong expansion capability is to avoid congestion or blocking. To quantify the disruption due to such blocking, examine the channel graphs for each input-output pair. We assume that the probability q_i of occupancy of each edge in a channel graph is known. The *blocking probability* of a channel graph is defined as the probability that every channel of that graph contains at least one occupied (blocked) edge. A channel graph with k stages is *superior* to another channel graph with k stages if the blocking probability of the former never exceeds that of the latter, independent of the occupancies for the E_i. There is no guarantee that constructing an interconnection network in which channel graphs are superior leads to an interconnection network with high expected throughput. Nevertheless, if the channel graphs have high blocking probability, this ensures poor throughput. Hence the design of superior channel graphs arises as a necessary step in interconnection network design.

9.2 The connection to designs

A network of diameter one is a 2-$(n, K, 1)$ covering; if we require in addition that each node i has a (disjoint) path to node $i + s$ mod n, the n pairs from the set $D_s = \{\{i, (i+s) \bmod n\} : 0 \leq i < n\}$ must appear in n distinct blocks. At first, this seems to be a complicated requirement, but a widely studied class of designs always has the desired property; we introduce them here. A set system $(V, \mathcal{B})$ with $V = \{0, 1, \ldots, n-1\}$ is *cyclic* if, whenever $\{b_1, \ldots, b_k\} \in \mathcal{B}$, $\{b_1 + 1, \ldots, b_k + 1\} \in \mathcal{B}$ (arithmetic modulo n is used). The *orbit* $\mathcal{O}(B)$ of a block B is the set $\{B + s \pmod{n} : 0 \leq s < n\}$; it is *full* when $|\mathcal{O}(B)| = n$. When all orbits are full, the set system is *full-cyclic*. It is easy to see that the pairs of D_s appear in at least n distinct blocks of a full-cyclic covering.

Any full-cyclic covering can then be used to design a barrel shifter. Each node finds the first orbit in which $\{0, s\}$ appears, say in block B. Node x now writes its value to the bus $B + x \pmod{n}$, and reads its value from the bus $B + x - s \pmod{n}$. In this way each node x reads the value node $x - s$ (mod n) wrote, and each communication path corresponds to a unique block in the orbit. Kilian, Kipnis and Leiserson [113] observed that to minimize the total number of buses and the number of buses incident at a node, the covering chosen is a *cyclic projective plane* (i.e. a projective plane which is cyclic).

The actual operation of a barrel shifter based on a cyclic projective plane is remarkably simple. To see this, consider the structure of cyclic projective planes. Since there are only n blocks, any two blocks B_1, B_2 satisfy $B_1 \equiv B_2 + s$ (mod n) for some $0 \leq s < n$. Consider a single block $B = \{b_1, \ldots, b_k\}$. Now for each element d, $1 \leq d < n$, $\{0, d\}$ appears in exactly one block. Hence B must contain exactly two elements b_i, b_j for which $b_j - b_i \equiv d \pmod{n}$. Every d, $1 \leq d < n$, is the difference of two elements of B; such a set B is a *difference*

Combinatorial Designs in Communications

set for $\{0, 1, \ldots, n-1\}$.

Using the difference set representation of the cyclic projective plane, the operation of a barrel shifter is straightforward. To shift a distance of s, each node finds the two elements b_i, b_j in the difference set with $b_j - b_i \equiv s$ (mod n). Node x then writes onto bus $x + b_i$, and reads from bus $x + b_j$.

When no cyclic projective plane on n elements exists, this very simple control logic can be retained nonetheless: This scheme requires only a set which covers all differences from 1 to $n-1$. Hence we can use a *difference cover*, in which each d, $1 \leq d < n$, is the difference of at least one pair of elements. Kilian, Kipnis and Leiserson [113] observed that they produce optimal barrel shifters of depth one. They also use difference covers to design 'permutation architectures', which realize permutations other than just cyclic shifts.

Let us turn to superconcentrators. Any depth-one network with $V_I \cap V_O = \emptyset$ can be equivalently written as a set system $(V_I, \mathcal{B})$, where $\mathcal{B} = \{\{v_i : \{v_i, v_o\} \in E\} : v_o \in V_O\}$. In this setting, an (n, α, β)-expander is a set system with n elements and n blocks, so that every set of α elements intersects at least β of the blocks. Intuitively, β is largest when the blocks intersect each other as little as possible. At the same time, however, for β to be large, each element must appear in a large number of blocks. To maximize the expansion, we choose to balance the block sizes, and balance the sizes of block intersections. Hence we consider symmetric designs.

Alon [108] proves that one class of symmetric designs, obtained from the points and hyperplanes of the projective geometries $\mathrm{PG}(d, q)$, provides good expansion properties: In the design from $\mathrm{PG}(d, q)$ on n elements, every set of α elements intersects $\beta \geq (\alpha n)/(\alpha + q - 1)$ blocks. Hence for all $\alpha = \mathrm{o}(n)$, $\alpha = \mathrm{o}(\beta)$; such a network is termed *highly expanding*. Moreover, Alon remarked that these expanders have essentially the smallest number of links of any network with equivalent expansion properties. Using projective geometries for expanders, Alon [108] established the existence of n-superconcentrators of depth three with $O(n^4/3)$ links; see [108] for further uses of the expanders and superconcentrators, and [112] for a similar use of symmetric designs.

In the same way that designs lead to desirable expansion properties, they also arise in the design of superior channel graphs. Chung [109, 111] used incidence graphs of block designs to determine connections between the second and third stage of a four-stage channel graph. She established that, among all channel graphs with the same numbers of nodes in each stage and the same numbers of interstage links, channel graphs arising from block designs are superior.

The design of interconnection networks employs design-theoretic tools in a number of ways. The use of designs to cover all pairs of nodes is prevalent in diameter one networks; on the other hand, the balanced intersection of blocks is shown to lead to high expansion factors, and hence to highly connecting networks.

10 Partial match queries on files

In this section, we explore an application of designs to file organization. We focus on the simpler applications of designs here, to somewhat specialized problems, and content ourselves with providing some references to the more general situations.

10.1 The application

A *file* is a collection of *records*; each record has a number of *attributes*, and we retrieve records by specifying their attributes. A primary requirement for any file organization is the support of *partial match queries*, in which values for some attributes are given and the remainder are unspecified. All records matching the values in the specified attributes are to be retrieved. Normally, records are relatively space-consuming objects; hence they are stored on a slower secondary storage device and an *accession number* records their address on this device. Our task, given a partial match query, is thus reduced to listing the relevant accession numbers.

We consider the situation in which there are n binary attributes. Moreover, we consider queries which request those records possessing certain attributes; the extension to the case in which we further stipulate that the records not have certain other attributes is not essentially more difficult. In a typical retrieval system, queries are relatively simple, in that they involve relatively few of the attributes. Hence, we first consider the case where partial match queries on up to t attributes must be supported, but queries on more than t attributes need not be.

The usual *inverted file system* creates a list of accession numbers for each attribute, and intersects these lists to reply to a partial match query. This requires the examination of very many accession numbers which do not form part of the final answer. At the other extreme, an *extended* inverted file system creates (in advance) a "bucket" of accession numbers for each partial match query. Redundancy is incurred in this scheme, but can be limited by placing an accession number in a bucket only when the partial match query is a maximal query which matches the record. The redundancy in storage pays off in retrieval, because in this scenario only the relevant accession numbers are examined. The impracticality of this approach arises from the very large number of buckets required, and a correspondingly large requirement for redundancy.

A compromise solution is to amalgamate many possible queries into a single bucket. Each bucket remains associated with a subset of the attributes, but may now contain information about many maximal partial match queries. The essential feature of the bucket subsets is that each query subset be contained in at least one bucket subset.

Rivest [124] remarked on a practical limitation, the large redundancy introduced by storing accession numbers many times. While this may be quite acceptable for small files, a very large difference in the sizes of records and their

Combinatorial Designs in Communications

accession numbers would be required, before the storage for buckets would be less than storage for the file itself. Nevertheless, combinatorial filing schemes remain useful when the cost of retrieving a record from the secondary storage is so large that we are unwilling to retrieve any record which may prove irrelevant to the query at hand. In the absence of such a prohibitive cost, however, we need only ensure that "most" of the records retrieved prove relevant; Rivest's scheme, which we explore next, has this property.

Rivest [124] considered partial match queries of any size on a file with n binary attributes; a query specifies records which do possess certain properties, do not possess certain others, and may or may not possess the remainder. In this case, records are placed in buckets; however, here the buckets partition the records, i.e. no redundancy is permitted. A record R is placed in a bucket M_i by evaluating a hash function h; if $h(R) = i$, then R is placed in M_i. The hash function is therefore a function which partitions all possible records into b buckets $M_1, \ldots, M_b$. To answer a partial match query Q, we determine (in a manner as yet unspecified) all buckets which could contain a record matching Q, and then linearly search all of the selected buckets. (If accession numbers rather than actual records are stored, we must access the secondary storage to retrieve all of the records in these buckets.)

Since data can be stored on many secondary storage devices, we can exploit parallelism if the records of interest lie on different disks. Suppose then that each bucket is associated with a disk. Examine the records which hash to the same bucket to determine when accesses must be made sequentially. We expect that two records meeting the requirements of a partial match query have many similar or equal attributes. This argues for the selection of a hash function which places records that are similar in many attributes in different buckets. Faloutsos and Metaxas [121] examine this problem under the assumption that attributes are binary (or can be made so by partitioning the set of values into 'high' and 'low', for example). Abdel-Ghaffar and El Abbadi [115] considered the case in which each attribute has p values (or, again, its range of values partitioned into p classes). The objective is to determine the largest value of d, given the numbers of records in a bucket (m), attributes (n) and values per attribute (p), so that we can partition the set of all possible records into buckets with no two records in the same bucket agreeing in more than $n - d$ attributes. Such a hash function has *maximum distance*.

10.2 The connection to designs

Now we are in a position to introduce combinatorial filing schemes. Let $A = \{a_1, \ldots, a_n\}$ be a set of attributes. Let $(A, \mathcal{B})$ be a t-covering, and write $\mathcal{B} = \{B_1, \ldots, B_m\}$. Each B_i has an associated list, or *bucket* M_i. Not all subsets of A appear in blocks of $\mathcal{B}$, but we are guaranteed that all t'-subsets with $t' \leq t$ are. A subset $A' \subseteq A$ which does appear may be a subset of many blocks; we write $f(A') = i$ if the "first" block containing the subset A' is B_i.

Now many subsets are associated with bucket M_i, and hence we partition this bucket into *subbuckets*; in particular, for each $A' \subset A$ with $f(A') = i$, we form a subbucket $M_{i,A'}$.

To enter a new record with attributes R, we place its accession number in subbucket $M_{i,A'}$ provided that $R \cap B_i = A'$ and $f(A') = i$. Each accession number thus appears in at most one subbucket of each bucket, but may appear in many different buckets. To answer a partial match query Q, we determine $i = f(Q)$ and only examine bucket M_i. The relevant accession numbers are then listed, each exactly once, by catenating all of the subbuckets $M_{i,A'}$ with $Q \subset A'$.

If only one bucket is used, this scheme reduces essentially to extended inverted filing. In fact, within each bucket, the scheme is like extended inverted filing, with one important difference. Subbuckets $M_{i,A'}$ exist even for sets A' which are too large to be partial match queries themselves; this eliminates redundancy within a bucket. The main advantage of first partitioning into buckets in this way is that the filing problems remaining within a bucket are intended to be of manageable size.

Two competing goals affect the selection of a t-covering to be used. First, the redundancy incurred by storing accession numbers in many buckets dictates that the t-covering should have few blocks; intuitively, fewer blocks lead to less redundancy. Second, larger blocks lead to more subbuckets per bucket, and hence leave larger filing problems within a bucket; intuitively, one prefers smaller blocks. The tradeoff between having few larger blocks or many smaller blocks is very application dependent. Ray-Chaudhuri [123] observed that, when $t = 2$, the first goal suggests the use of projective planes. If the second goal is taken into account, block designs with small block size are preferable [123]. When $t > 2$, the designs to be used are not as readily available, especially in view of the requirement that the index be 1. For $t \leq 5$, many suitable designs are known to exist, but as noted earlier, existence is far from settled in general.

A most profitable direction to extend this research was considered in [119, 123]. On each bucket, we can develop a second combinatorial filing scheme, and thereby develop an overall method which is multi-stage. To do this, on each block of a t-$(v, k, 1)$ design, we place a copy of a t-$(k, k', 1)$ design. The operation of the filing scheme is to first find the relevant block of the t-$(v, k, 1)$ design, and then within that block find the relevant block of the t-$(k, k', 1)$ design; this could naturally be repeated to form a filing scheme with any desired number of stages. In practical terms, the lack of known Steiner systems with large t and k limits the usefulness of this idea, however. Even when appropriate systems are known, it might be argued that the result is just a t-$(v, k', 1)$ design. Of course, it is such a design, but has the advantage that we need not search all blocks of the design in order to locate the relevant bucket. Two mappings would be employed here, one to locate the relevant block of the t-$(v, k, 1)$ design, and the second to locate within that subset the relevant block of the t-$(k, k', 1)$ design.

A second profitable direction is to generalize the scheme to handle multiple valued attributes. The extension of the design-theoretic approach to this problem has been studied by many authors, notably Bose and Koch [119], Bose et al. [118], Ghosh and Abraham [120], and Berman [117].

The essential ingredient in Rivest's scheme is the selection of the hash function. It must have two properties. We must be able to easily determine whether records in a given bucket could possibly match a given query. In addition, we want to examine as few buckets as possible (either on average or in the worst case). These decisions are not affected by the file itself. Rivest's main theorem here shows that if we have $b = 2^w$ buckets, to minimize the average number of buckets examined we choose a function which hashes a group of records to the same bucket if they are "close" in the following sense. For bucket M_i, there exist sets S_i^0 and S_i^1 for which all records which have attributes in S_i^0 set to 0 and attributes in S_i^1 set to 1, and no others, are hashed to M_i. Moreover, the number of specified attributes $|S_i^0 \cup S_i^1|$ is w. Hence the set of records hashed to bucket M_i can be easily encoded as an n-vector with entries 0, 1, $*$ containing w digits and $n - w$ $*$'s; we call these vectors *signatures* of the buckets. The asterisks denote "don't care" positions.

An easy example when $n > \log_2 b$ simply uses the first $\log_2 b$ bits of the record to determine the bucket. While this easy method has optimal average case performance, in the worst case it may require the examination of all buckets. What yields the best worst-case complexity? Rivest [124] addressed this question by considering a novel type of designs. An *associative block design* $\text{ABD}(n, w)$ is a $2^w \times n$ array with entries from 0, 1, $*$ so that

1. each row has w digits and $n - w$ $*$'s,

2. for every pair of rows, there is a column in which they contain different digits, and

3. every column contains the same number, $2^w(n - w)/n$, of $*$'s.

Conditions (1) and (2) ensure that the signatures (i.e. rows) form a partition of the file, which when used as a hash function delivers optimal average case performance. Condition (3) is designed to ensure that worst-case performance is also good.

Let us turn to maximum distance hash functions, following [121] and [115]. When we examine the records placed on one of the disks (buckets), treat the records as codewords in a p-ary code. This is a code of length n with m codewords, and distance as large as possible. The Singleton bound (see [7]) establishes that the distance d satisfies $d \leq n - \lceil \log_2(m) \rceil + 1$. When $m = p^\alpha$ and equality is met in the Singleton bound, the code is an MDS code, and then numerous connections with designs are well studied [125]. Cases when m is not a power of p pose challenging design-theoretic questions. In particular, when $p < m \leq p^2$, the Singleton bound permits the possibility of a code of distance $n-1$. A p-ary code of distance $n-1$ can be interpreted as a generalization of a transversal design as follows: Let X be a set of size np partitioned into n groups of p elements each. Then a *transversal packing* of type p^n is a collection of n-sets with one element from each group, so that no two of the n-sets intersect in more than one element (equivalently, no pair of elements occurs in more than one n-set). Existence of transversal packings of type p^n essentially asks for the maximum number of blocks (n-sets). When this maximum number of blocks is p^2, the packing is a transversal design and hence equivalent to a set of $n-2$ mutually orthogonal latin squares. However, there are numerous cases in which the required number n exceeds the block size of any, or any known, transversal design. In these cases, the application motivates the study of dense transversal packings. An equivalent formulation in terms of mutually orthogonal *partial* latin squares was given in [114].

In closing, we mention also the related notion of perfect hash families. Again, there is a useful connection with difference matrices and other designs [116].

11 Software testing

Testing is an important but expensive part of the software development process. This section describes how certain combinatorial designs are being used to reduce the number of tests needed in order to assure that a software product performs correctly in most reasonable instances.

11.1 The application

The problem is to design a test plan for a software system. For a moderate-sized system there can billions of possible test scenarios. So it is important to find a test plan that is not too large, yet tests for most of the interactions among the possible outputs in the modules of a software system. To do this, software developers have begun using combinatorial designs to test for these interactions. This usage is closely related to the use of combinatorial designs in the design of experiments.

To design a test plan, the tester identifies possible output values from each of the stages of the software system. An example from [128] comes from the testing of a telephone switch. In this example the stages are the *call type* with output values local, long distance and international; *billing type* with values either caller, collect or 800; *access parameter* with values ISDN, PBX or loop and the *status parameter* with values success, busy or blocked. Since each of the four stages has three values, there are $3^4 = 81$ different scenarios.

The goal is to reduce the number of test scenarios while still testing all of the two-way interactions. In [128] a test model with 13 scenarios is given that covers all possible two-way interactions. Larger examples can easily be constructed which show that the number of scenarios can be reduced substantially if only all the two-way interactions are covered. There is evidence that most errors that do occur in the development of a new software system can be found if the pairwise (and sometimes three-way) interactions are all tested.

A similar application arises in testing combinational logic circuits. Such a circuit implements a collection of functions from a set of n binary inputs to m binary outputs. When every function depends upon the value of at most d of the inputs, we want to generate a set of binary n-vectors which include every possible assignment of values to each subset of d inputs; this is a *universal test set*. Seroussi and Bshouti [135] discussed this application further and provided numerous references. In this case, the restriction to binary values is imposed by the nature of the circuits, but d-way interactions for $d > 2$ are typically of concern.

11.2 The connection to designs

A *covering array* $\mathrm{CA}(t, k, g)$ is an array with k rows with each cell filled with an element of an g-set S that has the property that given any set of t rows, every $t \times 1$ column vector with symbols from S occurs at least once. The parameter t is the *strength* of the covering array and when $t = 2$ a $\mathrm{CA}(2, k, g)$ is denoted $\mathrm{CA}(k, g)$. If the requirement is strengthened to require each pair exactly once, then a covering array is an *orthogonal array*. Just as orthogonal arrays are equivalent to transversal designs, covering arrays are equivalent to *transversal covers*.

The example in the previous section asks for a covering array with 4 rows and with $|S| = 3$. Such an array with exactly 13 columns exists; each column gives a test. In general, we can design a test plan for a software system with k stages each with g outputs by using a $\mathrm{CA}(k, g)$. This test plan would test for all possible two-way interactions of outputs from two different stages.

The first result of interest is a lower bound on the number of columns needed in a $\mathrm{CA}(k, g)$, to bound the number of runs needed to test the software system. To simplify the discussion we consider only two-way interactions (i.e. in the covering array, given any two rows, every ordered pair of symbols occurs in at least one column).

Poljak and Tuza [134] proved a bound for a partition problem in set theory concerning so called *qualitatively independent* partitions. That problem is essentially the dual of covering arrays and hence the Poljak-Tuza bound can be converted to a bound for covering arrays:

Theorem 11.1 *Let $c(k, g)$ denote the minimum number of columns needed in a $\mathrm{CA}(k, g)$. Then $c(k, g) \geq \frac{g}{2} \log_2 k$ as $k \to \infty$.*

Gargano, Korner and Vaccaro [133] provided constructions which prove that $c(k, g) \to \frac{g}{2} \log_2 k$ as $k \to \infty$. This asymptotic result is mainly of theoretical interest, and is not constructive. Subsequent work focused on finding specific constructions for given small values of k and g (see, for example,

[136, 137]). When $k \leq g+1$, we can use orthogonal arrays or, equivalently, sets of orthogonal latin squares. However, software testing applications typically have g small relative to k.

There is a nice construction in the case of $g = 2$. Let X be a set with $2n + 1$ points and index the rows of an array A by the subsets $Y \subset X$ where $|Y| = n$. There are exactly $\binom{2n+1}{n}$ rows. Then for $1 \leq i \leq 2n + 1$ place a 1 in row Y and column i if $i \in Y$; otherwise put a 0. It is not difficult to show that the array A is a covering array $CA(\binom{2n+1}{n}, 2)$ with exactly $2n + 1 = c$ columns. This is best possible as we can show (using Stirling's formula) that $c \sim \log_2 k$ as $n \to \infty$.

The next goal is to construct covering arrays with $g > 2$. Clearly, using the result for $g = 2$ we can asymptotically make a covering array for any $g > 3$ with $c = \binom{g}{2} \log_2 k$ columns. In [132] asymptotic constructions (for k large) are given for covering arrays with $c = s \log_2 k$ columns for some specific small values of $g \leq 13$. In particular, they showed that $s \leq 3.9$ when $g = 4$, that $s \leq 5.5$ when $g = 5$, and that $s \leq 8.9$ when $g = 6$.

The first recursive construction was given in [134]. There, the authors gave a construction that increases the number of rows in a covering array while leaving the value of g unchanged. This was strengthened in a paper by Cohen and Fredman [130]:

Theorem 11.2 *If there exist covering arrays* $CA(k_1, g)$ *and* $CA(k_2, g)$ *with* c_1 *and* c_2 *columns respectively, then there exists a covering array* $CA(k_1 k_2 + 1, g)$ *with* $c_1 + c_2 - g$ *columns.*

In [137] constructive techniques are given for covering arrays that are along the lines of some of the more well-known design-theoretic techniques. These include Wilson-type theorems using covering arrays with holes and the use of pairwise balanced designs. A table of explicit values of $n(k, g)$ for all $3 \leq g \leq 7$ and $2 \leq k \leq 50$ can be found in that paper.

In [128, 129, 131], the AETG system implemented at Bellcore to design efficient test sets is discussed. This system employs generalizations of covering arrays with the property that the number of symbols allowed in each row may differ. The success of their AETG system in finding such covering arrays has led to useful testing schemes.

Covering arrays for t-way interactions when $t > 2$ are treated in [126, 135, 136] and constructions when $t = 3$ are given in [127] using techniques developed initially for orthogonal arrays.

12 Disk layout and striping

This section examines the use of designs to balance access to a secondary storage device.

Combinatorial Designs in Communications 79

12.1 The application

We are again concerned with arrays of disks, as in Section 4. Here we concentrate on the application when erasure of any one disk can be corrected. We focus on balancing workload on individual disks in the disk array. In the simplest situation, if there are v disks, a *parity stripe* consists of $v-1$ blocks of data together with one *parity block* consisting of the modulo 2 sum of the data blocks. Placing one block from the stripe on each disk permits reconstruction if one disk fails. However, this reconstruction requires access to all remaining disks. Instead we can choose parity stripes to contain blocks on $k < v$ of the disks; then k is the *parity stripe size*. Although the overhead in storage increases from $\frac{1}{v}$ to $\frac{1}{k}$, only $k-1$ disks need be read in order to perform a reconstruction. This technique is *parity declustering*.

Parity declustering requires that we select the parity stripes, and the location of the parity block within each stripe, i.e. a *data layout*. We suppose that each of the v disks can store exactly s blocks. Altogether the vs blocks must be allocated either as data or parity blocks in stripes. Holland and Gibson [141] observed that in any layout, each stripe must have all blocks on different disks to permit reconstruction; parity blocks must occur in numbers as nearly equal as possible on each disk to balance disk workload; every two disks must be in a number of stripes together that is as nearly equal as possible, in order to balance disk workload resulting from reconstruction; and the task of determining where a block is stored in the disk array must be efficient.

12.2 The connection to designs

Muntz and Lui [143] suggested that stripes of the disk layout be treated as blocks in a design whose points are the disks. Then Holland and Gibson's first and third conditions are met by any block design. Even distribution of parity units requires that each block have a point designated to serve as the disk for the parity block, but the number of times each point serves as such a representative is to be as equal as possible for all points. Holland and Gibson [141] suggested repeating the blocks of the designs k times, where k is the block size, so that each point in a block can serve as the representative of one copy. Schwabe and Sutherland [144] established that such replication is not needed, proving via network flow techniques that occurrences of points as representatives can be equalized. Indeed, this result is implied by a stronger result of Levi [142].

In the connection with designs, the fourth requirement of Holland and Gibson is perhaps the most problematical. We cannot reasonably store the entire disk layout in array format and undertake table lookups whenever a block is to be mapped to a disk location. For this reason, effort has concentrated on properties of designs that make them quick to generate 'on the fly' [144]. This problem was addressed by Alvarez et al. [138]. They also examined disk layouts that permit reconstruction from multiple failures.

Mapping data to memory arises elsewhere in connection with designs, for example in the processing of partial match queries (Section 10), and in the mapping of matrix data to parallel memory modules. In the latter context, latin squares in which certain subarrays always contain distinct symbols arise [139, 140].

13 (t, m, s)-nets and numerical integration

Until this point, we have concentrated on connections focusing on communications, cryptography, and networking. This final section examines combinatorial tools that have found application to computation of definite integrals. This has had particular impact in the area of finance [146]. The principal current application is in the financial arena, where they are used to price exotic options. Nevertheless, the connection described affords a technique of general applicability. Indeed, Niederreiter [157] provides applications for the same combinatorial tools in pseudo-random number generation, which in turn are employed in cryptographic protocols.

13.1 The application

The Monte Carlo method is widely used in simulation and numerical computation, and applies particularly when a numerical approximation is needed to the value of a function whose exact value is not known or easily computed. Typically, a sample space of possible values to be explored is constructed. The dimensions of the sample space represent variables in the problem specification, and the range of values in each dimension represents the interval of possible values of the associated variable. Scaling and translating each variable so that its range is continuous from 0 to 1, and assuming that variables are treated independently of one another, we can take the sample space to be a unit cube in the associated number of dimensions. Typical Monte Carlo sampling selects points from the unit cube uniformly distributed in each dimension, performs a calculation using the values of the variables at each point, and aggregates the function values over all data points chosen.

When the unit cube has many dimensions, or the function to be evaluated is very irregular within the sample space, the Monte Carlo method often requires the selection of a number of data points which is too large to permit effective computation. The major obstacle is that dramatic local changes within the sample space are only detected by placing a sample point 'near' the change. While we want to preserve the random nature of the point spacing in the sample space, we also want to ensure that all small regions of the sample space receive their fair share of the sample points to be placed. To do this, the cube is partitioned into smaller parallelepipeds, and each parallelepiped is required to receive the correct share of points. The goal is to ensure that every parallelepiped, not just the chosen ones, receives a share of the points which

Combinatorial Designs in Communications

is at least approximately commensurate with its share of the total volume.

Let $\mathcal{I}_s$ be the unit cube of dimension s. Let $b \geq 2$ and $m \geq 1$ be integers, let $d_1, \ldots, d_s$ be nonnegative integers, and let $a_1, \ldots, a_s$ be integers satisfying $0 \leq a_i < b^{d_i}$. An *elementary interval in base b* in $\mathcal{I}_s$ is the parallelepiped of points $(x_1, \ldots, x_s)$ satisfying $a_i \leq x_i b^{d_i} < a_i + 1$. Let $s \geq 1$, $b \geq 2$ and $m \geq t \geq 0$ be integers. A (t, m, s)-*net in base b* is a multiset M of b^m points in $\mathcal{I}_s$ so that every elementary interval in base b in which $\sum_{i=1}^{s} d_i = m - t$ (i.e. the parallelepiped has volume b^{t-m}) contains precisely b^t points in M.

For a collection of σ points in the unit s-cube, the average number of points in a parallelepiped of volume v is σv. The *local discrepancy* of a particular parallelepiped is the difference between this average and the actual number of points found. The *discrepancy* is the maximum of the local discrepancies over all parallelepipeds. The main motivation for (t, m, s)-nets is that they provide point collections with "low" discrepancy. For certain elementary intervals, in fact, the discrepancy is guaranteed to be zero [157].

13.2 The connection to designs

Two (0,2,2)-nets in base 3 are shown in Figure 2. Each consists of $9 = 3^2$ points placed in a unit square. Lines have been drawn vertically and horizontally at multiples of $\frac{1}{3^2}$ to assist in the verification. Points are shown as black circles, but the actual point should be taken to be the centre of the circle shown.

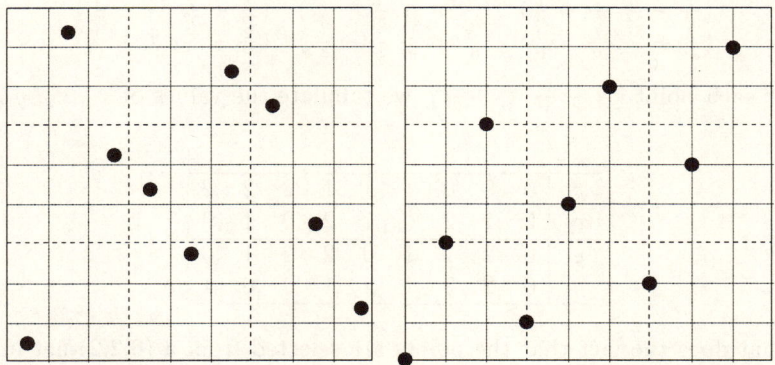

Figure 2: Two (0,2,2)-nets in base 3

To verify that each diagram is a (0,2,2)-net in base 3, we must check that every elementary interval of area $\frac{1}{3^2}$ contains exactly one point. There are three types of such elementary intervals in this case. The first type contains horizontal stripes of width 1 and height $\frac{1}{3^2}$, defined by an integer d satisfying $0 \leq d < 9$, including the points with $0 \leq x < 1$ and $\frac{d}{3^2} \leq y < \frac{d+1}{3^2}$. The second type contains vertical stripes of height 1 and width $\frac{1}{3^2}$, defined by an integer c satisfying $0 \leq c < 9$, including the points with $\frac{c}{3^2} \leq x < \frac{c+1}{3^2}$ and $0 \leq y < 1$. The third type contains subsquares of height $\frac{1}{3}$ and width $\frac{1}{3}$, defined by integers c and d satisfying $0 \leq c < 3$ and $0 \leq d < 3$, including the points with $\frac{c}{3^1} \leq x < \frac{c+1}{3^1}$ and $\frac{d}{3^1} \leq y < \frac{d+1}{3^1}$. There are nine of each type of elementary interval in the example given.

Now consider the (0,2,2)-net on the left. Within each small subsquare of side $\frac{1}{3^2}$, when a point is present, any relocation of that point within the small subsquare does not alter the fact that the diagram forms a (0,2,2)-net. Indeed we can encode each such small subsquare by indicating the (x, y) coordinate of its lower left corner, so that the coordinates are integer multiples of $\frac{1}{3^2}$. To specify the position of the point, it is sufficient to specify the coordinates of the lower left corner of the small subsquare in which it lies. For practical reasons, it may be reasonable to displace points from these coordinate positions. However, when finding (t, m, s)-nets in base b, we can always assume that points are placed at positions in the unit cube in which each coordinate is a multiple of $\frac{1}{b^m}$.

Carrying this further in our example, we can write the x and y coordinates which are multiples of $\frac{1}{3^2}$ in the form $\frac{x_1}{3^1} + \frac{x_2}{3^2}$ and $\frac{y_1}{3^1} + \frac{y_2}{3^2}$, where x_1, x_2, y_1, and y_2 are each one of the integers 0, 1, or 2 (and, in general, are nonnegative integers less than b). Let us then examine the coordinates of the nine points in the first (0,2,2)-net. They are

$$\left(\frac{0}{3^1} + \frac{0}{3^2}, \frac{0}{3^1} + \frac{0}{3^2}\right) \quad \left(\frac{0}{3^1} + \frac{1}{3^2}, \frac{2}{3^1} + \frac{2}{3^2}\right) \quad \left(\frac{0}{3^1} + \frac{2}{3^2}, \frac{1}{3^1} + \frac{2}{3^2}\right)$$
$$\left(\frac{1}{3^1} + \frac{0}{3^2}, \frac{1}{3^1} + \frac{1}{3^2}\right) \quad \left(\frac{1}{3^1} + \frac{1}{3^2}, \frac{0}{3^1} + \frac{2}{3^2}\right) \quad \left(\frac{1}{3^1} + \frac{2}{3^2}, \frac{2}{3^1} + \frac{1}{3^2}\right)$$
$$\left(\frac{2}{3^1} + \frac{0}{3^2}, \frac{2}{3^1} + \frac{0}{3^2}\right) \quad \left(\frac{2}{3^1} + \frac{1}{3^2}, \frac{1}{3^1} + \frac{0}{3^2}\right) \quad \left(\frac{2}{3^1} + \frac{2}{3^2}, \frac{0}{3^1} + \frac{1}{3^2}\right)$$

For each point $\left(\frac{x_1}{3^1} + \frac{x_2}{3^2}, \frac{y_1}{3^1} + \frac{y_2}{3^2}\right)$, we tabulate the values of x_1, x_2, y_1, and y_2:

x_1	0	0	0	1	1	1	2	2	2
x_2	0	1	2	0	1	2	0	1	2
y_1	0	2	1	1	0	2	2	1	0
y_2	0	2	2	1	2	1	0	0	1

What does the fact that the points are selected from a (0,2,2)-net imply? In each horizontal stripe which is an elementary interval, there is exactly one point. This requires that among the nine choices of (x_1, x_2), we find each ordered pair in which each element is 0, 1, or 2 exactly once. It can be verified that this condition is met in the table given. Two tuples $(g_1, \ldots, g_{b^s})$ and $(h_1, \ldots, h_{b_s})$ with elements from $0, 1, \ldots, b-1$ are called *orthogonal* when the ordered pairs (g_i, h_i) for $1 \leq i \leq b^s$ contain every ordered pair of elements from $0, 1, \ldots, b-1$ exactly b^{s-2} times. Hence, the requirement on horizontal stripes can be simply stated as a requirement that the rows x_1 and x_2 are orthogonal. In the same way, requiring a single point in each vertical stripe

Combinatorial Designs in Communications

which is an elementary interval is the same as requiring that rows y_1 and y_2 are orthogonal. Finally, the requirement that every $\frac{1}{3} \times \frac{1}{3}$ subsquare which is an elementary interval contains a single point is equivalent to the statement that rows x_1 and y_1 are orthogonal.

Selecting any four tuples of length 9 with elements 0, 1, or 2 and indexing them by x_1, x_2, y_1, and y_2, we find that when the rows x_1 and y_1 are orthogonal, the rows indexed by x_1 and x_2 are orthogonal, and the rows indexed by y_1 and y_2 are orthogonal, the process can be reversed to produce a (0,2,2)-net. Choosing these four tuples is not as difficult as it might seem. The reason is simple. If we choose the tuple indexed by y_2 to be the same as that indexed by x_1, and the tuple indexed by x_2 to be the same as that indexed by y_1, the requirements are met. In fact, then, we need only select two tuples in this case, to form:

x_1	0	0	0	1	1	1	2	2	2
x_2	0	2	1	1	0	2	2	1	0
y_1	0	2	1	1	0	2	2	1	0
y_2	0	0	0	1	1	1	2	2	2

Now using the entries in this table to determine locations of nine points again yields a (0,2,2)-net in base 3. In fact, it is the second (0,2,2)-net shown in Figure 2.

The natural question is to attempt to produce nets in higher dimensions. Our example of a (0,2,2)-net leads us to ask when we can form a $(0, 2, s)$-net in base 3, and in particular, what maximum value for s is achievable. With this in mind, we next describe a (0,2,4)-net in base 3. First form the array:

$$\begin{array}{ccccccccc} 0 & 0 & 0 & 1 & 1 & 1 & 2 & 2 & 2 \\ 0 & 1 & 2 & 0 & 2 & 1 & 0 & 2 & 1 \\ 0 & 1 & 2 & 1 & 0 & 2 & 2 & 1 & 0 \\ 0 & 1 & 2 & 2 & 1 & 0 & 1 & 0 & 2 \end{array}$$

This array has the important property that any two distinct rows are orthogonal. We shall construct coordinates for nine points in the unit 4-dimensional cube. Each point therefore needs four coordinates, each of which is written in the form $\frac{x_1}{3^1} + \frac{x_2}{3^2}$. We use the array given to determine the four coordinates of each point as follows. Each point corresponds to a column $(c_1, c_2, c_3, c_4)^{\mathrm{T}}$ of the array. The point in the net can then be written as

$$\left(\frac{c_1}{3^1} + \frac{c_2}{3^2}, \frac{c_2}{3^1} + \frac{c_3}{3^2}, \frac{c_3}{3^1} + \frac{c_4}{3^2}, \frac{c_4}{3^1} + \frac{c_1}{3^2}\right).$$

For example, the column $(1, 2, 0, 1)^{\mathrm{T}}$ corresponds to a point placed at $(\frac{5}{9}, \frac{6}{9}, \frac{1}{9}, \frac{4}{9})$.

In this way, we obtain a (0,2,4)-net in base 3. Niederreiter [157] established the equivalence between $(0,2,s)$-nets in base b and OA(s,b)s; indeed, we saw this correspondence earlier. The correspondence also can be used to establish nonexistence of certain (0,2,s)-nets in base b, by employing the observation that an OA(s,b) exists only if $s \leq b+1$ (see, for example, [2]). Equality can occur, and indeed is known to occur when b is a prime or a power of a prime number. Nevertheless, for certain choices of b, it is known that equality does *not* occur. Determining the largest value of s for which an OA(s,b) exists is a challenging open problem. When $b = 6$, for example, the largest value admitted for s is 3. When $b = 10$, the largest value for s is, at present, unknown. Mullen [153] discussed some remarkable connections with various combinatorial objects; see also [2] for results and existence tables for OA(s,b), given there in the equivalent formulation as mutually orthogonal latin squares.

Now let us examine the existence of (t, m, s)-nets in base b more generally. A (t, m, s)-net in base b gives a (u, n, r)-net in base b whenever $t \leq u \leq n \leq m$ and $1 \leq r \leq s$. The goal is to construct (t, m, s)-nets with $m - t$ "large", since they have stronger uniformity properties. When $m - t = 0$, any set of b^m points can be chosen. When $m - t = 1$, take the points $(i/b, \ldots, i/b)$ for $i = 0, 1, \ldots, b - 1$, each b^{m-1} times. Hence the cases of interest are for $m \geq t + 2$. Niederreiter [158] and Mullen and Whittle [156] showed that the existence of an $OA_{b^t}(2, s, b)$ is equivalent to the existence of a $(t, t + 2, s)$-net in base b. The equivalence mirrors that for $(0, 2, s)$-nets in base b that we have seen. When $m - t \geq 3$, one requires a more general combinatorial model. Lawrence [149] generalized the definition of orthogonal array as follows: A *cubical orthogonal array* $COA_\lambda(t, k, n)$ is a $t \times k \times \lambda n^t$ array $C = (c_{ijk})$ with elements from a set of size n, meeting a condition on certain subarrays as follows. A set $S \subseteq \{(x_i, y_i) : 1 \leq i \leq t,\ 1 \leq x_i \leq t,\ 1 \leq y_i \leq k\}$ is a *qualifying collection* of rows when $(x_i, y_i) \in S$ and $x_i > 1$ implies that $(x_i - 1, y_i) \in S$. Figure 3 shows three qualifying collections of rows with $t = 4$ and $k = 6$. The first is $\{(1,1), (2,1), (3,1), (4,1)\}$; the second is $\{(1,2), (1,4), (1,6), (2,6)\}$; and the third is $\{(1,3), (2,3), (3,3), (1,5)\}$. These are shown by different shading in the figure, and are meant only to illustrate the definition, and many more qualifying collections are present. Figure 3 depicts a 2-dimensional $t \times k$ 'slice' of the 3-dimensional COA.

The qualifying collections correspond exactly to the elementary intervals. When S is a qualifying collection of rows, define a $t \times \lambda n^t$ array $A = (a_{ik})$ by setting $a_{ik} = c_{x_i, y_i, k}$; then, A contains every $t \times 1$ column vector the same number λ of times. These are often called *generalized orthogonal arrays*, but we employ the adjective 'cubical' to suggest the three-dimensional structure of the array. Cubical OAs generalize orthogonal arrays in the following sense: In a $COA_\lambda(t, k, n)$ with $t \leq k$, the $k \times \lambda n^t$ array obtained by setting the first coordinate equal to 1 is an $OA_\lambda(t, k, n)$.

The construction of a (t, m, s)-net in base b from a $COA_{b^t}(m - t, s, b)$ proceeds as follows. For each p satisfying $1 \leq p \leq b^m$, consider the $(m - t) \times s$ array A_p whose (i, j) entry is the (i, j, p) entry of the COA. Each selection of p corresponds to one of the b^m points in the (t, m, s)-net, and A_p is used to

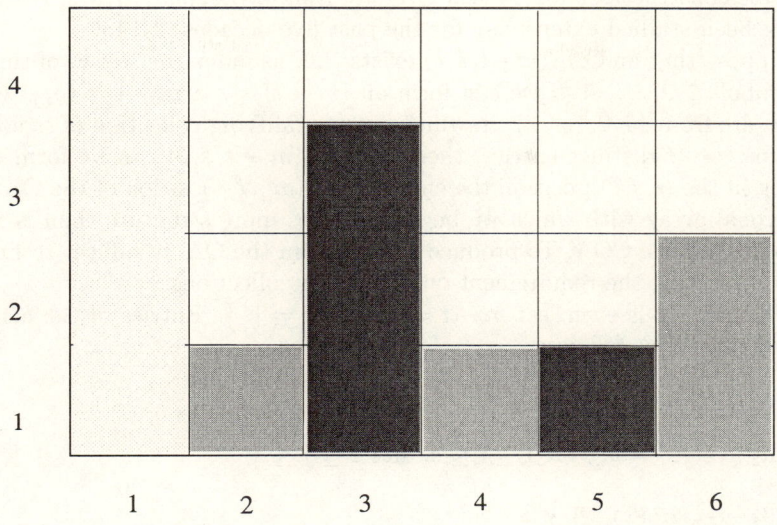

Figure 3: Qualifying collections

determine its position in the s-dimensional unit cube as follows. Denote the entry of A_p in position (i, j) by $a_{i,j}$, for $1 \leq i \leq m - t$ and $1 \leq j \leq s$. Each column of A_p determines the coordinate of the pth point of the net. The jth column gives the coordinate of the point in the jth dimension as

$$\frac{a_{1,j}}{b^1} + \frac{a_{2,j}}{b^2} + \cdots + \frac{a_{(m-t),j}}{b^{m-t}},$$

or equivalently, as

$$\sum_{i=1}^{m-t} \frac{a_{i,j}}{b^i}.$$

Carrying out this computation for each column j with $1 \leq j \leq s$ determines the position of the point in the s-dimensional unit cube completely. Of course, the position of the point can (and, to avoid bias, should) be displaced from the 'corner' by adding (random) values less than $\frac{1}{b^{m-t}}$ to each coordinate. This process is then repeated for each of the b^m points.

It may seem unnecessarily complicated to employ a cubical OA to construct a (t, m, s)-net. However, Lawrence [149] showed that the existence of a (t, m, s)-net in base b is *equivalent* to the existence of a $\text{COA}_{b^t}(m - t, s, b)$. Consequently, every construction of a (t, m, s)-net effectively relies upon the construction of some cubical orthogonal array.

For this reason, the primary question about the construction of (t, m, s)-nets is the construction of cubical OAs. We turn to this next. Most of the available constructions for cubical OAs are from orthogonal arrays, the latter having been studied extensively for the past five decades [2, 148].

Suppose that an $\text{OA}_{b^t}(m-t, \widehat{s}, b)$ exists. Let us index the $\widehat{s}$ rows of this OA by symbols $1, 2, \ldots, \widehat{s}$. If we can form an $(m - t) \times s$ array $A = (a_{i,j})$ whose entries are from $\{1, 2, \ldots, \widehat{s}\}$, in which every qualifying collection of size $m - t$ contains $m - t$ distinct entries, then a $\text{COA}_{b^t}(m - t, s, b)$ can be formed, by placing in the (i, j, ℓ) position the entry in the $(a_{i,j}, \ell)$ position of the OA. The orthogonal array with which we began provides more structure than is really needed to get the COA. To produce a COA from the OA, it suffices to find an array A meeting the requirement on qualifying collections.

When $m - t$ is even, set $m - t = 2e$ and $s = \lfloor \frac{\widehat{s}}{e} \rfloor$. Entries of the array A are determined as follows:

1. $a_{i,j} = s(i-1) + j$ for $1 \leq i \leq e$, and $1 \leq j \leq s$;

2. $a_{2e+1-i, j-1} = a_{i,j}$ for $1 \leq i \leq e$, and $2 \leq j \leq s$;

3. $a_{2e+1-i, s} = a_{i,1}$ for $1 \leq i \leq e$.

When $m - t$ is odd, set $m - t = 2e + 1$, and $s = \lfloor \frac{\widehat{s}-1}{e} \rfloor$. Entries of the array A are determined as follows:

1. $a_{i,j} = s(i-1) + j$ for $1 \leq i \leq e$, and $1 \leq j \leq s$;

2. $a_{e+1, j} = es + 1$ for $1 \leq j \leq s$;

3. $a_{2e+2-i, j-1} = a_{i,j}$ for $1 \leq i \leq e$, and $2 \leq j \leq s$;

4. $a_{2e+2-i, s} = a_{i,1}$ for $1 \leq i \leq e$.

For example, when $\widehat{s} = 14$ and $m - t = 5$, we find $s = 6$ and $e = 2$, and form the array

$$A = \begin{pmatrix} 2 & 3 & 4 & 5 & 6 & 1 \\ 8 & 9 & 10 & 11 & 12 & 7 \\ 13 & 13 & 13 & 13 & 13 & 13 \\ 7 & 8 & 9 & 10 & 11 & 12 \\ 1 & 2 & 3 & 4 & 5 & 6 \end{pmatrix}.$$

The symbol 14 is not used in this example, although the orthogonal array provides 14 rows. Using symbol 14 in place of some occurrences of 13 in the third row of A would result in further elementary intervals (with smaller volumes) containing their correct share of the points. There is no need in general to construct the COA explicitly to determine the point positions in the (t, m, s)-net. If an $\text{OA}_\lambda(5, 14, b)$ exists, each of the λb^5 columns determines the position of a point as follows. Let $(d_1, d_2, \ldots, d_{14})^\text{T}$ be a column of the OA. Then we use A to select rows from the OA, with the jth column of A determining the coordinate of the point in the jth dimension in the net.

The first coordinate, for example, is $\frac{d_1}{b^1} + \frac{d_7}{b^2} + \frac{d_{13}}{b^3} + \frac{d_8}{b^4} + \frac{d_2}{b^5}$. This provides a construction of COAs from OAs.

There are numerous other constructions for both COAs and (t, m, s)-nets. We do not enumerate them here, but refer the reader to [147, 150, 154]. See [151, 152] for the strongest general necessary conditions.

14 About things not said

The baker's dozen of applications described here provide compelling evidence that combinatorial design theory is finding many diverse connections to practical themes. These in turn suggest new and challenging research problems within design theory. We would be remiss, however, if we failed to point out that the thirteen topics discussed here are just a sample. Among other important applications are two-point sampling, derandomization, sequences with zero autocorrelation, and quorum systems. Somewhat further afield from our focus here, there are strong connections with sports tournaments and lotteries. See [2, 3, 9] for these and other interesting topics. Our goal has not been to be encyclopaedic, so we leave to the reader the enjoyable task of finding further unexpected applications of designs.

Acknowledgements

Research of the authors is supported by grants DAAG55-98-1-0272 from ARO (Colbourn) and RGPIN 203114-98 from NSERC Canada (Stinson).

References

0 Background

[1] Th. Beth, D. Jungnickel & H. Lenz, *Design Theory*, Cambridge University Press, Cambridge (1986).

[2] C. J. Colbourn & J. H. Dinitz (eds.), *CRC Handbook of Combinatorial Designs*, CRC Press, Boca Raton FL (1996).

[3] C. J. Colbourn & P. C. van Oorschot, Applications of combinatorial designs in computer science, *ACM Computing Surveys*, **21** (1989), 223–250.

[4] C. J. Colbourn & A. Rosa, *Triple Systems*, Oxford University Press, Oxford (1999).

[5] J. H. Dinitz & D. R. Stinson (eds.), *Contemporary Design Theory: A Collection of Surveys*, John Wiley & Sons, New York (1992).

[6] D. R. Hughes & F. C. Piper, *Projective Planes*, Springer Verlag, New York (1973).

[7] F. J. MacWilliams & N. J. A. Sloane, *The Theory of Error-Correcting Codes*, North-Holland, Amsterdam (1978).

[8] D. Raghavarao, *Constructions and Combinatorial Problems in Design of Experiments*, John Wiley & Sons, New York (1971).

[9] D. R. Stinson, Combinatorial designs and cryptography, in *Surveys in Combinatorics, 1993* (ed. K. Walker), Cambridge University Press, London (1993), pp. 257–287.

1 Optical orthogonal codes

[10] C. Argon & H. Farooq Ahmad, Optimal optical orthogonal code design using difference sets and projective geometry, *Optics Communications*, **118** (1995), 505–508.

[11] C. M. Bird & A. D. Keedwell, Design and applications of optical orthogonal codes—a survey, *Bulletin of the Institute for Combinatorics and its Applications*, **11** (1994), 21–44.

[12] S. Bitan & T. Etzion, On constructions for optimal optical orthogonal codes, *Lecture Notes in Computer Science*, **781** (1994), 111–125.

[13] M. Buratti, A powerful method for constructing difference families and optimal optical orthogonal codes, *Designs Codes and Cryptography*, **5** (1995), 13–25.

[14] K. Chen, G. Ge & L. Zhu, Starters and related codes, *Journal of Statistical Planning and Inference*, in press.

[15] F. R. K. Chung, J. A. Salehi & & V. K. Wei, Optical orthogonal codes: design, analysis and applications, *IEEE Transactions on Information Theory*, **35** (1989), 595–604. Correction: *IEEE Transactions on Information Theory*, **38** (1992), 1429.

[16] J. H. Dinitz & D. R. Stinson, Room squares and related designs, in *Contemporary Design Theory: A Collection of Surveys* (eds. J. H. Dinitz & D. R. Stinson), John Wiley & Sons, New York (1992).

[17] S. V. Maric & V. K. N. Lau, Multirate fiber-optic CDMA: System design and performance analysis, *Journal of Lightwave Technology*, **16** (1998), 9–17.

[18] S. V. Maric, O. Moreno & C. Corrada, Multimedia transmission in fiber-optic LANs using optical CDMA, *Journal of Lightwave Technology*, **14** (1996), 2149–2153.

[19] J. A. Salehi, Code division multiple-access techniques in optical fibre networks – part I: Fundamental principles, *IEEE Transactions on Information Theory*, **37** (1989), 824–833.

[20] G. C. Yang, Some new families of optical orthogonal codes for code-division multiple-access fibre-optic networks, *IEEE Transactions on Communications*, **142** (1995), 363–368.

[21] J. X. Yin, Some combinatorial constructions for optical orthogonal codes, *Discrete Mathematics*, **185** (1998), 201–219.

2 Synchronous multiple access to channels

[22] G. E. Atkin & H. P. Corrales, An efficient modulation/coding scheme for MFSK systems on bandwidth controlled channels, *IEEE Transactions on Sel. Areas Communications*, **7** (1989), 1396–1401.

[23] G. E. Atkin, D. A. Fares & H. P. Corrales, Coded multipulse position modulation in a noisy optical channel, *Microwave Optical Technology Letters*, **2** (1989), 336–340.

[24] C. J. Colbourn, Weakly union-free maximum packings, *Annals Combinatorics*, in press.

[25] C. J. Colbourn & S. Zhao, Maximum Kirkman signal sets for synchronous uni-polar multi-user communication systems, preprint, Computer Science, University of Vermont, 1998.

[26] D. A. Fares, Concatenated coding for multipulse signaling in noisy optical channels, *Microwave Optical Technology Letters*, **4** (1991), 359–361.

[27] D. A. Fares, W. H. Abul-Shohoud, N. A. Raslan & M. A. Nassef, $\delta_{\max}$ detection of multipulse signaling in noisy optical channels, *Microwave Optical Technology Letters*, **5** (1992), 269–273.

[28] S. Zhao, Application of BIBDs in MT-MFSK signal set design for multiplexing bursty sources, PhD thesis, University of Technology, Sydney, 1998.

[29] S. Zhao, K. W. Yates & K. Yasukawa, Application of Kirkman designs in joint detection multiple access schemes, *Proceedings of the International Symposium on Spread Spectrum Techniques and Applications*, **2** (1996), 857–861.

3 Group testing and superimposed codes

[30] D. J. Balding & D. C. Torney, Optimal pooling designs with error detection, *Journal of Combinatorial Theory, Series A*, **74** (1996), 131–140.

[31] D. J. Balding & D. C. Torney, The design of pooling experiments for screening a clone map, *Fungal Genetics and Biology*, **21** (1997), 302–307.

[32] T. Berger & J. W. Mandell, Bounds on the efficiency of two-stage group testing, preprint, Industrial Engineering, Cornell University, 1998.

[33] T. Berger, N. Mehravari, D. Towsley & J. Wolf, Random multiple-access communications and group testing, *IEEE Transactions on Communications*, **32** (1984), 769–778.

[34] W. J. Bruno, D. J. Balding, E. H. Knill, D. Bruce, C. Whittaker, N. Doggett, R. Stallings & D. C. Torney, Design of efficient pooling experiments, *Genomics*, **26** (1995), 21–30.

[35] Y. M. Chee, C. J. Colbourn & A. C. H. Ling, Weakly union-free twofold triple systems, *Annals Combinatorics*, **1** (1997), 215–225.

[36] D. Coppersmith & J. B. Shearer, New bounds for union-free families of sets, *Electronic Journal of Combinatorics*, **5** (1998), #R39.

[37] R. Dorfman, The detection of defective members of a large population, *Annals of Mathematical Statistics*, **14** (1943), 436–440.

[38] D. Z. Du & F. K. Hwang, *Combinatorial Group Testing and Its Applications*, World Scientific, Singapore (1993).

[39] A. D'yachkov, V. Rykov & A. M. Rashad, Superimposed distance codes, *Problems Control and Information Theory*, **18** (1989), 237–250.

[40] P. Erdős, P. Frankl & Z. Füredi, Families of finite sets in which no set is covered by the union of two others, *Journal of Combinatorial Theory, Series A*, **33** (1982), 158–166.

[41] P. Frankl & Z. Füredi, A new extremal property of Steiner triple systems, *Discrete Mathematics*, **48** (1984), 205–212.

[42] W. H. Kautz & R. R. Singleton, Nonrandom binary superimposed codes, *IEEE Transactions on Information Theory*, **10** (1964), 363–377.

[43] L. Riccio & C. J. Colbourn, An upper bound for disjunct matrices, preprint, Computer Science, University of Vermont, 1998.

[44] M. Ruszinkó, On the upper bound of the size of the r-cover-free families, *Journal of Combinatorial Theory, Series A*, **66** (1994), 302–310.

[45] D. R. Stinson, Tran van Trung & R. Wei, Secure frameproof codes, key distribution patterns, group testing algorithms, and related structures, *Journal of Statistical Planning and Inference*, in press.

[46] D. R. Stinson & R. Wei, Combinatorial properties and constructions of traceability schemes and frameproof codes, *SIAM Journal of Discrete Mathematics*, **11** (1998), 41–53.

[47] F. Vakil & M. Parnes, On the structure of a class of sets useful in non-adaptive group testing, *Journal of Statistical Planning and Inference*, **39** (1994), 57–69.

[48] J. K. Wolf, Born again group testing: multiaccess communications, *IEEE Transactions on Information Theory*, **IT-31** (1985), 185–191.

4 Erasure codes and information dispersal

[49] A. Albanese, J. Blömer, J. Edmonds, M. Luby & M. Sudan, Priority encoding transmission, *IEEE Transactions on Information Theory*, **42** (1996), 1737–1744.

[50] N. Alon, J. Edmonds & M. Luby, Linear time erasure codes with nearly optimal recovery, in *Proceedings of the 36th Annual Symposium on Foundations of Computer Science* IEEE, Piscataway NJ (1995), pp. 512–519.

[51] G. M. Amdahl, Validity of the single processor approach to achieving large scale computing capabilities, in *Proceedings of the 1967 Spring Joint Computer Conference* AFIPS, Washington DC **30** (1967), pp. 483–485.

[52] Y. M. Chee, C. J. Colbourn & A. C. H. Ling, Asymptotically optimal erasure-resilient codes for large disk arrays, preprint, Computer Science, University of Vermont, 1998.

[53] P. M. Chen, E. K. Lee, G. A. Gibson, R. H. Katz & D. A. Patterson, RAID: High-performance, reliable secondary storage, *ACM Computing Surveys*, **26** (1994), 145–185.

[54] P. Elias, Coding for two noisy channels, in *Information Theory: Third London Symposium*, Butterworth, London (1955), pp. 61–76.

[55] P. Erdős, Problems and results in combinatorial analysis, *Creation in Mathematics*, **9** (1976), 25.

[56] G. A. Gibson, *Redundant Disk Arrays: Reliable, Parallel Secondary Storage*, MIT Press, Cambridge, Mass. (1992).

[57] L. Hellerstein, G. A. Gibson, R. M. Karp, R. H. Katz & D. A. Patterson, Coding techniques for handling failures in large disk arrays, *Algorithmica*, **12** (1994), 182–208.

[58] H. Lefmann, P. Pudlák & P. Sacický, On sparse parity check matrices, *Designs Codes and Cryptography*, **12** (1997), 107–130.

[59] A. C. H. Ling, C. J. Colbourn, M. J. Grannell & T. S. Griggs, Construction techniques for anti-Pasch Steiner triple systems, preprint, Computer Science, University of Vermont, 1997.

[60] D. A. Patterson & J. L. Hennessy, *Computer Organization and Design: The Hardware/Software Interface*, Morgan Kaufmann, San Mateo, CA (1994).

[61] M. O. Rabin, Efficient dispersal of information for security, load balancing, and fault tolerance, *Journal of the Association for Computing Machinery*, **36** (1989), 335–348.

5 Threshold and ramp schemes

[62] G. R. Blakley, Safeguarding cryptographic keys, in *Proceedings of the National Computer Conference 1979, American Federation of Information Processing Societies Proceedings* 48, (1979), pp. 313–317.

[63] G. R. Blakley and C. Meadows, Security of ramp schemes, *Lecture Notes in Computer Science*, **196** (1985), 242–268.

[64] C. Blundo, A. De Santis & D. R. Stinson, On the contrast in visual cryptography schemes, *Journal of Cryptology*, in press.

[65] T. Hofmeister, M. Krause & H. U. Simon, Contrast-optimal k out of n secret sharing schemes in visual cryptography, *Lecture Notes in Computer Science*, **1276** (1997), 176–185.

[66] W.-A. Jackson & K. M. Martin, A combinatorial interpretation of ramp schemes, *Australasian Journal of Combinatorics*, **14** (1996), 51–60.

[67] K. M. Martin, Discrete Structures in the Theory of Secret Sharing, PhD Thesis, University of London, 1991.

[68] M. Naor & A. Shamir, Visual cryptography, *Lecture Notes in Computer Science*, **950** (1995), 1–12.

[69] A. Shamir, How to share a secret, *Communications of the ACM*, **22** (1979), 612–613.

[70] D. R. Stinson, Visual cryptography and threshold schemes, *Dr. Dobb's Journal*, **April** (1998), 36–43.

[71] D. R. Stinson & S. A. Vanstone, A combinatorial approach to threshold schemes, *SIAM Journal of Discrete Mathematics*, **1** (1988), 230–236.

6 Authentication codes

[72] E. N. Gilbert, F. J. MacWilliams & N. J. A. Sloane, Codes which detect deception, *Bell System Technical Journal*, **53** (1974), 405–424.

[73] K. Kurosawa, K. Okada, H. Saido & D. R. Stinson, New combinatorial bounds for authentication codes and key predistribution schemes, *Designs Codes and Cryptography*, **15** (1998), 87–100.

[74] J. L. Massey, Cryptography – a selective survey, in *Digital Communications* North-Holland, Amsterdam (1986), pp. 3–21.

[75] R. S. Rees and D. R. Stinson, Combinatorial characterizations of authentication codes II, *Designs Codes and Cryptography*, **7** (1996), 239–259.

[76] G. J. Simmons, A survey of information authentication, in *Contemporary Cryptology, The Science of Information Integrity* IEEE Press, Piscataway NJ (1992), pp. 379–419.

[77] D. R. Stinson, The combinatorics of authentication and secrecy codes, *Journal of Cryptology*, **2** (1990), 23–49.

[78] D. R. Stinson, Combinatorial characterizations of authentication codes, *Designs Codes and Cryptography*, **2** (1992), 175–187.

7 Resilient and correlation-immune functions

[79] C. H. Bennett, G. Brassard & J. M. Robert, How to reduce your enemy's information, *Lecture Notes in Computer Science*, **218** (1986), 468–476.

[80] J. Bierbrauer, K. Gopalakrishnan & D. R. Stinson, Orthogonal arrays, resilient functions, error correcting codes and linear programming bounds, *SIAM Journal of Discrete Mathematics*, **9** (1996), 424–452.

[81] P. Camion, C. Carlet, P. Charpin & N. Sendrier, On correlation-immune functions, *Lecture Notes in Computer Science*, **576** (1992), 86–100.

[82] B. Chor, O. Goldreich, J. Håstad, J. Friedman, S. Rudich & R. Smolensky, The bit extraction problem or t-resilient functions, *26th IEEE Symposium on the Foundations of Computer Science*, (1985), 396–407.

[83] J. Friedman, On the bit extraction problem, *33rd IEEE Symposium on the Foundations of Computer Science*, (1992), 314–319.

[84] K. Gopalakrishnan & D. R. Stinson, Three characterizations of non-binary correlation-immune and resilient functions, *Designs Codes and Cryptography*, **5** (1995), 241–251.

[85] T. Siegenthaler, Correlation-immunity of nonlinear combining functions for cryptographic applications, *IEEE Transactions on Information Theory*, **30** (1984), 776–780.

[86] D. R. Stinson, Resilient functions and large sets of orthogonal arrays, *Congressus Numerantium*, **92** (1993), 105–110.

[87] D. R. Stinson, On some methods for unconditionally secure key distribution and broadcast encryption, *Designs Codes and Cryptography*, **12** (1997), 215–243.

8 Multidrop networks

[88] B. E. Aupperle & J. F. Meyer, Fault-tolerant BIBD networks, in *Proceedings of the Eighteenth International Symposium on Fault Tolerant Computing* IEEE, Piscataway NJ (1988), pp. 306–311.

[89] S. Ball, A. Blokhuis & F. Mazzocca, Maximal arcs in Desarguesian planes of odd order do not exist, *Combinatorica*, **17** (1997), 31–41.

[90] A. Barlotti, Su $\{k-n\}$-archi di un piano lineare finito, *Bolletino della Unione Matematica Italiana*, **11** (1956), 553–556.

[91] E. Berkovich & S. Berkovich, A combinatorial architecture for instruction-level parallelism, *Microprocessors and Microsystems*, **22** (1998), 23–31.

[92] J. C. Bermond, J. Bond & S. Djelloul, Dense bus networks of diameter 2, in *Interconnection Networks and Mapping and Scheduling Parallel Computations* (eds. D. F. Hsu, A. L. Rosenberg & D. Sotteau), American Mathematics Society, Providence RI (1994), pp. 9–16.

[93] J. C. Bermond, J. Bond, M. Paoli & C. Peyrat, Graphs and interconnection networks: diameter and vulnerability, in *Surveys in Combinatorics 1983* (ed. E. K. Lloyd), Cambridge University Press, Cambridge (1983), pp. 1–30.

[94] J. C. Bermond, J. Bond & J. F. Saclé, Large hypergraphs of diameter 1, in *Graph Theory and Combinatorics* (ed. B. Bollobás), Academic Press, London (1984), pp. 19–28.

[95] J.-C. Bermond & F. Ö. Ergincan, Bus interconnection networks, *Discrete Applied Mathematics*, **68** (1996), 1–15.

[96] C. J. Colbourn, Projective planes and congestion-free networks, preprint, Computer Science, University of Vermont, 1998.

[97] R. H. F. Denniston, Some maximal arcs in finite projective planes, *Journal of Combinatorial Theory*, **6** (1969), 317–319.

[98] W. W. M. Dai, Y. Kajitani & Y. Hirata, Optimal single hop multiple bus networks, in *Proceedings of the 1993 IEEE International Symposium on Circuits and Systems* IEEE, Piscataway NJ (1993), pp. 2541–2544.

[99] Z. Füredi, Maximum degree and fractional matchings in uniform hypergraphs, *Combinatorica*, **1** (1981), 155–162.

[100] J. W. P. Hirschfeld, *Projective Geometries Over Finite Fields*, second edition, Oxford University Press, Oxford (1998).

[101] J. W. P. Hirschfeld & L. Storme, The packing problem in statistics, coding theory and finite projective spaces, *Bulletin of the Belgian Mathematics Society Simon Stevin*, in press.

[102] M. D. Mickunas, Using projective geometry to design bus connection networks, in *Proceedings of the Workshop on Interconnection Networks for Parallel and Distributed Processing* (1980), pp. 47–55.

[103] D. Wiedemann, Cyclic difference covers through 133, *Congressus Numerantium*, **90** (1992), 181–185.

[104] R. Yao, T. Chen & T. Kang, An investigation of multibus multiprocessor systems, *Acta Electronica Sinica*, **18** (1990), 125–127.

[105] B. Yener, Y. Ofek & M. Yung, Combinatorial design of congestion-free networks, *IEEE/ACM Transactions on Networking*, **5** (1997), 989–1000.

[106] S. Q. Zheng, Sparse hypernetworks based on Steiner triple systems, in *Proceedings of the 1995 International Conference on Parallel Processing* IEEE, Piscataway NJ (1995), pp. I.92–I.95.

[107] S. Q. Zheng, An abstract model for optical interconnection networks, in *Parallel Computing Using Optical Interconnections* (eds. K. Li, Y. Pan & S. Q. Zheng), Kluwer Academic, Norwell, MA (1998), pp. 139–162.

9 Channel graphs and interconnection networks

[108] N. Alon, Expanders, sorting in rounds and superconcentrators of limited depth, *Proceedings of the Seventeenth ACM Symposium on the Theory of Computing*, (1985), 98–102.

[109] F. R. K. Chung, Zone-balanced networks and block designs, *Bell System Technical Journal*, **57** (1978), 2957–2981.

[110] F. R. K. Chung, On concentrators, superconcentrators, generalizers and nonblocking networks, *Bell System Technical Journal*, **58** (1979), 1765–1777.

[111] F. R. K. Chung, On switching networks and block designs II, *Bell System Technical Journal*, **59** (1980), 1165–1173.

[112] A. Ghafoor, T. R. Bashkow & I. Ghafoor, Bisectional fault-tolerant communication architecture for supercomputer systems, *IEEE Transactions on Computers*, **38** (1989), 1425–1446.

[113] J. Kilian, S. Kipnis & C. E. Leiserson, The organization of permutation architectures with bussed interconnections, *IEEE Transactions on Computers*, **39** (1990), 1346–1358.

10 Partial match queries on files

[114] K. A. S. Abdel-Ghaffar, On the number of mutually orthogonal partial latin squares, *Ars Combinatoria*, **42** (1996), 259–286.

[115] K. A. S. Abdel-Ghaffar and A. El Abbadi, Optimal disk allocation for partial match queries, *ACM Transactions on Database Systems*, **18** (1993), 132–156.

[116] M. Atici, S. S. Magliveras, D. R. Stinson & W.-D. Wei, Some recursive constructions for perfect hash functions, *Journal of Combinatorial Designs*, **4** (1996), 353–363.

[117] G. Berman, The application of difference sets to the design of a balanced multiple-valued filing scheme, *Information and Control*, **32** (1976), 128–138.

[118] R. C. Bose, C. T. Abraham & S. P. Ghosh, File organization of records with multiple-valued attributes for multi-attribute queries, in *Combinatorial Mathematics and Its Applications* (eds. R. C. Bose & T. A. Dowling), UNC Press, Chapel Hill (1969), pp. 277–297.

[119] R. C. Bose & G. G. Koch, The design of combinatorial information retrieval systems for files with multiple-valued attributes, *SIAM Journal on Applied Mathematics*, **17** (1969), 1203–1214.

[120] S. P. Ghosh & C. T. Abraham, Application of finite geometry in file organization for records with multiple-valued attributes, *IBM Journal of Research and Development*, **12** (1968), 180–187.

[121] C. Faloutsos & D. Metaxas, Declustering using error-correcting codes, *Proceedings of the ACM Symposium on Principles of Database Systems*, (1989), 253–258.

[122] T. Fujiwara, M. Ito, T. Kasami, M. Kataoka & J. Okui, Performance analysis of disk allocation method using error-correcting codes, *IEEE Transactions on Information Theory*, **37** (1991), 379–384.

[123] D. K. Ray-Chaudhuri, Combinatorial information retrieval systems for files, *SIAM Journal on Applied Mathematics*, **16** (1968), 973–992.

[124] R. L. Rivest, Partial-match retrieval algorithms, *SIAM Journal on Computing*, **5** (1976), 19–50.

[125] V. D. Tonchev, Codes, in *CRC Handbook of Combinatorial Designs* (eds. C. J. Colbourn & J. H. Dinitz), CRC Press, Boca Raton FL (1996), pp. 517–542.

11 Software testing

[126] N. Alon, Explicit construction of exponential sized families of k-independent sets, *Discrete Mathematics*, **58** (1986), 191–193.

[127] M. A. Chateauneuf, C. J. Colbourn & D. L. Kreher, Covering arrays of strength three, *Designs Codes and Cryptography*, in press.

[128] D. M. Cohen, S. R. Dalal, M. L. Fredman & G. C. Patton, The AETG system: an approach to testing software based on combinatorial design, *IEEE Transactions on Software Engineering*, **23** (1997), 437–444.

[129] D. M. Cohen, S. R. Dalal, J. Parelius & G. C. Patton, The combinatorial design approach to automatic test generation, *IEEE Software*, **13** (1996), 83–88.

[130] D. M. Cohen & M. L. Fredman, New techniques for designing qualitatively independent sets, *Journal of Combinatorial Designs*, in press.

[131] S. R. Dalal & C. L. Mallows, Factor-covering designs for testing software, *Technometrics*, **40** (1998), 234–243.

[132] L. Gargano, J. Körner & U. Vaccaro, Qualitative independence and Sperner problems for directed graphs, *Journal of Combinatorial Theory, Series A*, **61** (1992), 173–192.

[133] L. Gargano, J. Körner & U. Vaccaro, Sperner capacities, *Graphs and Combinatorics*, **9** (1993), 31–46.

[134] S. Poljak & Z. Tuza, On the maximum number of qualitatively independent partitions, *Journal of Combinatorial Theory, Series A*, **51** (1989), 111–116.

[135] G. Seroussi & N. H. Bshouty, Vector sets for exhaustive testing of logic circuits, *IEEE Transactions on Information Theory*, **34** (1988), 513–522.

[136] N. J. A. Sloane, Covering arrays and intersecting codes, *Journal of Combinatorial Designs*, **1** (1993), 51–63.

[137] B. Stevens & E. Mendelsohn, New recursive methods for transversal covers, *Journal of Combinatorial Designs*, in press.

12 Disk layout and striping

[138] G. A. Alvarez, W. A. Burkhard, L. J. Stockmeyer & F. Cristian, Declustered disk array architectures with optimal and near-optimal parallelism, in *Proceedings of the 25th ACM/IEEE International Symposium on Computer Architecture* IEEE, Piscataway NJ (1998), pp. 109–120.

[139] C. J. Colbourn & K. E. Heinrich, Conflict-free access to parallel memories, *Journal of Parallel and Distributed Computing*, **14** (1992), 193–200.

[140] K. Heinrich, K. Kim & V. K. Prasanna Kumar, Perfect Latin squares, *Discrete Applied Mathematics*, **37/38** (1992), 281–286.

[141] M. Holland & G. A. Gibson, Architectures and algorithms for on-line failure recovery in redundant disk arrays, *Journal of Parallel Distributed Databases*, **2** (1994),

[142] F. W. Levi, *Finite Geometrical Systems*, University of Calcutta, Calcutta (1942).

[143] R. Muntz & J. Lui, Performance analysis of disk arrays under failure, in *Proceedings of the Conference on Very Large Data Bases* (1990), pp. 162–173.

[144] E. J. Schwabe & I. M. Sutherland, Improved parity-declustered layouts for disk arrays, *Journal of Computer and System Sciences*, **53** (1996), 328–343.

[145] J. E. Thomas, S. J. Schwartz & W. A. Burkhard, Disk array declustering using resolvable designs, preprint, Computer Science, University of California at San Diego, 1997.

13 (t, m, s)-nets and numerical integration

[146] P. P. Boyle, M. Broadie & P. Glasserman, Monte Carlo methods for security pricing, *Journal of Economic Dynamics Control*, in press.

[147] A. T. Clayman, K. M. Lawrence, G. L. Mullen, H. Niederreiter & N. J. A. Sloane, Updated tables of parameters of (t, m, s)-nets, *Journal of Combinatorial Designs*, in press.

[148] A. S. Hedayat, N. J. A. Sloane & J. Stufken, *Orthogonal Arrays: Theory and Practice*, in press.

[149] K. M. Lawrence, A combinatorial characterization of (t, m, s)-nets in base b, *Journal of Combinatorial Designs*, **4** (1996), 275–293.

[150] K. M. Lawrence, A. Mahalanabis, G. L. Mullen & W. C. Schmid, Construction of digital (t, m, s)-nets from linear codes, in *Finite Fields and their Applications* (eds. S. Cohen & H. Niederreiter), Cambridge University Press, Cambridge (1996), pp. 189–208.

[151] W. J. Martin & D. R. Stinson, A generalized Rao bound for ordered orthogonal arrays and (t, m, s)-nets, preprint, Mathematics, University of Winnipeg, 1997.

[152] W. J. Martin & D. R. Stinson, Association schemes for ordered orthogonal arrays and (t,m,s)-nets, preprint, Mathematics, University of Winnipeg, 1997.

[153] G. L. Mullen, A candidate for the "next Fermat problem", *Mathematics Intelligencer*, **17** (1995), 18–22.

[154] G. L. Mullen, A. Mahalanabis & H. Niederreiter, Tables of (t, m, s)-net and (t, s)-sequence parameters, *Lecture Notes in Statistics*, **106** (1995), 58–86.

[155] G. L. Mullen & W. C. Schmid, An equivalence between (t, m, s)-nets and strongly orthogonal hypercubes, *Journal of Combinatorial Theory, Series A*, **76** (1996), 164–174.

[156] G. L. Mullen & G. Whittle, Point sets and sequences with small discrepancy, *Monatshefte Mathematik*, **113** (1992), 265–273.

[157] H. Niederreiter, Point sets and sequences with small discrepancy, *Monatshefte Mathematik*, **104** (1987), 273–337.

[158] H. Niederreiter, Orthogonal arrays and other combinatorial aspects in the theory of uniform point distributions in unit cubes, *Discrete Mathematics*, **106/107** (1992), 361–367.

[159] C. R. Rao, Factorial experiments derivable from combinatorial arrangements of arrays, *Journal of the Royal Statistical Society*, **9** (1947), 128–139.

Computer Science, University of Vermont
Burlington, VT 05405, U.S.A.
Charles.Colbourn@uvm.edu

Mathematics and Statistics, University of Vermont
Burlington, VT 05405, U.S.A.
Jeff.Dinitz@uvm.edu

Combinatorics and Optimization, University of Waterloo
Waterloo, Ontario, CANADA N2L 3G1
dstinson@cacr.math.uwaterloo.ca

Random Walks on Combinatorial Objects

Martin Dyer and Catherine Greenhill

Summary Approximate sampling from combinatorially-defined sets, using the Markov chain Monte Carlo method, is discussed from the perspective of combinatorial algorithms. We also examine the associated problem of discrete integration over such sets. Recent work is reviewed, and we re-examine the underlying formal foundational framework in the light of this. We give a detailed treatment of the coupling technique, a classical method for analysing the convergence rates of Markov chains. The related topic of perfect sampling is examined: in perfect sampling, the goal is to sample exactly from the target set. We conclude with a discussion of negative results in this area: these are results which imply that there are no polynomial time algorithms of a particular type for a particular problem.

1 Introduction

The focus of this paper is approximate sampling and approximate counting (or approximate integration), using the Markov chain Monte Carlo (MCMC) method, and viewed from the perspective of combinatorial algorithms. There has been much work in this area in recent years, some of which we survey below in Section 4. We illustrate this work with a closer examination of one particular technique which has proved successful recently, that of *coupling*. This is a classical method from applied probability, but its application in this area has involved some new insights.

Formal foundations for work in this area were provided in the seminal paper of Jerrum, Valiant and Vazirani [50]. However, the subject seems subsequently to have outgrown the framework it provided. The present paper makes a modest attempt to update the situation.

We begin, in Section 2, by fixing notation which we use throughout. In Section 3 we offer some formal definitions of the central concepts, following [50] and [73]. A review of recent developments in the areas of approximate sampling, approximate counting and perfect sampling is given in Section 4.

The coupling method is described in Section 5, and is illustrated on a simple Markov chain for independent sets in graphs. This example forms a running theme throughout the paper. A variant of coupling which has been employed since 1997, called *path coupling*, is presented in Section 6.

Much attention has recently been paid to the topic of "exact" or "perfect" sampling using Markov chains. In Section 7, we explore some aspects of perfect sampling, in particular the relationship between perfect and approximate sampling.

Another area of recent interest has been in showing that the MCMC (or any) technique fails on certain problems. As an illustration, we conclude by discussing two typical negative results in Section 8. (Some others are included in the survey of Section 4.)

2 Notation and preliminaries

Throughout $\mathbb{N} = \{0, 1, 2, \ldots\}$, $\mathbb{N}_+ = \mathbb{N} \setminus \{0\}$, $\mathbb{Q}_+ = \{q \in \mathbb{Q} \colon q > 0\}$, and $[n] = \{1, 2, \ldots, n\}$ for $n \in \mathbb{N}_+$.

Let $\mathcal{M}$ be a Markov chain on finite state space Ω, with transition matrix P, i.e. if X_t is the state at time t,

$$\Pr(X_t = \sigma \mid X_{t-1} = \omega) = P(\omega, \sigma) \qquad (t = 1, 2, \ldots),$$

which we also write as P_σ^ω. We assume that $\mathcal{M}$ is ergodic, and therefore has unique stationary distribution π. In most cases of interest, $\mathcal{M}$ is *reversible*, i.e.

$$\pi(\omega) P(\omega, \sigma) = \pi(\sigma) P(\sigma, \omega) \qquad (\forall \omega, \sigma \in \Omega). \tag{1}$$

The central role of reversible chains in applications rests on the fact that π can be deduced from (1). If $\mu \colon \Omega \to \mathbb{R}$ satisfies (1), then it determines π up to normalisation. In fact, we usually *design* the chain to satisfy (1). Without reversibility, there is no apparent method of determining π, other than to explicitly construct the transition matrix, an exponential time (and space) computation in our setting.

If $p_0(\omega) = \Pr(X_0 = \omega)$, then $p_t(\sigma) = \sum_\omega p_0(\omega) P^t(\omega, \sigma)$ is the distribution at time t. As a measure of convergence, the natural choice in this context is *variation distance*,

$$\mathrm{d_{TV}}(p_t, \pi) = \tfrac{1}{2} \sum_{\omega \in \Omega} |p_t(\omega) - \pi(\omega)| = \max_{A \subseteq \Omega} (p_t(A) - \pi(A)). \tag{2}$$

[Some authors, e.g. Lindvall [59], define this without the $\tfrac{1}{2}$.] The *mixing time* of the chain is then

$$\tau(\varepsilon) = \max_{p_0} \min_t \{\mathrm{d_{TV}}(p_t, \pi) \leq \varepsilon\},$$

and it is easy to show that the maximum occurs when $X_0 = \omega_0$, with probability one, for some state ω_0.

For further information on Markov chains, see [2].

Throughout the paper, we refer to the well-known complexity classes P, NP and #P. For their definitions and further information, see [64].

3 A computational framework

Some foundations for approximate counting and uniform sampling were set out in the influential paper of Jerrum, Valiant and Vazirani [50]. However, the ideas of that paper have rather more generality than the results contained within it. Also the paper concentrates its attention on the class of *self-reducible* problems, as introduced by Schnorr [72]. The restrictive definition of self-reducibility makes this concept difficult to apply, as we discuss below. For these reasons, we re-examine these foundations, to reflect the substantial body of subsequent work which has used the ideas, rather than the results, of [50]. The reader who is more interested in techniques and applications might prefer to skim this material, at least on first reading.

The sample spaces we have in mind are sets of combinatorial objects. However, in order to discuss the computational complexity of generation, it is necessary to consider a sequence of instances of increasing size. We therefore work within the following formal framework, which extends the ideas of Jerrum, Valiant and Vazirani [50]. The models of computation are the Turing Machine (TM) for deterministic computations and the Probabilistic Turing Machine (PTM) for randomized computations. (A PTM is a TM with a source of uniform and independent random bits.) The following definition generalises that of [50] to allow non-uniform distributions, and is closely related to that of Sinclair [73, pp. 86–87] for the same purpose. We must confine ourselves to some class of distributions which are "easily described", from a computational viewpoint, in large instances. We identify this below with a class of unnormalised measures which we call "weight functions".

Let Σ be a fixed alphabet of at least two symbols, and $W: (\Sigma^*)^2 \to \mathbb{N}$ be such that, for some polynomial b, $W(\sigma, \omega) = 0$ unless $|\omega| \leq b(|\sigma|)$. Moreover $W(\sigma, \omega)$ must be computable in time polynomial in $|\sigma|$ whenever $W(\sigma, \omega) > 0$. (If the TM for W may ignore part of its input, this implies that W is *always* computable in polynomial time.) Let us call W a *weight function*. Here σ may be thought of as an encoding of an instance of some combinatorial problem, and the ω of interest are encodings of the structures we wish to generate.

Let $\Omega_\sigma = \{\omega : W(\sigma, \omega) > 0\}$. Then the sequence of discrete probability spaces determined by W is $(\Omega_\sigma, \pi_\sigma)$, where π_σ is the *density*

$$\pi_\sigma(\omega) = W(\sigma, \omega)/Z(\sigma), \quad \text{with} \quad Z(\sigma) = \sum_{\omega' \in \Omega_\sigma} W(\sigma, \omega')$$

being the corresponding *normalising function*. It is easy to see that the class of normalising functions so defined is essentially Valiant's [76] class #P. The definition implies that, for some fixed $c \in \mathbb{N}$, $|\Omega_\sigma| \leq Z(\sigma) \leq 2^{|\sigma|^c}$. If $Z(\sigma) = 0$, then $\Omega_\sigma = \emptyset$ and π_σ is the unique (improper) measure on Ω_σ.

In our definition, two distinct weight functions may define the same sequence of spaces. Therefore let us say weight functions W_1, W_2 are *equivalent* if there exists $\kappa: \Sigma^* \to \mathbb{Q}_+$ so that $W_2(\sigma, \omega) = \kappa(\sigma) W_1(\sigma, \omega)$ ($\forall \sigma, \omega \in \Sigma^*$). Then there is a bijection between sequences of probability spaces $(\Omega_\sigma, \pi_\sigma)$ and equivalence classes of weight functions. Thus, if we write $\widetilde{W}$ for the equivalence class containing W, we may identify it with the sequence $(\Omega_\sigma, \pi_\sigma)$.

Example 3.1 Let $G = (V, E)$ be a (simple) graph, with maximum degree Δ and $|V| = n$, and let $\mathcal{I}(G)$ denote the collection of independent sets in G. It is known [38, 75] to be #P-complete to compute $|\mathcal{I}(G)|$ exactly for $\Delta \geq 3$, and is easily shown to be in P for $\Delta < 3$. For given $\lambda > 0$, the *hard-core gas model* (see for example [4]) samples from a density where independent sets of size s have probabilities proportional to λ^s. We use this problem as a running example. In our setting, let $\lambda = r/q$ for integers r, q. Then $W(\sigma, \omega) = 0$ unless σ encodes a graph, and ω an independent set of some size s in G. If $W(\sigma, \omega) \neq 0$, then $W(\sigma, \omega) = q^n \lambda^s = r^s q^{(n-s)} \in \mathbf{N}_+$. ∎

We insist that sample spaces are discrete, and weight functions are integer valued. Computationally, discrete spaces are essential. If we wish to work with continuous spaces, then approximations must be made to some predetermined number of bits. The same is true if we are interested in real-valued densities (as in some statistical applications). However, the effect of such approximations can be absorbed into the variation distance of the sampling procedure. The reader may still wonder why we require W to have codomain $\mathbf{N}$ rather than $\mathbf{Q}$, which would seem more natural. This is because we use unnormalised measures, and we wish to avoid the following technical difficulty. In a large sample space it is possible to specify polynomial size rationals for the unnormalised measure which result in exponential size rationals for the probabilities. An example is the set $[2^n]$, with the measure assigning probability proportional to $1/i$ to $i \in [2^n]$. (See [73, pp. 27–28] for details, where this example is used for a slightly different purpose.) In such spaces there is no possibility of *exact* sampling in sub-exponential expected time, and we must accept approximations. We prefer not to deal with these anomalous spaces, but to insist that these approximations be made explicit. Thus, in this example we could use weights $\lfloor K/i \rfloor$ for some suitably large integer K.

A *fully polynomial approximate sampler* (which we shorten to *good sampler*) for $(\Omega_\sigma, \pi_\sigma)$ is a PTM which, on inputs σ and $\varepsilon \in \mathbf{Q}_+$ ($0 < \varepsilon \leq 1$), outputs $\omega \in \Sigma^*$, according to a measure μ_σ satisfying $\mathrm{d}_{\mathrm{TV}}(\mu_\sigma, \pi_\sigma) \leq \varepsilon$, in time bounded by a bivariate polynomial in $|\sigma|, \log \varepsilon^{-1}$. We allow $\omega \notin \Omega_\sigma$. If $\Omega_\sigma = \emptyset$, the algorithm does not terminate within its time bound. However, this can be detected, and we may construct a polynomial time algorithm which terminates either with a random ω or a proof that Ω_σ is empty.

A good sampler essentially coincides with the *almost uniform generator* of Jerrum, Valiant and Vazirani [50]. Their definition is given in terms of a relation $\mathcal{R}$ rather than a function, so is clearly contained within ours by restricting $W: \Sigma^* \to \{0, 1\}$. However, the following converse is easily proved. Let $(\Omega_\sigma, \pi_\sigma) = \widetilde{W}$ and $\Omega'_\sigma = \{(\omega, i) : \omega \in \Omega_\sigma, 1 \leq i \leq W(\sigma, \omega)\}$. Then there is a good sampler for $(\Omega_\sigma, \pi_\sigma)$ if and only if there is an almost uniform generator for $\mathcal{R} = \{(\sigma, (\omega, i)) : (\omega, i) \in \Omega'_\sigma\}$. So our definition is reducible to that of [50], but it seems more natural to work directly with non-uniform distributions.

Our real interest here is in combinatorial Markov chains, which we define as follows. Let $M: (\Sigma^*)^3 \to \mathbf{N}$ and define

$$\mathcal{R}_\sigma = \{(\omega, \omega') : M(\sigma, \omega, \omega') > 0\}, \quad \Omega_\sigma = \{\omega : \exists \omega' \text{ with } (\omega, \omega') \in \mathcal{R}_\sigma\}.$$

Random Walks on Combinatorial Objects

Let M have the following properties.

(a) There is a polynomial b such that $|\omega|, |\omega'| \leq b(|\sigma|)$ if $M(\sigma, \omega, \omega') > 0$, and M is computable in time polynomial in $|\sigma|$ whenever $M(\sigma, \omega, \omega') > 0$.

(b) There exist constants $K(\sigma) \in \mathbf{N}_+$, of polynomial size, such that
$$\sum_{\omega' \in \Sigma^*} M(\sigma, \omega, \omega') = K(\sigma) \qquad (\forall \omega \in \Omega_\sigma).$$

(c) The transitive closure of $\mathcal{R}_\sigma$ is $\Omega_\sigma \times \Omega_\sigma$, and for some ω, $(\omega, \omega) \in \mathcal{R}_\sigma$.

(d) Writing $M_\omega(\sigma, \omega') = M(\sigma, \omega, \omega')$ ($\omega \in \Sigma^*$), it follows from (a) that M_ω is a weight function. We require that there is a good sampler for $\widetilde{M_\omega}$ ($\forall \omega$).

We call M a *density matrix*, and associate with it a sequence of Markov chains $\mathcal{M}_\sigma = (\Omega_\sigma, P_\sigma)$, with transition matrices
$$P_\sigma(\omega_1, \omega_2) = M(\sigma, \omega_1, \omega_2)/K(\sigma) \qquad (\omega_1, \omega_2 \in \Omega_\sigma).$$

Properties (a) and (c) ensure that $\mathcal{M}_\sigma$ is finite and ergodic. Property (d) ensures that we can efficiently simulate $\mathcal{M}_\sigma$ to a close approximation for any given number of steps. Property (b) ensures that polynomial powers of the transition matrix cannot generate rationals of superpolynomial size, and hence the state probabilities at any polynomial time cannot be rationals of superpolynomial size. We include this property since we do not wish to preclude exact generation using Markov chains. In any case, this condition can always be satisfied to any desired approximation, and is usually satisfied naturally. There is little loss in restricting $K(\sigma)$ to be a power of 2. If any such $K(\sigma)$ exist, it is easy to show that there is a chain with the same stationary distribution and K a power of 2, simply by increasing the "self-loop" probability on all states. Since we are interested in the stationary distribution, we can use this slightly slower chain. Thus we may insist on K being a power of 2 where convenient.

Density matrices M_1, M_2 are *equivalent* if there exists $\kappa \colon \Sigma^* \to \mathbb{Q}_+$ such that $M_2(\sigma, \omega, \omega') = \kappa(\sigma) M_1(\sigma, \omega, \omega')$ for all $\sigma, \omega, \omega' \in \Sigma^*$. We can identify the equivalence class $\widetilde{M}$ with the sequence $\mathcal{M}_\sigma$. We say that $\mathcal{M}_\sigma$ is a *rapidly mixing Markov chain* if its mixing time $\tau_\sigma(\varepsilon)$ is bounded by a polynomial in $|\sigma|, \log \varepsilon^{-1}$.

If $\mathcal{M}_\sigma$ is a Markov chain sequence, let π_σ denote the stationary distribution of $\mathcal{M}_\sigma$. Then, if W is a weight function, $\mathcal{M}_\sigma$ is a *Monte Carlo Markov chain* (MCMC) for $\widetilde{W}$ if both $\widetilde{W}, \mathcal{M}_\sigma$ determine the same sequence of probability spaces $(\Omega_\sigma, \pi_\sigma)$. (This slight overloading of the MCMC abbreviation should not cause confusion.) The usual way to establish this is by reversibility, i.e. if $W(\sigma, \omega) M(\sigma, \omega, \omega') = W(\sigma, \omega') M(\sigma, \omega', \omega)$ for all $\sigma \in \Sigma^*$ and $\omega, \omega' \in \Omega_\sigma$. Clearly we have a good sampler for $\widetilde{W}$ if $\mathcal{M}_\sigma$ is a rapidly mixing Markov chain.

Example 3.2 Continuing Example 3.1, a possible MCMC for this problem is:

INSERT/DELETE CHAIN

   Let T be a time bound and $X_0 \in \mathcal{I}(G)$ arbitrary.
   Let X_t be the current independent set.
(a)   Choose $v \in V$ uniformly at random.
(b)   (Delete)   If $v \in X_t$, set $X_{t+1} \leftarrow X_t \setminus \{v\}$ with probability $1/(1+\lambda)$.
      (Insert)   If $v \notin X_t$, set $X_{t+1} \leftarrow X_t \cup \{v\}$ with probability $\lambda/(1+\lambda)$,
                 if this is an independent set.
(c)   Set $t \leftarrow t+1$. If $t < T$, return to (a).

To express this in our formalism, $M(\sigma,\omega,\omega') = 0$ unless σ encodes a graph and ω,ω' encode independent sets which differ in at most one vertex of G. Otherwise, as $\lambda = r/q$, $M(\sigma,\omega,\omega') = q$ corresponds to a possible deletion from ω, and $M(\sigma,\omega,\omega') = r$ to a possible insertion. Then $M(\sigma,\omega,\omega)$ is chosen to ensure $\sum_{\omega'} M(\sigma,\omega,\omega') = K(\sigma) = n(r+q)$. It is easy to see that M satisfies our conditions and, using (1), that the weight function W from Example 3.1 determines its stationary distribution. ∎

One of the main applications of sampling is to *approximate integration*. In our setting this means estimating $Z(\sigma)$ to some specified relative error. In the important case where W is a characteristic function, we call the approximate integration problem *approximate counting*. Specifically, a *fully polynomial randomized approximation scheme* (fpras) for $Z(\sigma)$ is a PTM which on input σ, ϵ outputs $\hat{Z}$ so that

$$\Pr(1/(1+\epsilon) \leq \hat{Z}/Z \leq 1+\epsilon) \geq \tfrac{3}{4},$$

and which runs in time polynomial in $|\sigma|$ and $1/\epsilon$.

The theoretical foundations for approximate uniform sampling and counting were laid in [50]. There, an equivalence between sampling and approximate counting was shown for the class of so-called self-reducible problems [72]. Subsequent work has exposed several disadvantages of this class as a general framework. First, whether a problem is self-reducible depends strongly on the encoding. The problem may have to be re-encoded in an unnatural manner in order to render it self-reducible. An example is adding an edge to a graph, where we may have to encode a graph by its complement to have self-reducibility. Furthermore, many natural reductions employed do not satisfy the self-reducibility criteria. An example is adding an arbitrary ordered pair to a finite partial order (and forming the appropriate closure), which occurs in one approach to the problem of approximately counting linear extensions. (See [78, pp. 301–302].) More seriously, some problems to which the techniques of [50] have been applied, do not seem to be self-reducible under any re-encoding. Examples are the volume approximation problem of [26], and approximately counting contingency tables [25, 31].

Random Walks on Combinatorial Objects

We therefore propose a modified framework. We emphasise that the techniques are those of [50], and we wish only to emphasise the extent of their applicability. Let size: $\Sigma^* \to \mathbb{N}$ be such that size(σ) is polynomially bounded in $|\sigma|$, and if size$(\sigma') <$ size(σ) then $|\sigma'|$ is polynomially bounded in $|\sigma|$. If size$(\sigma) = 0$, we call the problem a *base problem*. For the class of base problems, we assume the existence of a good sampler and a fpras for $Z(\sigma)$.

For all σ, let $\Xi(\sigma)$ be a polynomial time computable set such that

(a) size$(\xi) <$ size(σ) $(\forall \xi \in \Xi)$.

(b) There exist polynomial time computable constants $k_\xi(\sigma) \in \mathbb{Q}_+$ and injections $\phi_\xi(\sigma): \Omega_\xi \to \Omega_\sigma$ $(\forall \xi \in \Xi)$, such that

$$k_\xi W(\xi, \omega) \leq W(\sigma, \phi_\xi(\omega)) \qquad (\forall \omega \in \Omega_\xi).$$

Both $\phi_\xi(\omega)$ and $\phi_\xi^{-1}(\omega)$ must be computable in polynomial time, given $\omega \in \Omega_\xi$ and $\omega \in \Omega_\sigma$, respectively.

(c) For some $\zeta \in \Xi$, $Z(\sigma)/(k_\zeta(\sigma)Z(\zeta))$ is polynomially bounded in $|\sigma|$.

If $\widetilde{W}$ satisfies these conditions, we call the problem *self-contractible*. Summing over $\omega \in \Omega_\xi$ and using the injectivity of ϕ_ξ shows that (b) implies $k_\xi Z(\xi) \leq Z(\sigma)$ $\forall \xi \in \Xi$. Now, suppose we have a good sampler for $\widetilde{W}$. Then, following [50], we may estimate $k_\xi Z(\xi)/Z(\sigma)$ by rejection sampling. We sketch the method, ignoring details concerning the allowable closeness of intermediate approximations, for which we refer the reader to [50]. We sample ω from $W(\sigma, \cdot)$, and accept with probability $k_\xi W(\xi, \phi^{-1}(\omega))/W(\sigma, \omega)$ if $\phi^{-1}(\omega) \neq \emptyset$. The overall acceptance probability is

$$\sum_{w \in \phi(\Omega_\xi)} \frac{k_\xi W(\xi, \phi^{-1}(\omega))}{W(\sigma, \omega)} \frac{W(\sigma, \omega)}{Z(\sigma)} = \frac{k_\xi Z(\xi)}{Z(\sigma)}.$$

Moreover, from (c) there is some $\zeta \in \Xi$ such that we can estimate this ratio to sufficient relative accuracy in polynomial time. Since size$(\zeta) <$ size(σ), we may repeat this process with ζ replacing σ. Then, letting $\sigma_0 = \sigma$, $\sigma_1 = \zeta$, ..., we may iterate until size$(\sigma_r) = 0$. Now $|\sigma_i|$ is polynomially bounded in $|\sigma|$ for all $i = 0, 1, \ldots, r$. For σ_r we can approximate $Z(\sigma_r)$ in polynomial time. Then we may multiply estimates together to approximate

$$Z(\sigma_r) \prod_{i=1}^{r} \frac{Z(\sigma_{i-1})}{k_{\sigma_i}(\sigma_{i-1}) Z(\sigma_i)} = \frac{Z(\sigma)}{\prod_{i=1}^{r} k_{\sigma_i}(\sigma_{i-1})}$$

to the required relative error, and hence $Z(\sigma)$.

Example 3.3 Consider the independent set problem of Example 3.1. We take size(G) to be its number of edges, $|E|$. The set Ξ contains the n subgraphs G_v given by deleting a particular nonisolated vertex v. The injections ϕ_{G_v} map independent set X in G_v to X in G. The k_{G_v} are all q. It is easy to check that this satisfies (a),(b) above. Moreover, since the ranges of the ϕ_{G_v} cover $\mathcal{I}(G)$ provided G has an edge, it follows that there exists a v with $Z(G)/qZ(G_v) \leq n$ if $|E| > 0$. Hence (c) is also satisfied. If $\text{size}(G) = |E| = 0$, then we may sample from the independent sets directly (or see Example 5.2) and efficiently compute the normalising function. Thus the problem is self-contractible, and it follows that a fpras exists for the normalising function if we have a good sampler. It is shown in Example 6.1 that this is true for $\lambda \leq 2/(\Delta - 2)$. This is not a self-reducibility reduction, though (a different) one does exist for this problem. (See Example 3.4.) ∎

A converse result may be obtained under rather stronger conditions. First suppose that the base problems are such that $Z(\sigma)$ may be determined *exactly*, and suppose that (b) and (c) are strengthened to

(b)′ There exist polynomial time computable constants $k_\xi(\sigma) \in \mathbb{Q}_+$ and injections $\phi_\xi(\sigma) \colon \Omega_\xi \to \Omega_\sigma$ $(\forall \xi \in \Xi)$, such that
$$k_\xi W(\xi, \omega) = W(\sigma, \phi_\xi(\omega)) \qquad (\forall \omega \in \Omega_\xi).$$
Both $\phi_\xi(\omega)$ $(\omega \in \Omega_\xi)$ and $\phi_\xi^{-1}(\omega)$ $(\omega \in \Omega_\sigma)$ must be computable in polynomial time.

(c)′ The sets $\phi_\xi(\Omega_\xi)$ form a partition of Ω_σ.

Let us call such a problem *self-partitionable*. Clearly (b)′ implies (b). Also, from (b)′ and (c)′, since

$$\sum_{\xi \in \Xi} k_\xi Z(\xi) = \sum_{\xi \in \Xi} \sum_{\omega \in \Omega_\xi} k_\xi W(\xi, \omega) = \sum_{\xi \in \Xi} \sum_{\omega \in \Omega_\xi} W(\sigma, \phi_\xi(\omega))$$
$$= \sum_{\omega \in \Omega_\sigma} W(\sigma, \omega) = Z(\sigma), \qquad (3)$$

and the polynomial size of Ξ now implies (c). We sketch the generation procedure, skipping details. Suppose we can estimate $Z(\sigma)$ by $\hat{Z}(\sigma)$ within relative error ϵ to high enough probability. We branch to $\xi \in \Xi$ with probability $k_\xi \hat{Z}(\xi)/(1+\epsilon)\hat{Z}(\sigma)$. If the total of these probabilities over $\Xi(\sigma)$ is more than 1 we "fail", i.e. we abandon this whole sampling "trial". If the total is less than 1, as we would expect, then we fail with the (small) unassigned probability. Otherwise we repeat, getting $\sigma = \sigma_0, \sigma_1, \ldots, \sigma_r$ until $\text{size}(\sigma_r) = 0$, in which case $\hat{Z}(\sigma_r) = Z(\sigma_r)$. Then we generate ω' from $W(\sigma_r, .)$ with small enough variation distance that all probabilities have relative error at most ϵ' for some very small ϵ'. Then ω is determined from $\omega_{i-1} = \phi_{\sigma_i}(\omega_i)$ $(i = 1, \ldots, r)$, with $\omega_0 = \omega$, $\omega_r = \omega'$. Then (with high probability) the probability that ω is generated is within relative error ϵ' of

$$\frac{k_{\sigma_1}\hat{Z}(\sigma_1)}{(1+\epsilon)\hat{Z}(\sigma_0)} \frac{k_{\sigma_2}\hat{Z}(\sigma_2)}{(1+\epsilon)\hat{Z}(\sigma_1)} \cdots \frac{k_{\sigma_r}Z(\sigma_r)}{(1+\epsilon)\hat{Z}(\sigma_{r-1})} \frac{W(\sigma_r, \omega_r)}{Z(\sigma_r)} = \frac{W(\sigma, \omega)}{(1+\epsilon)^r \hat{Z}(\sigma)},$$

after an easy induction. This is equivalent to the desired weight function. Provided that ϵ, ϵ' are sufficiently small, the failure probability and the variation distance can be kept small on a single trial. Then we may output an arbitrary ω if we fail after some large enough number of trials. Hence the overall variation distance is small. It follows that for self-partitionable problems, approximate integration and good sampling are equivalent. It is easy to see that self-reducible [50] problems are self-partitionable, but the converse is not necessarily true. An example is the volume approximation problem of [26]. (See [27] for some details.)

Example 3.4 Consider further the independent set problem of Example 3.3. We take size(G) to be its number of vertices, n. Choose a fixed vertex $v \in V$. Then the set Ξ contains two subgraphs, G_0 given by deleting v, and G_1 given by deleting v and all its neighbours. The injection ϕ_{G_0} maps independent set X in G_0 to X in G, and ϕ_{G_1} maps X in G_1 to $X \cup \{v\}$ in G. We take $k_{G_0} = q$, $k_{G_1} = r$. This satisfies (a), (b)$'$, (c)$'$ above. If size$(G) = 0$, then there is only one independent set. Thus the problem is self-partitionable. This reduction is a self-reducibility reduction if the vertex v is chosen to be the lowest numbered vertex, but not if it is chosen to be (say) the vertex of largest degree, which is a logical choice. ■

We can show that approximate integration implies good sampling under rather weaker conditions than self-partitionability. We do not develop this here, however, since we have no example of a problem satisfying these conditions which is not self-partitionable. In any case, the usual direction in applications is to go from sampling to integration.

4 Review

This section presents a brief review of developments in the areas of approximate sampling and approximate integration during the last couple of years. There are several excellent survey articles which introduce the methods and results in this area, including Welsh [78], Jerrum and Sinclair [49] and Jerrum [48]. Therefore we do not give an exhaustive survey here, but concentrate on more recent results from the theoretical computer science community. There is a very extensive body of work on the use of Markov chain Monte Carlo in statistics and statistical mechanics. We do not have space to survey that work here, but simply direct the reader to [8,74] for further information. We discuss the closely related area of perfect sampling in a separate subsection below.

There are only a few approaches to proving rapid mixing in Markov chains. Much of this paper involves new variations of the classical method of coupling. However, we start this review section by considering results obtained by other methods.

Some of the quantities used to bound the mixing time of a Markov chain are the conductance, spectral gap and log-Sobolev constant of the chain. Using the conductance approach, Kannan, Tetali and Vempala [55] proved rapid convergence of Markov chains for generating bipartite graphs and tournaments. By bounding the spectral gap, Chung, Graham and Yau [17] proved pseudo-polynomial convergence of a Markov chain for contingency tables, when the row and column sums are large enough. This result was improved by Dyer, Kannan and Mount [31]. Their chain mixes in polynomial time when the row and column sums are large enough (the threshold being lower than in [17]). This result was also proved by bounding the spectral gap of the chain.

Diaconis and Saloff-Coste [21] proved rapid mixing of a Markov chain for sampling generating sets of abelian groups, by comparing the log-Sobolev constants of two Markov chains. They used a comparison theorem from an earlier paper [22]. Other methods for relating the mixing times of Markov chains by comparing their spectral gap or log-Sobolev constant have been proposed by Chung and Graham [16], Randall and Tetali [70] and the authors [29]. Further results on bounding the spectral gap can be found in Guattery, Leighton and Miller [40] and in Chung and Yau [18].

Another approach used to bound the mixing time is Dobrushin uniqueness [23]. This approach was used by Salas and Sokal [71] in work on graph colourings, and by Peinado and Lengauer [65] to sample structures which arise in computational chemistry. The relationship between Dobrushin uniqueness and path coupling (see below) was explored in [9].

Madras and Randall [63] gave a method for analyzing the product of two Markov chains, and applied their results to a Markov chain for colourings on a grid.

One area where the Markov chain Monte Carlo approach has been very successful is in estimating the volume of convex bodies (see [52] for an overview). Typically, the mixing rates of random walks in convex bodies are analysed using isoperimetric inequalities (see, for example [26]). The state of the art is the $O^*(n^5)$ volume approximation algorithm of Kannan, Lovász and Simonovits [54] (where $O^*(\cdot)$ is notation which hides logarithmic factors). Kannan and Li [53] showed how to sample according to the multivariate normal density, and relate this to volume computation. In [56], Kannan and Vempala determined when the volume of a convex polytope is a good estimate for the number of lattice points in the polytope.

Now let us consider results which were proved using coupling. As stated in Section 2, the problem of sampling independent sets is a running example in this paper. Let λ denote the parameter used in the weight function, and let Δ denote the maximum degree of a given graph. Luby and Vigoda [62] described a Markov chain for independent sets which is rapidly mixing for the following values of λ, Δ: when $\Delta = 3$ the chain is rapidly mixing for $\lambda \leq 1$, and when $\Delta \geq 4$ the chain is rapidly mixing for $\lambda \leq 1/(\Delta - 3)$. Propp and Wilson [69] adapted the work of [62] to show how to sample perfectly in expected polynomial time for these values of λ, Δ.

Bubley, Dyer and Jerrum [14] used coupling to give a new approach to sampling points in a convex body. Also, Wilson [79] used coupling to prove upper and lower bounds on the mixing time for lozenge tiling and card shuffling Markov chains. Hernek [44] described a Markov chain for two-rowed contingency tables and, using coupling, showed that the chain has pseudopolynomial mixing time.

Another application of coupling is the popular area of perfect sampling. We review perfect sampling separately, in Section 4.1 below.

A variant of traditional coupling which is described in Section 6 is path coupling, introduced in [10]. (Bubley's thesis [9] contains much work on coupling and path coupling, most of which we mention here.) Using path coupling, the authors [29] proved rapid mixing of a new chain for independent sets, whenever $\lambda \leq 2/(\Delta - 2)$ and $\Delta \geq 3$. (This result is outlined in Example 6.1, and was independently discovered by Luby, Mitzenmacher and Vigoda [77].)

Another result obtained using path coupling is an $O(n^3 \log n)$ bound on the mixing time for a Markov chain for sampling linear extensions of a partial order, due to Bubley and Dyer [12]. This chain is similar to the combinatorial chain of Karzanov and Khachiyan [57], but uses a different distribution to choose the position to update in the partial order. Recently, Wilson [79] built on this result to prove an $O(n^3 \log n)$ upper bound for the mixing time of the Karzanov–Khachiyan chain. He also outlined how one may obtain an $\Omega(n^3 \log n)$ lower bound for the mixing time of the Karzanov–Khachiyan chain (or for the chain given in [12]).

A new Markov chain for two-rowed contingency tables was described by the authors in [28]. The transitions of this chain are very simple: select two columns of the table uniformly at random, and replace this 2×2 submatrix by another 2×2 matrix with the same row and column sums, chosen uniformly at random. Using path coupling, the mixing time of this chain is shown to be polynomial.

The Potts model is a generalisation of graph colouring which arises in statistical physics. The antiferromagnetic Potts model generalises proper colourings, while in the ferromagnetic Potts model configurations are favoured if many vertices are coloured with the same colour as their neighbour. The Swendsen–Wang process is a well-known method for sampling from the ferromagnetic q-state Potts model. It was mildly conjectured that this process might be rapidly mixing on any graph. This conjecture has been disproved by two results which we state below. On the other hand, Cooper and Frieze [19] proved two positive results. Using path coupling, they showed that Swendsen–Wang mixes rapidly for graphs with small maximum degree, for small enough values of the "coupling constant" K. For the special case of trees, they showed that the mixing time is linear in the number of vertices, for any value of K. The latter result was obtained using classical coupling.

In the limit, the antiferromagnetic Potts model describes proper colourings of graphs. The simple Markov chain for graph colourings described independently by [47] and [71] is rapidly mixing if there are at least 2Δ colours, where

Δ is the maximum degree of the graph. This result was improved in a paper by the authors [30], which describes a Markov chain which has better bounds on the mixing time for regular graphs, and when fewer than 3Δ colours are used. The latter chain is also rapidly mixing whenever at least 2Δ colours are used. Indeed, it was thought that 2Δ colours might be needed in order to achieve rapid mixing. This was disproved by Bubley and the authors in [13] where a rapidly mixing Markov chain for 5-colourings of graphs with maximum degree three was described. The coupling was constructed by solving a large number of transportation problems. The result was extended to give a rapidly mixing Markov chain for counting 7-colourings of triangle-free 4-regular graphs.

A variation of path coupling, called delayed path coupling, was used by Czumaj et al. [20] to prove rapid mixing for a Markov chain for generating random permutations in parallel.

We now describe some negative results which have been obtained. Gore and Jerrum [36] investigated the behaviour of the Swendsen–Wang process on the complete graph. They showed that for certain values of the "coupling constant" K, the process has exponential mixing time. Cooper and Frieze [19] extended this result to the random graph $G_{n,p}$, showing that there are critical values of K for which (with high probability) the Swendsen–Wang process requires exponential time to mix. Jerrum and Goldberg [35] proved that the so-called "Burnside process" for sampling from unlabelled structures does not always mix rapidly. Two negative results for independent sets were established by Dyer, Frieze and Jerrum [25]. These results are described in Section 8.

Finally, we mention some results in related areas, which do not directly use Markov chain Monte Carlo. There are many results concerning random walks on expander graphs, of which the paper of Broder, Frieze and Upfal [7] is an example. Gore et al. [37] gave an algorithm for sampling words from a context-free language, while Frieze et al. [34] presented an algorithm for generating Hamiltonian cycles in random regular graphs.

4.1 Review of perfect sampling

There has been much interest recently in algorithms for sampling perfectly from a given distribution. We now briefly review results in this area. Further aspects of perfect sampling are considered in Section 7 below.

The first paper in this area was by Propp and Wilson [67]. This paper introduced the approach known as coupling from the past (CFTP). This method is particularly efficient if the chain is *monotone*. In such a chain the state space forms a lattice with respect to a partial order which is preserved stochastically by the transitions of the chain. For our purposes we may take this as meaning that there exists a coupling of the chain under which all transitions preserve the ordering. (See, for example, [33] for a more formal definition.)

The idea of monotone CFTP is to run two copies of the chain from some time in the past, one starting from the top element of the state space and one from the bottom element. If at time zero both copies of the chain have coalesced, then the state of the chain at time zero is distributed according to the stationary distribution. Otherwise, the procedure is repeated from a time twice as far into the past. In this monotone setting, the expected running time of the algorithm can be bounded in terms of the mixing time of the chain. Let T be a random variable denoting the running time of the algorithm, and suppose that the partial order has height r. Then

$$\mathbf{E}(T) \leq 2\tau(e^{-1})(1 + \log(r)), \tag{4}$$

where $\tau(\varepsilon)$ is the mixing time of the Markov chain [67]. Weaker variations of CFTP been proposed for other situations, for example antimonotone [42] and bounding chains [41, 45].

Fill [33] proposed an alternative algorithm for perfect sampling, known as Fill's algorithm. His approach is based on rejection sampling, where the coin toss used to decide whether to reject the output is performed by running the reversed chain in a clever way. Fill's algorithm is interruptible, in the sense that the impatient user does not bias the output by prematurely terminating a run. Like CFTP, Fill's algorithm works best in the monotone situation.

Monotonicity is a very strong condition. In particular, it is possible to obtain very good experimental estimates for the mixing time in the monotone situation without any use of CFTP (see Johnson [51, Section 2b]).

We now review the main results obtained in the area of perfect sampling. Again, our interest is focused on those papers with a theoretical computer science slant. For a wider bibliography, see the web site [80].

We first consider applications of monotone CFTP. Felsner and Wernisch [32] showed how to perfectly sample random linear extensions of two-dimensional partially ordered sets, while Kim, Shor and Winkler applied monotone CFTP to independent sets in bipartite graphs, as cited in [69]. (An alternative approach to the latter problem is described in Example 7.2 below.)

Propp and Wilson [67] applied monotone CFTP to several problems, including sampling lattice paths uniformly at random, sampling permutations uniformly, sampling from the Gibbs distribution in the ferromagnetic Ising model, sampling from the ferromagnetic q-state Potts model if $q \geq 1$, and sampling lozenge tilings uniformly at random. In most applications the expected running time is not analyzed, although this is polynomial whenever the Markov chain in question is rapidly mixing, by (4).

Now let us consider *antimonotone* systems. Here there is a partial order which is reversed by transitions of the Markov chain. This concept builds on the work of Kendall [58] on repulsive point-processes. Häggström and Nelander [42] showed how to apply CFTP to several systems, including independent sets in graphs, the antiferromagnetic Ising model, and the random cluster model when $q < 1$ (see [42] for details).

The concept of a *bounding chain* generalises both monotone and antimonotone chains. Huber [45] used bounding chains to obtain a uniform sampling algorithm for graph colourings, which has expected polynomial time if at least $(\Delta + 1)^2$ colours are used, where Δ is the maximum degree of the graph. By modifying the analysis of [11], Huber obtained an expected polynomial time uniform sampler for sink free orientations of a graph. Very recently, Huber [46] analysed the Dyer-Greenhill chain for independent sets [29] using bounding chains, to give an algorithm for perfect sampling in expected polynomial time whenever $\lambda \leq 2/(\Delta - 2)$, $\Delta \geq 3$.

There are two results on performing perfect sampling in a Markov chain where the transition probabilities are unknown. The first such result was given by Lovász and Winkler [60]. This was improved upon by Propp and Wilson [68], who presented an algorithm with universal randomized stationary stopping time which is within a constant factor of optimal. These algorithms are related to the generation of random spanning trees in a weighted directed graph. Aldous [1] and Broder [6] independently discovered an algorithm for generating random spanning trees of undirected graphs uniformly at random. Propp and Wilson [68] gave faster and more general algorithms for perfect generation of spanning trees.

We return to the subject of perfect sampling in Section 7 below.

5 Coupling

As indicated in Section 4, recent work in constructing polynomial time sampling methods has centred around Markov chain Monte Carlo algorithms, with coupling as the proof technique. We now describe this idea, as far as is relevant to finite Markov chains. For further information, see [59].

A *coupling* $\mathcal{C}(\mathcal{M})$ for $\mathcal{M}$ is a stochastic process (X_t, Y_t) on Ω^2 such that each of X_t, Y_t is marginally a copy of $\mathcal{M}$,

$$\begin{aligned} \Pr(X_t = \sigma_1 \mid X_{t-1} = \omega_1) &= P(\omega_1, \sigma_1), \\ \Pr(Y_t = \sigma_2 \mid Y_{t-1} = \omega_2) &= P(\omega_2, \sigma_2), \end{aligned} \quad (\forall t > 0). \tag{5}$$

The following simple but powerful inequality, due to Doeblin [24], then follows easily from these definitions.

Lemma 5.1 (Coupling Lemma) *Let X_t, Y_t be a coupling for $\mathcal{M}$ such that Y_0 has the stationary distribution π. Then, if X_t has distribution p_t,*

$$d_{\mathrm{TV}}(p_t, \pi) \leq \Pr(X_t \neq Y_t). \tag{6}$$

Proof Suppose $A_t \subseteq \Omega$ maximizes in (2). Then, since Y_t has distribution π,

$$\begin{aligned} d_{\mathrm{TV}}(p_t, \pi) &= \Pr(X_t \in A_t) - \Pr(Y_t \in A_t) \\ &\leq \Pr(X_t \in A_t, Y_t \notin A_t) \leq \Pr(X_t \neq Y_t). \quad \blacksquare \end{aligned}$$

It is important to remember that the Markov chain Y_t is simply a proof construct, and X_t the chain we actually observe. We also require that $X_t = Y_t$ implies $X_{t+1} = Y_{t+1}$, since this makes the right side of (6) nonincreasing. Then the earliest epoch T at which $X_T = Y_T$ is called *coalescence*, making T a random variable. A *successful coupling* is such that $\lim_{t \to \infty} \Pr(X_t \neq Y_t) = 0$. Clearly we are only interested in successful couplings.

A coupling is a *Markovian coupling* if the process $\mathcal{C}(\mathcal{M})$ is a Markov chain on Ω^2. Griffeath [39] showed that there always exists a *maximal coupling*, which gives equality in (6). (An easier proof is given in [66].) This maximal coupling is in general non-Markovian, and is seemingly not constructible without knowing p_t ($t = 1, 2, \ldots$). But coupling has little algorithmic value if we already know p_t. More generally, it seems difficult to prove mixing properties of non-Markovian couplings in our setting. Therefore we restrict attention to Markovian couplings, at the (probable) cost of sacrificing equality in (6). Burdzy and Kendall [15] investigated efficient Markovian couplings.

Let $\mathcal{C}(\mathcal{M})$ be a Markovian coupling, with Q its transition matrix, i.e. the probability of a joint transition from (ω_1, ω_2) to (σ_1, σ_2) is $Q_{\sigma_1 \sigma_2}^{\omega_1 \omega_2}$. The precise conditions required of Q are then

$$Q_{\sigma_1 \sigma_2}^{\omega \omega} \neq 0 \quad \text{implies} \quad \sigma_1 = \sigma_2 \quad (\forall \omega \in \Omega), \qquad (7)$$

$$\sum_{\sigma_2 \in \Omega} Q_{\sigma_1 \sigma_2}^{\omega_1 \omega_2} = P_{\sigma_1}^{\omega_1} \quad (\forall \omega_2 \in \Omega), \qquad \sum_{\sigma_1 \in \Omega} Q_{\sigma_1 \sigma_2}^{\omega_1 \omega_2} = P_{\sigma_2}^{\omega_2} \quad (\forall \omega_1 \in \Omega). \qquad (8)$$

Here (7) implies equality after coalescence, and (8) implies the marginals are copies of $\mathcal{M}$. Our goal is to design Q so that $\Pr(X_t \neq Y_t)$ quickly becomes small. We need only specify Q to satisfy (8) for $\omega_1 \neq \omega_2$. The other entries are completely determined by (7) and (8).

To prove rapid mixing using coupling, it is usual to map $\mathcal{C}(\mathcal{M})$ to a process on $\mathbf{N}$ by defining a function $\psi : \Omega^2 \to \mathbf{N}$ such that $\psi(\omega_1, \omega_2) = 0$ implies $\omega_1 = \omega_2$. We call this a *proximity function*. Then $\Pr(X_t \neq Y_t) \leq \mathbf{E}(\psi(X_t, Y_t))$, by Markov's inequality, and we need only show that $\mathbf{E}(\psi(X_t, Y_t))$ converges quickly to zero.

Example 5.2 Continuing Example 3.2, as an illustration we analyse the INSERT/DELETE chain in the case when $\Delta = 0$, i.e. $E = \emptyset$. Then the qualification in step (b) of its description becomes redundant. We use this example merely to fix ideas. There are, of course, many other (possibly simpler) ways of approaching this easy case.

At time t, let X_t, Y_t be the states occupied by our two copies of the chain. Consider the following coupling, where for convenience we drop temporarily the subscript t:

Choose the same $v \in V$ uniformly at random in both X, Y.
(i) If $v \in X \cap Y$, delete in both X, Y with probability $1/(1 + \lambda)$,
no move in both X, Y with probability $\lambda/(1 + \lambda)$.
(ii) If $v \notin X \cup Y$, insert in both X, Y with probability $\lambda/(1 + \lambda)$,
no move in both X, Y with probability $1/(1 + \lambda)$.
(iii) If $v \in X \setminus Y$, delete in X, no move in Y, with probability $1/(1 + \lambda)$,
insert in Y, no move in X, with probability $\lambda/(1 + \lambda)$.
(iv) If $v \in Y \setminus X$, insert in X, no move in Y, with probability $\lambda/(1 + \lambda)$,
delete in Y, no move in X, with probability $1/(1 + \lambda)$.

It is easy to see that this coupling has the correct marginals for both X_t, Y_t, though they are far from independent. To monitor convergence, let $H(X, Y)$ be the *Hamming distance* between X, Y, i.e. $H(X, Y) = |X \oplus Y|$, where $\oplus$ denotes symmetric difference. Hamming distance clearly satisfies the requirements of a proximity function, and $H(X, Y) \leq n$.

In cases (i) and (ii) of the coupling we have $H(X_{t+1}, Y_{t+1}) = H(X_t, Y_t)$, whereas in cases (iii) and (iv) we have $H(X_{t+1}, Y_{t+1}) = H(X_t, Y_t) - 1$. However, the probability that case (iii) or (iv) occurs is simply $H(X_t, Y_t)/n$. Hence

$$\mathbf{E}(H(X_{t+1}, Y_{t+1}) \mid (X_t, Y_t)) = H(X_t, Y_t) - H(X_t, Y_t)/n = (1 - 1/n)H(X_t, Y_t),$$

from which it follows by induction that

$$\mathbf{E}(H(X_t, Y_t)) = (1 - 1/n)^t H(X_0, Y_0) \leq n(1 - 1/n)^t \leq n e^{-t/n}.$$

But now, by (6) and the non-negativity and integrality of H,

$$\mathrm{d_{TV}}(p_t, \pi) \leq \Pr(X_t \neq Y_t) \leq \mathbf{E}(H(X_t, Y_t)) \leq n e^{-t/n},$$

from which it follows easily that $\tau(\varepsilon) \leq n \log(n/\varepsilon)$. Thus we have established rapid mixing of the chain for this simple case.

Using a more clever coupling, the mixing time of this chain can be improved by a constant factor. We omit the details since our interest is purely illustrative. ∎

6 Path coupling

A major difficulty with coupling is that we are obliged to specify it, and show improvement in the proximity function, for every pair of states. The idea of *path coupling* [10], where applicable, can be a major saving in this respect. We describe the approach below.

Recall that a *quasi-metric* satisfies the conditions for a metric except possibly the symmetry condition. Any metric is a quasi-metric, but a simple example of a quasi-metric which is not a metric is directed edge distance in a digraph.

Suppose we have a relation $S \subseteq \Omega^2$ such that S has transitive closure Ω^2, and suppose that we have a proximity function defined for all pairs in S, i.e. $\psi: S \to \mathbb{N}$. Then we may lift ψ to a quasi-metric $\delta(\omega, \omega')$ on Ω as follows. For each pair $(\omega, \omega') \in \Omega^2$, consider the set $\mathcal{P}(\omega, \omega')$ of all sequences

$$\omega = \omega_1, \omega_2, \ldots, \omega_{r-1}, \omega_r = \omega' \quad \text{with} \quad (\omega_i, \omega_{i+1}) \in S \quad (i = 1, \ldots, r-1). \tag{9}$$

Then we set

$$\delta(\omega, \omega') = \min_{\mathcal{P}(\omega, \omega')} \sum_{i=1}^{r-1} \psi(\omega_i, \omega_{i+1}). \tag{10}$$

It is easy to prove that δ is a quasi-metric. We call a sequence minimizing (10) *geodesic*. We now show that, without any real loss, we may define the (Markovian) coupling only on pairs in S. Such a coupling is a called a path coupling. We give a detailed development below. Clearly $S = \Omega^2$ is always a relation whose transitive closure is Ω^2, but path coupling is only useful when we can define a suitable S which is "much smaller" than Ω^2. A relation of particular interest is $\mathcal{R}_\sigma$ from Section 3, but this is not always the best choice.

As in Section 5, we use σ (or σ_i) to denote a state obtained by performing a single transition of the chain from the state ω (or ω_i). Let P_σ^ω denote the probability of a transition from state ω to state σ in the Markov chain, and let $Q_{\sigma\sigma'}^{\omega\omega'}$ denote the probability of a joint transition from (ω, ω') to (σ, σ'), where $(\omega, \omega') \in S$, as specified by the path coupling. Since this coupling has the correct marginals, we have

$$\sum_{\sigma' \in \Omega} Q_{\sigma\sigma'}^{\omega\omega'} = P_\sigma^\omega, \qquad \sum_{\sigma \in \Omega} Q_{\sigma\sigma'}^{\omega\omega'} = P_{\sigma'}^{\omega'} \qquad (\forall (\omega, \omega') \in S). \tag{11}$$

We extend this to all pairs $(\omega, \omega') \in \Omega^2$, as follows. For each pair, fix a sequence $(\omega_1, \omega_2, \ldots, \omega_r) \in \mathcal{P}(\omega, \omega')$. We do not assume the sequence is geodesic here, or indeed the existence of any proximity function, but this is our eventual purpose. The implied global coupling $\bar{Q}_{\sigma_1\sigma_r}^{\omega_1\omega_r}$ is then defined along this sequence by successively conditioning on the previous choice. Using (11), this can be written explicitly as

$$\bar{Q}_{\sigma_1\sigma_r}^{\omega_1\omega_r} = \sum_{\sigma_2 \in \Omega} \sum_{\sigma_3 \in \Omega} \cdots \sum_{\sigma_{r-1} \in \Omega} Q_{\sigma_1\sigma_2}^{\omega_1\omega_2} \frac{Q_{\sigma_2\sigma_3}^{\omega_2\omega_3}}{P_{\sigma_2}^{\omega_2}} \cdots \frac{Q_{\sigma_{r-1}\sigma_r}^{\omega_{r-1}\omega_r}}{P_{\sigma_{r-1}}^{\omega_{r-1}}}. \tag{12}$$

Summing (12) over σ_r or σ_1, and again applying (11), causes the right side to successively simplify, giving

$$\sum_{\sigma_r \in \Omega} \bar{Q}_{\sigma_1\sigma_r}^{\omega_1\omega_r} = P_{\sigma_1}^{\omega_1} \quad (\forall \omega_r \in \Omega), \qquad \sum_{\sigma_1 \in \Omega} \bar{Q}_{\sigma_1\sigma_r}^{\omega_1\omega_r} = P_{\sigma_r}^{\omega_r} \quad (\forall \omega_1 \in \Omega). \tag{13}$$

Hence the global coupling satisfies (8), as we would anticipate from the properties of conditional probabilities.

Now suppose the global coupling is determined by geodesic sequences. We bound the expected value of $\delta(\sigma_1, \sigma_r)$. This is

$$\begin{aligned}
\mathbf{E}(\delta(\sigma_1, \sigma_r)) &= \sum_{\sigma_1} \cdots \sum_{\sigma_r} \delta(\sigma_1, \sigma_r) \frac{Q^{\omega_1 \omega_2}_{\sigma_1 \sigma_2} Q^{\omega_2 \omega_3}_{\sigma_2 \sigma_3} \cdots Q^{\omega_{r-1} \omega_r}_{\sigma_{r-1} \sigma_r}}{P^{\omega_2}_{\sigma_2} \cdots P^{\omega_{r-1}}_{\sigma_{r-1}}} \\
&\leq \sum_{\sigma_1} \cdots \sum_{\sigma_r} \sum_{i=1}^{r-1} \delta(\sigma_i, \sigma_{i+1}) \frac{Q^{\omega_1 \omega_2}_{\sigma_1 \sigma_2} Q^{\omega_2 \omega_3}_{\sigma_2 \sigma_3} \cdots Q^{\omega_{r-1} \omega_r}_{\sigma_{r-1} \sigma_r}}{P^{\omega_2}_{\sigma_2} \cdots P^{\omega_{r-1}}_{\sigma_{r-1}}} \\
&= \sum_{i=1}^{r-1} \sum_{\sigma_1} \cdots \sum_{\sigma_r} \delta(\sigma_i, \sigma_{i+1}) \frac{Q^{\omega_1 \omega_2}_{\sigma_1 \sigma_2} Q^{\omega_2 \omega_3}_{\sigma_2 \sigma_3} \cdots Q^{\omega_{r-1} \omega_r}_{\sigma_{r-1} \sigma_r}}{P^{\omega_2}_{\sigma_2} \cdots P^{\omega_{r-1}}_{\sigma_{r-1}}} \\
&= \sum_{i=1}^{r-1} \sum_{\sigma_i} \sum_{\sigma_{i+1}} \delta(\sigma_i, \sigma_{i+1}) Q^{\omega_i \omega_{i+1}}_{\sigma_i \sigma_{i+1}},
\end{aligned} \quad (14)$$

where we have used the triangle inequality for a quasi-metric and the same observation as that leading from (12) to (13).

Suppose we can find $\beta \leq 1$, such that, for all $(\omega, \omega') \in S$,

$$\mathbf{E}(\delta(\sigma, \sigma')) = \sum_{\sigma} \sum_{\sigma'} \delta(\sigma, \sigma') Q^{\omega \omega'}_{\sigma \sigma'} \leq \beta \delta(\omega, \omega'). \quad (15)$$

Then, from (14), (15) and (10) we have

$$\mathbf{E}(\delta(\sigma_1, \sigma_r)) \leq \sum_{i=1}^{r-1} \beta \delta(\omega_i, \omega_{i+1}) = \beta \sum_{i=1}^{r-1} \delta(\omega_i, \omega_{i+1}) = \beta \delta(\omega_1, \omega_r). \quad (16)$$

Thus we can show (15) for every pair, merely by showing that this holds for all pairs in S. To apply path coupling to a particular problem, we must find a relation S and proximity function ψ so that this is possible. In particular we need $\delta(\omega, \omega')$ for $(\omega, \omega') \in S$ to be easily deducible from ψ.

Suppose that Ω has *diameter* D, i.e. $\delta(\omega, \omega') \leq D$ for all $\omega, \omega' \in \Omega$. Then, if $\beta < 1$, a similar calculation to that at the end of Example 5.2 gives

$$\mathrm{d}_{\mathrm{TV}}(p_t, \pi) \leq \varepsilon \quad \text{for} \quad t \geq \ln(D\varepsilon^{-1})/(1-\beta). \quad (17)$$

This bound is polynomial even when D is exponential in the problem size. It is also possible to prove a bound when $\beta = 1$, provided we know the quasi-metric cannot "get stuck". Specifically, we need an $\alpha > 0$ (inversely polynomial in the problem size) such that, in the above notation,

$$\Pr(\delta(\sigma, \sigma') \neq \delta(\omega, \omega')) \geq \alpha \quad (\forall \omega, \omega' \in \Omega). \quad (18)$$

Observe that it is not sufficient simply to establish (18) for pairs in S. However, the structure of the path coupling can usually help in proving it. In this case, we can show that

$$\mathrm{d}_{\mathrm{TV}}(p_t, \pi) \leq \varepsilon \quad \text{for} \quad t \geq \lceil eD^2/\alpha \rceil \lceil \ln(\varepsilon^{-1}) \rceil. \quad (19)$$

This is most easily shown using a martingale argument. We omit the proof, but see [61] for details. Here we need D to be polynomial in the problem size.

Random Walks on Combinatorial Objects   119

Example 6.1 We continue Example 5.2, by analysing a modification of the INSERT/DELETE chain. We show that this is rapidly mixing if $\lambda \leq 2/(\Delta - 2)$. The reader may observe that the analysis would be difficult to tackle directly using coupling.

INSERT/DELETE/DRAG CHAIN

Let X_t be the current independent set.
(a) Choose $v \in V$ uniformly at random.
(b) (Delete) If $v \in X_t$, $X_{t+1} \leftarrow X_t \setminus \{v\}$ with probability $1/(1+\lambda)$.
    (Insert) If $v \notin X_t$ and $X_t \cup \{v\}$ is an independent set,
    $X_{t+1} \leftarrow X_t \cup \{v\}$ with probability $\lambda/(1+\lambda)$.
    (Drag) If $v \notin X_t$, but v has a *unique* neighbour $u \in X_t$,
    $X_{t+1} \leftarrow X_t \cup \{v\} \setminus \{u\}$ with probability $\lambda/4(1+\lambda)$.
(c) Set $t \leftarrow t+1$, go to (a).

This chain has the same stationary distribution as the INSERT/DELETE chain, since the DRAG move is symmetric. The set S contains all pairs in $\mathcal{I}(G)$ at unit Hamming distance. The proximity function on S is Hamming distance. This lifts to Hamming distance globally, which is a metric.

For a pair $(X, Y) \in S$, let w denote the unique vertex in which they differ, and without loss assume $X \setminus Y = \{w\}$. Let X', Y' denote the evolved states after one step. We then use the following path coupling. We interleave the analysis with its statement.

(a) Choose the same v uniformly at random in both X, Y.

(b) If $v \neq w$ or a neighbour of w, perform the same step in both X, Y.
    Clearly then $\delta(X', Y') = \delta(X, Y) = 1$.

(c) If $v = w$, then
    INSERT in Y, do nothing in X with probability $\lambda/(1+\lambda)$.
    DELETE in X, do nothing in Y with probability $1/(1+\lambda)$.
    In either case we have $\delta(X', Y') = 0$.

(d) If v is a neighbour of w which has no neighbour in Y,
    DRAG in X, INSERT in Y with probability $\lambda/4(1+\lambda)$.
    Do nothing in X, INSERT in Y with probability $3\lambda/4(1+\lambda)$.
    Do nothing in X or Y with probability $1/(1+\lambda)$.
    In the first event, we have $X' = Y'$, but in the second, the Hamming distance increases by 1. Thus (conditional on v)
    $$\mathbf{E}(\delta(X', Y')) = 1 + \frac{-1 \times \lambda}{4(1+\lambda)} + \frac{1 \times 3\lambda}{4(1+\lambda)} = 1 + \frac{\lambda}{2(1+\lambda)}.$$

(e) If v is a neighbour of w which more than one neighbour in Y, do nothing in both X, Y. Clearly $\delta(X', Y') = 1$.

(f) If v is a neighbour of w which has a unique neighbour in Y, DRAG in Y, do nothing in X with probability $\lambda/4(1+\lambda)$.
Do nothing in either X or Y with probability $1 - \lambda/4(1+\lambda)$.
In the first event, the Hamming distance increases by 2. Thus (conditional on v)

$$\mathbf{E}(\delta(X',Y')) = 1 + \frac{2\times \lambda}{4(1+\lambda)} = 1 + \frac{\lambda}{2(1+\lambda)}.$$

Now suppose w has d_1, d_2, d_3 neighbours as in (d), (e) and (f) above. Then, since $d_1 + d_2 + d_3 \leq \Delta$,

$$\mathbf{E}(\delta(X',Y')) = 1 - \frac{1}{n} + \frac{d_1 \lambda}{2n(1+\lambda)} + \frac{d_3 \lambda}{2n(1+\lambda)} \leq 1 + \frac{1}{n}\left(\frac{\Delta\lambda}{2(1+\lambda)} - 1\right),$$

and $\mathbf{E}(\delta(X',Y')) \leq \delta(X,Y) = 1$ when $\Delta\lambda/2(1+\lambda) \leq 1$, i.e. $\lambda \leq 2/(\Delta-2)$. ∎

7 Perfect sampling

Perfect sampling was reviewed in Section 4.1 above. We now collect some results and observations on this idea. First we present an easy application of path coupling to monotone couplings. This is illustrated using independent sets in bipartite graphs.

We then attempt to relate the notions of perfect sampling and good sampling, by showing how a rapidly mixing Markov chain can be transformed into an expected polynomial time perfect sampler. This shows that good sampling is at least as hard as expected polynomial-time perfect sampling.

Before presenting these results, we give two definitions of perfect sampling. These definitions seem to encompass all reasonable possibilities. Let Ω be a state space and π a distribution on Ω. We say that an algorithm $\mathcal{A}$ is a *weak perfect sampler* for π if $\mathcal{A}$ outputs $i \in \Omega$ with probability π_i, and runs in expected polynomial time. The well known coupling from the past (CFTP) algorithm [67] is an example of a weak perfect sampler. Here we cannot exclude the possibility that there exists a state $i \in \Omega$ which the algorithm $\mathcal{A}$ cannot output in polynomial time. There are results where a weak perfect sampler has the additional property that

$$\Pr(T > k\,\mathbf{E}(T)) < (1/4)^k \qquad (k \in \mathbb{N}), \qquad (20)$$

or similar (see, for example, [45]). This guarantees that the perfect sampler could be turned into a good sampler, and may be of some comfort, but it does not imply that perfect sampling is achievable within any fixed polynomial time. If perfect sampling is the goal here, there seems little to be gained by showing that something like (20) holds. Even if the algorithm terminates in an acceptable time, we can only claim that the sample produced is perfectly distributed if we were truly prepared to wait forever for it to terminate. Otherwise the sample is only approximate, and the variation distance is determined by exactly how long we would have been prepared to wait. This point is illustrated by the following example.

Example 7.1 Consider a random walk on the integers $[n]$ with transitions from state i given as follows: with probability $1/2$ move to state $\min(i+1,n)$, otherwise move to state $\max(i-1,1)$. (This generalises an example used by Propp and Wilson [67] for a different purpose.) This chain is monotone under the obvious linear ordering of the states, and the stationary distribution is uniform. A monotone coupling for this chain can be defined as follows. From a given state (i,j), move to $(\min(i+1,n),\min(j+1,n))$ with probability $1/2$, otherwise move to $(\max(i-1,1),\max(j-1,1))$. The expected time for coalescence to occur, from initial state $(1,n)$, is $O(n^2)$. However, for all states i, j and times t, the probability that the coupling is in state (i,j) at time t is of the form $x2^{-t}$, for some integer x. If n is not a power of 2, therefore, we cannot guarantee perfect sampling within *any* bounded amount of time, using monotone CFTP. The difficulty here is not due to the use of the (possibly restrictive) PTM model of computation. Even if we move to the more powerful oracle coin model [73, p. 18], the conclusion still holds. ∎

Propp and Wilson [67] referred to this phenomenon as "user impatience", but we emphasise that how impatient *we might have been* is still relevant even when the algorithm terminates in an acceptable time. Since it is unlikely that we were really prepared to wait forever, this seems an immense difficulty for this form of perfect sampling. The situation here is in marked contrast to an expected polynomial time *optimization* algorithm where, if we wish, we may stop and restart the algorithm at any point without adverse consequences.

By contrast, an algorithm $\mathcal{A}$ is called a *strong perfect sampler* for π if the following conditions hold:

(a) The output of $\mathcal{A}$ is either the symbol $\perp$ or an element of Ω.

(b) The probability that $i \in \Omega$ is output, conditional on an element of Ω being output, is equal to π_i.

(c) The probability that $\perp$ is output is at most $\frac{1}{2}$.

(d) The running time of the algorithm is polynomial.

Fill's algorithm [33] is an example of a strong perfect sampler. (Fill used the term "interruptible".) Our definition of a strong perfect sampler for a distribution π generalises the definition of a uniform generator, given in [50]. Here we can truly claim that the output is a perfect sample, provided it is not $\perp$. Conditions (c) and (d) guarantee that it is likely we do not have to wait too long for this to occur. We do not, as with the weak perfect sampler, have to reason about what we might have done on other possible executions of the algorithm to assert this. It seems therefore that, where available, this form of perfect sampling is markedly superior. Huber [45, 46] used bounding chains to turn rapidly mixing Markov chains into weak perfect samplers, by applying CFTP. With a little modification, one can instead turn these algorithms into strong perfect samplers by applying Fill's algorithm [33].

A strong perfect sampler can easily be turned into a good sampler. Simply run the strong perfect sampler to completion. If the output is an element of Ω, then return this element. If the output is the symbol $\bot$, then return a fixed element of Ω, chosen arbitrarily beforehand. The variation distance between this distribution and the desired distribution is bounded above by the probability that the strong perfect sampler outputs the symbol $\bot$. We can ensure that the variation distance is at most ε, in time increased only by a factor $O(\log \varepsilon^{-1})$, simply by running the sampler repeatedly. Therefore strong perfect sampling is at least as hard as good sampling.

Good samplers exist for many problems, but few are known to possess strong perfect samplers. Can strong perfect sampling algorithms generally be found where good samplers exist? We conjecture that this is not generally possible, and that strong perfect sampling is harder than good sampling. Less ambitiously, can we devise a strong perfect sampler for all matchings in a graph, or for the linear extensions of a partial order? These questions are all open.

In Section 7.2, we show that good sampling is at least as hard as weak perfect sampling. One may also ask whether good samplers can be constructed whenever weak perfect sampling is possible. We present a very restrictive model of computation where weak perfect sampling is possible, but with high probability good sampling is not achievable. This shows that good sampling is harder than weak perfect sampling. Ideally, we would like to prove a similar result in a more realistic model of computation.

Bellare, Goldreich and Petrank [3] showed how to perform strong perfect sampling of NP-witnesses, using an NP-oracle. It follows that, if P = NP, we can perform strong perfect sampling for any relation in NP. Conversely, if strong perfect sampling is impossible for any problem in NP, then P $\neq$ NP.

7.1 Monotone path coupling

Suppose we have a Markov chain $\mathcal{M}$ on a state space Ω. Further suppose that there exists a partial order $\leq$ on Ω, possessing unique top and bottom elements. Let a path coupling $(\omega, \omega') \mapsto (\sigma, \sigma')$ be defined with respect to a subset S of Ω^2 and a proximity function ψ. We call this a *monotone path coupling* if the following conditions hold.

(a) Whenever $(\omega, \omega') \in \Omega^2$ such that $\omega < \omega'$, we can form a strictly increasing geodesic sequence $\omega = \omega_1, \omega_2, \ldots, \omega_r = \omega'$ between ω and ω'.

(b) If $(\omega, \omega') \in S$ and $\omega \leq \omega'$ then $\sigma \leq \sigma'$.

If these conditions are satisfied, then the implied global coupling is also monotone. An example of a monotone path coupling is given below.

Example 7.2 Consider the INSERT/DELETE/DRAG chain of Example 6.1, restricted to bipartite graphs. Let $G = (V, E)$ be a bipartite graph with vertex bipartition $V = V_1 \cup V_2$. A natural partial order on $\mathcal{I}(G)$ is defined by $X \leq Y$ if and only if

$$X \cap V_1 \subseteq Y \cap V_1 \quad \text{and} \quad Y \cap V_2 \subseteq X \cap V_2.$$

This partial order has a top element V_1 and a bottom element V_2. We now show that the coupling described in Example 6.1 is monotonic with respect to this partial order.

Suppose $X, Y \in \mathcal{I}(G)$ satisfy $X \leq Y$. Then $X \setminus Y \subseteq V_2$ and $Y \setminus X \subseteq V_1$. We can easily form a strictly increasing geodesic sequence between X and Y. Thus it suffices to show that the coupling is monotonic for pairs at distance one apart.

Let $X, Y \in \mathcal{I}(G)$ satisfy $H(X, Y) = 1$ and $X \leq Y$. Without loss of generality, we can assume that $X = Y \cup \{w\}$ for some $w \in V_2$. Let v be the vertex chosen uniformly at random in the step (a) of the coupling. If a vertex is either added to or deleted from both X and Y then the resulting pair (X', Y') satisfies $X' \leq Y'$. Therefore we need only check monotonicity when v satisfies condition (c), (d) or (f) of Example 6.1. However, it is easy to see that the coupling defined in each of these cases is monotone. For example, suppose that v satisfies condition (f). Then v is a neighbour of w which has a unique neighbour u in Y. Since $w \in V_2$ we have $v \in V_1$ and $u \in V_2$. Here $(X', Y') = (X, (Y \setminus \{u\}) \cup \{v\})$ or $(X', Y') = (X, Y)$. In the second case there is nothing to prove, while in the first case we have

$$X' = X < Y < Y \setminus \{u\} < (Y \setminus \{u\}) \cup \{v\} = Y'.$$

This shows that the coupling is monotone, and the other cases are proved similarly.

Suppose now that $\lambda \leq 2/(\Delta - 2)$ and $\Delta \geq 3$. Since the coupling is monotone, we can use a slight modification of Fill's algorithm [33] to give a strong perfect sampler for independent sets in bipartite graphs. For the necessary details on Fill's algorithm, see [33].

7.2 Transforming an approximate sampler into a weak perfect sampler

This section shows that weak perfect sampling is not difficult to achieve if we have a good MCMC sampler. The idea is simple: with very high probability, we output the result of running the Markov chain for a polynomial number of steps, just as we would in approximate sampling. Then with exponentially small probability, we perform an exponential amount of computation and output some state. The computation ensures that the output is distributed exactly according to the stationary distribution. The probability is chosen small enough that the expected cost of computation is polynomial.

Let $\mathcal{M}$ be a combinatorial Markov chain with transition matrix P, and state space Ω. Let $N = |\Omega|$ and suppose that $n = \lceil \log_2 N \rceil$. We assume that the transitions of $\mathcal{M}$ from state i are performed by a strong perfect sampler, for each i. Let δ be a parameter, the value of which is fixed later, such that $0 < \delta \leq \frac{1}{2}$. Let f, g be polynomial-time computable integers such that $0 < f/g \leq \pi_i$ for all $i \in \Omega$. We can certainly find such f, g: for example, let $f = 1$ and $g = 2^{|\sigma|^c}$, in the notation of Section 2. (We do not assume that the exact values of the π_i are known, although in almost all applications they will be known up to some normalising constant.) Define ε by

$$\varepsilon = \frac{\delta f}{(1-\delta)g},$$

for this particular value of δ. Let $t = \tau(\varepsilon)$, where $\tau(\cdot)$ is the mixing time of $\mathcal{M}$. Fix some start state, which we denote by 1, and let p_t denote the distribution of the Markov chain after t steps, starting from initial state 1.

Let r be the map defined by

$$\delta r_i = \pi_i - (1-\delta) p_t(i).$$

Then

$$\delta r_i = (1-\delta)(\pi_i - p_t(i)) + \delta \pi_i \geq -\varepsilon(1-\delta) + \delta \pi_i = \delta(\pi_i - f/g) \geq 0.$$

Hence r_i is nonnegative for all $i \in \Omega$. Since $\sum_{i \in \Omega} r_i = 1$, the map r is a probability distribution on Ω. Consider the algorithm given in pseudocode in Figure 1. The probability that this algorithm outputs $i \in \Omega$ is given by

$$(1-\delta) p_t(i) + \delta r_i = \pi_i.$$

Hence the algorithm performs perfect sampling.

We now describe an implementation of this algorithm with expected running time which is polynomial in n. We also check that the algorithm requires only a polynomial amount of space.

In order to calculate r_i exactly, we must be able to calculate the t-step probabilities $p_t(i)$. We now show how to do this. Without loss of generality, assume that t is a power of two, by rounding up if necessary. For all $\ell \geq 0$ and $i, j \in \Omega$ let $q^{(\ell)}(i,j)$ be defined by

$$q^{(\ell)}(i,j) = \Pr\left(X_{2^\ell} = j \mid X_0 = i\right).$$

Random Walks on Combinatorial Objects

> Begin.
>   with probability $(1 - \delta)$ do
>   run $\mathcal{M}$ for t steps from initial state 1 and output the final state;
>   otherwise
>   choose $i \in \Omega$ with probability r_i and output i;
>   end;
> End.

Figure 1: A perfect sampling algorithm

Then $q^{(0)}(i,j) = P_{ij}$ for all $i, j \in \Omega$. Moreover,

$$\begin{aligned}
q^{(\ell)}(i,j) &= \Pr\left(X_{2^\ell} = j \mid X_0 = i\right) \\
&= \sum_{k \in \Omega} \Pr\left(X_{2^{(\ell-1)}} = k \mid X_0 = i\right) \cdot \Pr\left(X_{2^{(\ell-1)}} = j \mid X_0 = k\right) \\
&= \sum_{k \in \Omega} q^{(\ell-1)}(i,k) \cdot q^{(\ell-1)}(k,j)
\end{aligned}$$

for all $i, j \in \Omega$ and $\ell \geq 0$. Now $2^s P$ is integral, for some polynomially bounded computable integer s, by definition of a weight matrix. Let $b_\ell = 2^\ell s$ for $\ell \geq 0$. Using induction, it is not difficult to show that $2^{b_\ell} q^{(\ell)}(i,j)$ is integral for all i, $j \in \Omega$ and $\ell \geq 0$.

> Begin.
>   if $\ell = 0$ then
>   return $2^s P_{ij}$;
>   else
>   sum := 0;
>   for $k := 1$ to N do
>   sum := sum + $Q(i, k, \ell - 1) \cdot Q(k, j, \ell - 1)$;
>   endfor;
>   return sum;
>   endif;
> End.

Figure 2: The procedure $Q(i, j, \ell)$

This leads to the recursive procedure $Q(i, j, \ell)$ for calculating $2^{b_\ell} q^{(\ell)}(i,j)$, given in Figure 2. Let $R(t)$ denote the running time of $Q(i, j, \log_2(t))$. We now bound $R(t)$ from above. Let c be a constant which is equal to the maximum of $R(1)$ and the cost of the overheads involved in a call to the procedure Q.

Then $R(2t) \leq c + 2^{n+1} R(t)$ for all t. Using induction we can show that

$$R(t) \leq c \frac{(2t)^{n+1} - 1}{2^{n+1} - 1} < 2c\, t^{n+1}$$

for $t = 2^\ell$, $\ell \geq 0$. Therefore $R(t) = O(t^{n+1})$ for all t, proving that the running time is polynomial.

Now consider the storage requirement of this procedure. It takes at most $b_\ell + 1$ bits to store $2^{b_\ell} q^{(\ell)}(i,j)$. At the ℓth level, the procedure needs $2(b_{\ell-1}+1)$ bits to store $Q(i,k,\ell-1) Q(k,j,\ell-1)$ and at most $n + b_\ell + 2$ bits to store the cumulative sum of these. Hence the total storage requirement is at most

$$\sum_{\ell=1}^{\log_2(t)} (2_b \ell + n + 4) < (n+4) \log_2(t) + 4st.$$

This is polynomial in n, as required.

Finally, we show how the procedure $Q(i,j,\ell)$ can be used to sample from the distribution r. Let $b = st = b_{\log_2(t)}$, so $b = b_\ell$ where $\ell = \log_2(t)$. Suppose that $\delta = 2^{-k}$ for some polynomially bounded integer k. We divide the interval $\{0, \ldots, 2^b Z - 1\}$ into N subintervals, where the ith subinterval has length $2^b Z r_i$. We can express $2^b Z r_i$ as

$$2^b Z\, r_i = 2^{k+b} M_i - Z(2^k - 1)\, Q(1, i, \ell).$$

This shows that $2^b Z r_i$ is an integer. Let $d = \lceil \log_2(Z) \rceil$. To sample from the distribution r, we repeatedly choose $X \in \{0, \ldots, 2^{d+b} - 1\}$ as a string of $d + b$ random bits, until $X < 2^b Z$. We output i if and only if X lies in the ith subinterval. The procedure is given in pseudocode in Figure 3 below.

The space requirements of this procedure are polynomial, as is easily checked. With overheads ignored, the expected value of the running time is

$$R(t) \sum_{i=1}^{N} i\, r_i \leq N R(t).$$

This completes the description of the implementation of the perfect sampling algorithm. It remains only to show that the expected running time is polynomial. Let T be a random variable which is the running time of the perfect sampler. Then the expected value of T satisfies

$$\mathbf{E}(T) \leq (1 - \delta) t + \delta N R(t).$$

Clearly this expression is polynomial if and only if $\delta N R(t)$ is polynomial. Since $\mathcal{M}$ is rapidly mixing, there are known polynomials p, q such that

$$t \leq p(n) q(\log(\varepsilon^{-1})).$$

Random Walks on Combinatorial Objects

```
Begin.
  repeat
    generate X ∈ {0,...,2^(d+b) − 1} as a string of d + b random bits;
  until X < 2^b Z;
  cM := 2^(b+k);
  cQ := (2^k − 1)Z;
  Msum := 0;
  Qsum := 0;
  i := 0;
  while Msum − Qsum ≤ X do
    i := i + 1;
    Msum := Msum + cM · M_i;
    Qsum := Qsum + cQ · Q(1, i, t);
  endwhile;
  return i;
End.
```

Figure 3: A procedure for sampling from the distribution r

Therefore
$$\delta NR(t) \leq \frac{g}{f\varepsilon^{-1}}(2p(n))^{n+1}q(\log(\varepsilon^{-1}))^{n+1},$$
writing δ in terms of ε. Now
$$\lim_{\varepsilon \to 0} \frac{\log(\varepsilon^{-1})^c}{\varepsilon^{-1}} = 0$$
for any constant c. Therefore, by choosing ε small enough, we can make the quantity $\delta NR(t)$ as small as we like. In particular, let $\varepsilon = n^{-Kn}$ where $K = \deg(p) + 2\deg(q) + 1$. Then it is not difficult to show that $\delta NR(t) \leq g/fn^K$, which is polynomial.

One technical point remains. If $\varepsilon = n^{-Kn}$ then δ is unlikely to be equal to some negative power of two. This can be easily addressed. Let $\delta = 2^{-k}$, where
$$k = \lceil \log_2(f/(\varepsilon g) + 1) \rceil,$$
and let $\varepsilon' = f\delta/(g(1-\delta))$. Then
$$\varepsilon' = \frac{f}{g(2^k - 1)} \leq \varepsilon,$$
so ε' is small enough to ensure that $\delta NR(t)$ is polynomial. Moreover, it is not difficult to show that
$$\varepsilon' \geq (2\varepsilon^{-1} + g/f)^{-1},$$

which is not doubly exponentially small. So all requirements are satisfied if we use ε' instead of ε.

We have shown how to transform an efficient approximate sampler into a weak perfect sampler. Moreover, by increasing t if necessary we can assume that $p_t(i) > 0$ for all $i \in \Omega$. This ensures that there does not exist a state i which cannot be output in polynomial time.

A possible criticism of our algorithm is that, in order to claim that the sample is perfect, the user must really have been prepared to carry out a huge amount of computation in the second phase, even though this occurs with exponentially small probability. However, other perfect sampling methods such as CFTP suffer from precisely this problem, even in the monotone case, as illustrated in Example 7.1 above. On the other hand, this algorithm does have one advantage over CFTP. It runs in deterministic polynomial space, whereas CFTP may require unbounded space to be available if truly perfect samples are to be output.

We make one final remark on the topic of perfect sampling. Consider the following highly restrictive model of computation. An algorithm has access to a tape, which stores the details of a distribution ρ on the set $[2^n]$. The distribution has the form

$$\rho_1 = 1 - 2^{-n}, \qquad \rho_j = 2^{-n}, \qquad \rho_i = 0 \text{ otherwise,}$$

where $j \in \{2^{n-1}+1, \ldots, 2^n\}$ is selected uniformly at random beforehand. The tape head is positioned at cell 1, and the tape can only be read sequentially. In this model, weak perfect sampling is possible, as follows. With probability $1 - 2^{-n}$, output 1. Otherwise, read through the tape and output j when you find it. This procedure has expected running time $\mathbf{E}(T) \leq 2$. However, no algorithm can be guaranteed to perform good sampling, since otherwise we may demand variation distance at most 3^{-n}. But in polynomial time the algorithm can only check a polynomial number of values of ρ_i, and so the algorithm is not be able to find j. The best the algorithm can do is guess. The probability that it guesses correctly is $2^{-(n-1)}$. Whenever it guesses incorrectly, the total variation distance between its output and ρ is greater than $2^{-(n+1)}$, which is greater than 3^{-n}. Hence with probability $1 - 2^{-(n-1)}$ the algorithm fails, uniformly over all choices of j.

This shows that good sampling is harder than weak perfect sampling, at least in this very severe model of computation. It would be very interesting to know whether a similar result can be proved for some more realistic model of computation.

8 Negative results

There are many problems for which good sampling or integration do not seem possible. For these problems, we seek proofs of impossibility. If the decision problem of determining whether there is an ω at which $W(\sigma, \omega) > 0$

is NP-complete, then clearly this is immediate since (presumably) we cannot access the sample space in polynomial time. Otherwise, the "magnification" technique from [50] exploits the existence of an embedded NP-complete problem to show that we can only access a negligible portion of the sample space. See [78, pp. 295–296] for a discussion. In more complex situations, we may still be able to show that good sampling or integration implies a (randomized or deterministic) polynomial time algorithm for some NP-complete problem. These are the most satisfactory types of results, but in other situations we are less ambitious, and simply focus on proving that the MCMC approach is unlikely to succeed. Here we attempt to show that some natural MCMC algorithm, or class of algorithms has mixing time superpolynomial in problem size.

We do not review all results in this area. These may be found in the work cited in Section 4 or in [78, pp. 295–296]. Instead, we examine two typical results which have been obtained recently for the problem of sampling or counting in $\mathcal{I}(G)$. These correspond to taking $\lambda = 1$ in the measure defined in Example 3.1. It follows from Example 6.1 that a good sampler is available for $\lambda = 1$ if $\Delta \leq 4$. This implies the existence of a good approximate counter for $\mathcal{I}(G)$. The question then arises as to whether $\Delta \leq 4$ is an artefact of the analysis, or is there some constant Δ for which sampling (or approximate counting) is impossible? This question was partially answered by Luby and Vigoda [62], who showed that there *exists* a constant value for which approximate counting is impossible, unless P = NP. This could be made quantitative, using more recent results of Berman and Karpinski [5], but the best Δ obtainable in this way seems to be around 8000. However, Dyer, Frieze and Jerrum [25] adopted a different (though related) approach, and proved the following stronger result.

Theorem 8.1 *If $\Delta \geq 25$ then there is no polynomial time algorithm which can approximate the logarithm of the number of independent sets in a Δ-regular graph to within relative error 10^{-6} unless* P = NP *(under randomized reductions).*

(By "randomized reductions", we mean that the algorithm used to transform an instance of problem A to problem B may be randomized.)

Theorem 8.1 clearly also rules out the possibility of any good sampler when $\Delta \geq 25$. We omit the proof, but simply remark that the result rests on recent deep work in the complexity of approximate optimization. In particular, it relies on work of Håstad [43].

However, there is a clear gap between what is known to be possible ($\Delta \leq 4$) and this impossibility result ($\Delta \geq 25$). In this region, the following weaker result is proved in [25].

Theorem 8.2 *Let $\Delta \geq 6$ and $b(n) \leq 0.35n$. There exists a family of bipartite graphs $G_0(n)$ on $n + n$ vertices, and maximum degree Δ, with the following property. Any Markov chain $\mathcal{M}$ on $\mathcal{I}(G_0)$ which changes the status of at most $b(n)$ vertices at each step has mixing time $\Omega(e^{\gamma n})$, for some constant $\gamma \geq 0$.*

Again we omit the proof, remarking only that it relies principally on random graph methods.

If $\Delta \geq 6$, Theorem 8.2 rules out any MCMC good sampler which extends (say) INSERT/DELETE/DRAG, but changes any constant number of vertices at one step. It rules out even faster moving chains. The only possibility is a chain which could change the status of very many vertices at a single step. However, it is difficult to see how a step of such a chain could be implemented in polynomial time.

The reader may observe that Theorem 8.2 still leaves an intriguing gap. What can be said for $\Delta = 5$? At the time of writing this remains open, but we conjecture that a good sampler does exist in this case.

Acknowledgements

We thank Mark Jerrum, Mark Huber and Russ Bubley for commenting on a draft of this paper and suggesting some improvements. This work was sponsored by the ESPRIT Working Group RAND2.

References

[1] D. Aldous, A random walk construction of uniform spanning trees and uniform labelled trees, *SIAM Journal on Discrete Mathematics*, **3** (1990), 450–465.

[2] D. Aldous & J. Fill, *Reversible Markov chains and random walks on graphs*, Monograph in preparation, available from http://www.stat.berkeley.edu/users/aldous/book.html, Department of Statistics, University of California, Berkeley.

[3] M. Bellare, O. Goldreich & E. Petrank, Uniform generation of NP-witnesses using an NP-oracle, *Electronic Colloquium on Computational Complexity*, (1998), Report number TR98-032.

[4] J. van den Berg & J. E. Steif, Percolation and the hard-core lattice gas model, *Stochastic Processes and their Applications*, **49** (1994), 179–197.

[5] P. Berman & M. Karpinski, On some tighter inapproximability results, preprint, Department of Computer Science, University of Bonn, 1998.

[6] A. Broder, Generating random spanning trees, in *30th Annual Symposium on Foundations of Computer Science,* IEEE, Los Alimitos (1989), pp. 442–447.

[7] A. Broder, A. Frieze & E. Upfal, Static and dynamic path selection on expander graphs: a random walk approach, in *29th Annual Symposium on Theory of Computing,* ACM, New York (1997), pp. 531–539.

[8] S. Brooks (Administrator), MCMC Preprint Service, Web site, http://www.stats.bris.ac.uk/MCMC.

[9] R. Bubley, Randomized algorithms: approximation, generation and counting, PhD thesis, University of Leeds, 1998.

[10] R. Bubley & M. Dyer, Path coupling: a technique for proving rapid mixing in Markov chains, in *38th Annual Symposium on Foundations of Computer Science,* IEEE, Los Alimitos (1997), pp. 223–231.

[11] R. Bubley & M. Dyer, Graph orientations with no sink and an approximation for a hard case of #SAT, in *8th Annual Symposium on Discrete Algorithms,* ACM–SIAM, New York–Philadelphia (1997), pp. 248–257.

[12] R. Bubley & M. Dyer, Faster random generation of linear extensions, in *9th Annual Symposium on Discrete Algorithms,* ACM–SIAM, New York–Philadelphia (1998), pp. 350–354.

[13] R. Bubley, M. Dyer & C. Greenhill, Beating the 2Δ bound for approximately counting colourings: a computer-assisted proof of rapid mixing, in *9th Annual Symposium on Discrete Algorithms,* ACM–SIAM, New York–Philadelphia (1998), pp. 355–363.

[14] R. Bubley, M. Dyer & M. Jerrum, An elementary analysis of a procedure for sampling points in a convex body, *Random Structures and Algorithms,* **12** (1998), 213–235.

[15] K. Burdzy & W. Kendall, Efficient Markovian couplings: examples and counterexamples, Technical Report 331, Department of Statistics, University of Warwick, 1998.

[16] F. Chung & R. Graham, Random walks on generating sets for finite groups, *Electronic Journal of Combinatorics,* **4** (1997), Research paper 7.

[17] F. Chung, R. Graham & S. Yau, On sampling with Markov chains, *Random Structures and Algorithms,* **9** (1996), 55–77.

[18] F. Chung & S. Yau, Eigenvalue inequalities for graphs and convex subgraphs, *Communications in Analysis and Geometry,* **5** (1997), 575–623.

[19] C. Cooper & A. Frieze, Mixing properties of the Swendsen–Wang process on classes of graphs, (Preprint), Department of Mathematical Sciences, University of North London, 1998.

[20] A. Czumaj, P. Kanarek, M. Kutyłowski & K. Loryś, Delayed path coupling and generating random permutations via distributed stochastic processes, in *10th Annual Symposium on Discrete Algorithms,* ACM–SIAM, New York–Philadelphia (1999), pp. 271–280.

[21] P. Diaconis & L. Saloff-Coste, Walks on generating sets of abelian groups, *Probability Theory and Related Fields*, **105** (1996), 393–421.

[22] P. Diaconis & L. Saloff-Coste, Logarithmic Sobolev inequalities for finite Markov chains, *The Annals of Applied Probability*, **6** (1996), 695–750.

[23] R. L. Dobrushin, The description of a random field by means of conditional probabilities and conditions of its regularity, *Theory of Probability and its Applications*, **13** (1968), 197–224.

[24] W. Doeblin, Exposé de la théorie des chaines simples constantes de Markov à un nombre fini d'états, *Revue Mathématique de l'Union Interbalkanique*, **2** (1933), 77–105.

[25] M. Dyer, A. Frieze & M. Jerrum, On counting independent sets in sparse graphs, (Preprint), School of Computer Studies, University of Leeds, 1998.

[26] M. Dyer, A. Frieze & R. Kannan, A random polynomial-time algorithm for approximating the volume of convex bodies, *Journal of the ACM*, **38** (1991), 1–17.

[27] M. E. Dyer & A. M. Frieze, Computing the volume–a case where randomization provably helps, in *Probabilistic Combinatorics and its Applications* (ed. B. Bollobás), *Proceedings of Symposia in Applied Mathematics*, 44, American Mathematical Society, Providence, RI (1991), pp. 123–170.

[28] M. Dyer & C. Greenhill, A genuinely polynomial-time algorithm for sampling two-rowed contingency tables, in *25th International Colloquium on Automata, Languages and Programming*, Springer, Berlin (1998), pp. 339–350.

[29] M. Dyer & C. Greenhill, On Markov chains for independent sets, (Preprint), School of Computer Studies, University of Leeds, 1997.

[30] M. Dyer & C. Greenhill, A more rapidly mixing Markov chain for graph colourings, *Random Structures and Algorithms*, **13** (1998), 285–317.

[31] M. Dyer, R. Kannan & J. Mount, Sampling contingency tables, *Random Structures and Algorithms*, **10** (1997), 487–506.

[32] S. Felsner & L. Wernisch, Markov chains for linear extensions: the two-dimensional case, in *8th Annual Symposium on Discrete Algorithms*, ACM–SIAM, New York–Philadelphia (1997), pp. 239–247.

[33] J. A. Fill, An interruptible algorithm for perfect sampling via Markov chains, *Annals of Applied Probability*, **8** (1998), 131–162.

[34] A. Frieze, M. Jerrum, M. Molloy, R. Robinson, N. Wormald, Generating and counting Hamiltonian cycles in random regular graphs, *Journal of Algorithms*, **21** (1996), 176–198.

[35] L. Goldberg & M. Jerrum, The "Burnside process" converges slowly, (Preprint), Department of Computer Science, University of Edinburgh, 1998.

[36] V. Gore & M. Jerrum, The Swendsen–Wang process does not always mix rapidly, in *29th Annual Symposium on Theory of Computing*, ACM, New York (1997), pp. 674–681.

[37] V. Gore, M. Jerrum, S. Kannan, Z. Sweedyk & S. Mahaney, A quasi-polynomial-time algorithm for sampling words from a context-free language, *Information and Computation*, **134** (1997), 59–74.

[38] C. Greenhill, The complexity of counting colourings and independent sets in sparse graphs and hypergraphs, (Preprint), School of Computer Studies, University of Leeds, 1998.

[39] D. Griffeath, A maximal coupling for Markov chains, *Zeitschrift für Wahrscheinlichkeitstheorie und verwandte Gebeite*, **31** (1975), 95–106.

[40] S. Guattery, T. Leighton & G. Miller, The path resistance method for bounding λ_2 of a Laplacian, in *8th Annual Symposium on Discrete Algorithms*, ACM–SIAM, New York–Philadelphia (1997), pp. 201–210.

[41] O. Häggström & K. Nelander, On exact simulation of Markov random fields, using coupling from the past, (Preprint), Department of Mathematical Statistics, Chalmers University of Technology, 1997.

[42] O. Häggström & K. Nelander, Exact sampling from anti-monotone systems, *Statistica Neerlandica*, in press.

[43] J. Håstad, Some optimal inapproximability results, in *29th Annual Symposium on Theory of Computing*, ACM, New York (1997), pp. 1–10.

[44] D. Hernek, Random generation of $2 \times n$ contingency tables, *Random Structures and Algorithms*, **13** (1998), 71–79.

[45] M. Huber, Exact sampling and approximate counting techniques, in *30th Annual Symposium on Theory of Computing*, ACM, New York (1998), pp. 31–40.

[46] M. Huber, Exact random sampling from independent sets, (Preprint), School of Operations Research and Industrial Engineering, Cornell University, 1998.

[47] M. Jerrum, A very simple algorithm for estimating the number of k-colorings of a low-degree graph, *Random Structures and Algorithms*, **7** (1995), 157–165.

[48] M. Jerrum, Mathematical foundations of the Markov chain Monte Carlo method, in *Probabilistic Methods for Algorithmic Discrete Mathematics* (eds. M. Habib, C. McDiarmid, J. Ramirez-Alfosin & B. Reed), Springer-Verlag, Berlin (1998), pp. 116–165.

[49] M. Jerrum & A. Sinclair, The Markov chain Monte Carlo method: an approach to approximate counting and integration, in *Approximation Algorithms* (ed. D. S. Hochbaum), PWS Publishing Company, Boston (1996), pp. 482–520.

[50] M. R. Jerrum, L. G. Valiant & V. V. Vazirani, Random generation of combinatorial structures from a uniform distribution, *Theoretical Computer Science*, **43** (1986), 169–188.

[51] V. E. Johnson, Studying convergence of Markov chain Monte Carlo algorithms using coupled sample paths, *Journal of the American Statistical Association*, **91** (1996), 154–166.

[52] R. Kannan, Markov chains and polynomial time algorithms, in *35th Annual Symposium on Foundations of Computer Science,* IEEE, Los Alimitos (1994), pp. 656–671.

[53] R. Kannan & G. Li, Sampling according to the multivariate normal density, in *36th Annual Symposium on Foundations of Computer Science,* IEEE, Los Alimitos (1996), pp. 204–212.

[54] R. Kannan, L. Lovász & M. Simonovits, Random walks and an $O^*(n^5)$ volume algorithm, *Random Structures and Algorithms*, **11** (1997), 1–50.

[55] R. Kannan, P. Tetali & S. Vempala, Simple Markov chain algorithm for generating bipartite graphs and tournaments, in *8th Annual Symposium on Discrete Algorithms,* ACM–SIAM, New York–Philadelphia (1997), pp. 193–200.

[56] R. Kannan & S. Vempala, Sampling lattice points, in *29th Annual Symposium on Theory of Computing,* ACM, New York (1997), pp. 696–700.

[57] A. Karzanov & L. Khachiyan, On the conductance of order Markov chains, *Order*, **8** (1991), 7–15.

[58] W. Kendall, Perfect simulation for the area-interaction point process, in *Probability towards 2000*, Springer, Berlin (1998), pp. 218–234.

[59] T. Lindvall, *Lectures on the Coupling Method*, Wiley-Interscience, New York (1992).

[60] L. Lovász & P. Winkler, Exact mixing in an unknown Markov chain, *Electronic Journal of Combinatorics*, **2** (1995), Research paper 15.

[61] M. Luby, D. Randall & A. Sinclair, Markov chain algorithms for planar lattice structures, in *36th Annual Symposium on Foundations of Computer Science*, IEEE, Los Alimitos (1995), pp. 150–159.

[62] M. Luby & E. Vigoda, Approximately counting up to four, in *29th Annual ACM Symposium on Theory of Computing*, ACM, New York (1997), pp. 682–687.

[63] N. Madras & D. Randall, Factoring graphs to bound mixing rates, in *37th Annual Symposium on Foundations of Computer Science*, IEEE, Los Alimitos (1996), pp. 194–203.

[64] C. H. Papadimitriou, *Computational Complexity*, Addison-Wesley, Reading, MA (1994).

[65] M. Peinado & T. Lengauer, Random generation of embedded graphs and an extension to Dobrushin uniqueness, in *30th Annual Symposium on Theory of Computing*, ACM, New York (1998), pp. 176–185.

[66] J. W. Pitman, On coupling of Markov chains, *Zeitschrift für Wahrscheinlichkeitstheorie und verwandte Gebeite*, **35** (1976), 315–322.

[67] J. G. Propp & D. B. Wilson, Exact sampling with coupled Markov chains and applications to statistical mechanics, *Random Structures and Algorithms*, **9** (1996), 223–252.

[68] J. G. Propp & D. B. Wilson, How to get a perfectly random sample from a generic Markov chain and generate a random spanning tree of a directed graph, *Journal of Algorithms*, **27** (1998), 170–217.

[69] J. Propp & D. Wilson, Coupling from the past: a user's guide, in *Microsurveys in Discrete Probability* (eds. D. Aldous & J. Propp), American Mathematical Society, to appear (1998), pp. 181–192.

[70] D. Randall & P. Tetali, Analyzing Glauber dynamics by comparison of Markov chains, in *3rd Latin American Symposium on Theoretical Informatics*, Springer, Campinas, Brazil (1998), pp. 292–304.

[71] J. Salas & A. Sokal, Absence of phase transition for antiferromagnetic Potts models via the Dobrushin uniqueness theorem, *Journal of Statistical Physics*, **86** (1997), 551–579.

[72] C. Schnorr, Optimal algorithms for self-reducible problems, in *3rd International Colloquium on Automata, Languages and Programming*, Edinburgh University Press, Edinburgh (1976), pp. 322–337.

[73] A. J. Sinclair, *Algorithms for Random Generation and Counting*, Birkhäuser, Boston (1993).

[74] A. Sokal, Introduction to Monte Carlo, Postscript file, available from http://math.nyu.edu/faculty/goodman/teaching/Monte_Carlo/Sokal.ps

[75] S. Vadhan, The complexity of counting in sparse, regular and planar graphs, preprint, Laboratory for Computer Science, MIT, 1997.

[76] L. G. Valiant, The complexity of computing the permanent, *Theoretical Computer Science*, **8** (1979), 189–201.

[77] E. Vigoda, personal communication, University of Berkeley, 1997.

[78] D. Welsh, Approximate counting, in *Surveys in Combinatorics 1997* (ed. R. A. Bailey), Cambridge University Press, Cambridge (1997), pp. 287–323.

[79] D. B. Wilson, Mixing times of lozenge tiling and card shuffling Markov chains, (Preprint), Department of Mathematics, MIT, 1997.

[80] D. B. Wilson (Maintainer), Perfectly random sampling with Markov chains, Web site, http://dimacs.rutgers.edu/~dbwilson/exact.html.

School of Computer Studies
University of Leeds
Leeds
LS2 9JT
United Kingdom
dyer@scs.leeds.ac.uk
csg@scs.leeds.ac.uk

Bose–Burton Type Theorems for Finite Projective, Affine and Polar Spaces

Klaus Metsch

Summary We study the following two problems in finite projective, affine and polar spaces of dimension d. Given integers $0 \leq s, t \leq d$ what is the cardinality of the smallest set T consisting of subspaces of dimension t with the property that every subspace of dimension s is incident with at least one element of T? Likewise, what is the cardinality of the largest set T consisting of subspaces of dimension t with the property that every subspace of dimension s is incident with at most one element of T? Bose and Burton solved the case $t = 0$ of the first question for projective spaces in 1966. We survey the known results, present some open problems, and prove new results for finite polar spaces.

1 Introduction

Consider a finite projective, affine or polar space of dimension d and two integers s and t with $0 \leq s, t \leq d$ and the following problems.

Problem 1: What is the smallest cardinality of a set T of t-subspaces such that every s-subspace is incident with at least one element of T?

Problem 2: What is the largest cardinality of a set T of t-subspaces such that every s-subspace is incident with at most one element of T?

Much attention has been paid to particular cases of both problems in the last 30 years. This article gives a survey and several new results concerning Problem 1. Some results concerning Problem 2 are mentioned too.

It is also quite natural to determine the extremal structures occurring in the solutions of Problems 1 and 2. Sometimes much attention has been paid to the second or even third smallest (largest) structures. This is related to the so-called extension problems. We also give some information on these topics.

The first result concerning Problem 1 was probably given by Bose and Burton [7] in 1966. They showed that a set T of points of $\mathrm{PG}(d,q)$ that meets every s-subspace has at least $(q^{d+1-s} - 1)/(q - 1)$ points with equality if and only if T is the point set of a subspace of dimension $d - s$.

It seems to be hopeless to answer Problem 1 in general for projective (or other) spaces. Apart from the three trivial cases $s = t$ and $s = d$ and $t = d$, an answer for $\mathrm{PG}(d,q)$ is known in the case $d \leq s + \frac{s}{t} - 1$, in the case $d = 2s$ and $t = 1$ and in the case $d = 2s + 1 = 5$ and $t = 1$. We report on these results in Section 2. We also describe some examples that might have the correct cardinality for special values of s, t and d but for which no proof has yet been given.

In Section 3 some variations of Problem 1 are given and several open problems are posed. Problem 2 for projective spaces is briefly discussed in Section 4. Not much is known for affine spaces. One important result proved by Jamison can be found in Section 5.

In the remaining sections polar spaces are studied. Of particular interest here are ovoids and spreads, which have been extensively studied by various authors. Since 1995, good progress has been made. If the dimension of the finite polar space is d, the question of the existence of ovoids and spreads corresponds to the cases $(s,t) = (d,0)$ and $(s,t) = (0,d)$ respectively in Problems 1 and 2. We report the state of knowledge in Section 5.

Problem 1 has been completely solved for $t = 0$ and arbitrary s for the elliptic polar spaces $Q^-(2n+1,q)$ in [30]. Not much more is known for polar spaces. Some easy results can be obtained from the existence of spreads.

In Sections 6 and 7 we solve Problem 1 in the case $t = 0$ and $s = 1$ for all finite classical polar spaces except for the symplectic polar spaces $W(2n+1,q)$ when q is odd. In other words, we determine the smallest number of points that *block* all lines of the polar space. For the hyperbolic space $O^+(2n+1,q)$, the unitary space $U(2n+1,q)$ and the parabolic space $Q(2n,q)$, this is quite easy. A counting argument immediately gives the correct answer. When the dimension of the polar space is large enough, the extremal configurations can be characterized. The result is as follows.

Theorem 1.1 *For the hyperbolic polar spaces $Q^+(2n+1,q)$, the unitary polar spaces $U(2n+1,q)$ and the parabolic polar spaces $Q(2n+2,q)$, $n \geq 1$, the smallest sets T of points that meet every line have respectively cardinality $|Q(2n,q)|$, $|U(2n,q)|$ and $|Q^-(2n+1,q)|$.*
For $n \geq 2$ the only examples for T are the point sets of the polar space that lie in a non-tangent hyperplane of the ambient projective space, where the hyperplane has to be an elliptic one in the parabolic case.

The ambient projective space is the projective space in which a polar space is defined by a form of the appropriate type; see Section 6.

The elliptic case is more difficult. However, as mentioned above, it was settled in [30]. The smallest set of points that blocks all lines in $Q^-(2n+1,q)$ has size $q|Q^-(2n-1,q)|$. Given a point P of the polar space and the tangent hyperplane H at P in the ambient projective space $\mathrm{PG}(2n+1,q)$, the set consisting of the points of $Q^-(2n+1,q)$ different from P in H is an example of that size. There are no other examples. The unitary case $U(2n,q)$ is even more difficult. However, we prove the following analogous result in Section 8.

Theorem 1.2 *The smallest set T of points that meets every line in the unitary polar space $U(2n,q)$, $n \geq 3$, has cardinality $q|U(2n-2,q)|$.*

Every example for T of this cardinality is obtained from a tangent hyperplane H of the ambient projective space by taking all points of $U(2n, q)$ in H except the pole $H^\perp$.

For $n = 2$ the problem is open, that is, it is not known how many points are needed to block all lines of $U(4, q)$. Certainly one needs more than $q^2\sqrt{q} + 1$, since $U(4, q)$ has no ovoids (see Section 6), and $q|U(2, q)| = q^2\sqrt{q} + q$ points are sufficient (use the same example as for $n \geq 3$).

The polar spaces $W(2n+1, q)$ of symplectic type behave in a different way. Consider $W(2n + 1, q)$ defined by a symplectic polarity in $\mathrm{PG}(2n + 1, q)$. All points of $\mathrm{PG}(2n + 1, q)$ belong to $W(2n + 1, q)$. Of course, a hyperplane of $\mathrm{PG}(2n + 1, q)$ meets all lines of $W(2n + 1, q)$, but at least for even q there exist smaller sets. In fact, if q is even, then $W(2n + 1, q)$ is isomorphic to $Q(2n + 2, q)$ and we have found above the minimum cardinality for a set of points that meets every line. For odd q, I cannot solve the problem.

The hyperbolic space $Q^+(2n+1, q)$ is the only finite classical polar space for which the ambient projective space $\mathrm{PG}(2n+1, q)$ has a subspace of codimension two that meets every maximal subspace of $Q^+(2n + 1, q)$ in a subspace of codimension two. This property is responsible for the following result, which is the solution to Problem 1 in the case $t = 0$ and $s = 2$. The proof is given in Section 6.

Theorem 1.3 *If T is a set of points of $Q^+(2n+1, q)$ that meets every plane, then $|B| \geq |Q^-(2n - 1, q)|$. For $n \geq 3$ equality holds if and only if B consists of the points of $Q^+(2n + 1, q)$ that lie in an elliptic subspace of codimension two of the ambient projective space $\mathrm{PG}(2n + 1, q)$.*

Here a subspace of $\mathrm{PG}(2n + 1, q)$ of codimension two is called *elliptic* if it meets $Q^+(2n + 1, q)$ in a $Q^-(2n - 1, q)$.

2 Blocking configurations for projective spaces

Throughout this paper, we consider q as a fixed prime power and we use the following notation

$$\Theta_i := \frac{q^{i+1} - 1}{q - 1} = 1 + q + q^2 + \ldots + q^i.$$

Thus Θ_i is the number of points of the finite projective space $\mathrm{PG}(i, q)$.

This section reports in which cases the answer to Problem 1 is known in projective spaces. We repeat the question.

How many subspaces are necessary in a collection T of t-dimensional subspaces of $\mathrm{PG}(d, q)$ such that every s-dimensional subspace of $\mathrm{PG}(d, q)$ is incident with at least one element of T?

Let us here first state the result of Bose and Burton, which is the answer to Problem 1 in the case $t = 0$.

Theorem 2.1 *If T is a set of points of $\mathrm{PG}(d,q)$ that meets every s-subspace, then $|T| \geq \Theta_{d-s}$ with equality if and only if T consists of the points in a subspace of dimension $d-s$.*

Proof Clearly the set of points of a subspace of dimension $d-s$ meets every s-subspace. Conversely, suppose that T is a set of at most Θ_{d-s} points that meets every s-subspace.

Each line lies on the same number c_1 of subspaces of dimension $s-1$. Each point lies on the same number c_0 of subspaces of dimension $s-1$ where $c_0 \Theta_{s-2} = c_1 \Theta_{d-1}$. If P is a point not in T, then some subspace S of dimension $s-1$ on P misses T, since otherwise a simple double counting argument would give $c_0 \cdot 1 \leq |T| c_1$, which is not true.

Since S lies in Θ_{d-s} subspaces of dimension s, it follows that $|T| = \Theta_{d-s}$ and every s-subspace on S meets T in exactly one point. This implies that every line on P meets T in at most one point. Then T must be a subspace, since this holds for all points P outside T. Clearly T has dimension $d-s$. ∎

Also the case $s = 0$ can be settled. In this case, Problem 1 asked for the smallest t-*cover*, that is for the smallest set of t-subspaces that cover all points of the space $\mathrm{PG}(d,q)$. Since $\mathrm{PG}(d,q)$ has Θ_d points, a t-cover necessarily has at least $\lceil \Theta_d/\Theta_t \rceil$ elements. Beutelspacher [3] showed that this bound is sharp.

Result 2.2 *The space $\mathrm{PG}(d,q)$ admits a t-cover of size $\lceil \Theta_d/\Theta_t \rceil$.*

Eisfeld [16] gave the following structural information on t-covers T obtaining the bound of Result 2.2. If $d = k(t+1) + r$ with $0 \leq r \leq t$, then there is a subspace U of dimension $t - r - 1$ such that every point outside U is covered exactly once and every point of U is covered exactly $q^{r+1} + 1$ times by elements of T.

Since projective spaces are self-dual, we can restrict Problem 1 to the case when $0 \leq t \leq s \leq d$. So far we have settled the case when $t = 0$ and the case when $s = d$ (the dual of $s = 0$). The case $s = t$ is trivial.

There is one other easy case, in which Problem 1 can be answered. Suppose that $t+1$ is a divisor of $d+1$ where $t < d$. Then Θ_d/Θ_t is an integer and the previous result shows that the point set of $\mathrm{PG}(d,q)$ can be partitioned by subspaces of dimension t. Such a partition is called a t-*spread*. It is well known that there are so called *geometric* t-*spreads* T, which by definition have the following property: If U and V are two members of T, then the elements of T that are contained in the $(2t+1)$-space $\langle U, V \rangle$ form a t-spread of $\langle U, V \rangle$. It is known that geometric t-spreads have the property that every subspace of dimension $d + 1 - (d+1)/(t+1)$ contains at least one element of T (see, for example, Result 2.3).

Geometric t-spreads provide blocking configurations of minimum cardinality. To see this, consider $\mathrm{PG}(d',q)$ as a subspace of $\mathrm{PG}(d,q)$ with $d \geq d'$ and suppose that $\mathrm{PG}(d',q)$ admits a geometric t-spread T, so that $t+1$ divides $d'+1$.

Then every subspace of $\mathrm{PG}(d',q)$ of dimension $s' := d' + 1 - (d' + 1)/(t + 1)$ contains an element of T. This implies that every subspace of dimension $s := s' + d - d'$ contains an element of T. This example in fact is optimal and unique as was noticed in [4]. Changing the notation of [4], the result can be stated as follows:

Result 2.3 *Suppose T is a set of t-subspaces of $\mathrm{PG}(d,q)$ such that every s-subspace contains an element of T. If*

$$d \leq s + \frac{s}{t} - 1,$$

then

$$|T| \geq \frac{q^{(t+1)(d+1-s)} - 1}{q^{t+1} - 1}$$

with equality if and only if T is a geometric t-spread in a subspace of dimension $(d + 1 - s)(t + 1) - 1$.

We come to a more complex example. Put $x := \max\{0, \lceil d - s - \frac{s}{t} + 1 \rceil\}$. Then $d - x \leq s + \frac{s}{t} - 1$. Consider a subspace V of dimension $x - 1$ and a subspace U of dimension $x + (d - x + 1 - s)(t + 1) - 1$ with $V \subseteq U$. Consider also a set F of subspaces R of dimension $t + x$ with $V \subseteq R \subseteq U$ that form a geometric t-spread in the quotient geometry U/V. Let T_F be the set consisting of all t-subspaces that are skew to V and that are contained in some member of F. Then every s-subspace that is skew to V contains at least one element of T_F.

So far, this collection of t-subspaces looks promising. In fact, for every subspace Z of dimension $d - x = s + \frac{s}{t} - 1$ that is skew to V, the elements of T_F in Z form a geometric t-spread in the subspace $U \cap Z$ of dimension $(d - x + 1 - s)(t + 1) - 1$ and Result 2.3 shows that this is the optimal solution to block the s-subspaces of Z. The question is: Is T_F contained in an optimal set T for $\mathrm{PG}(d,q)$? At least in some cases, this is true. For example when $x = 0$, then T_F already blocks all s-subspaces; in fact, it is just the example of Result 2.3. But there are more cases.

Consider for example the case $x = t = 1$, that is, $d = 2s$. Then V is a point, U is the whole space, F is a set of planes that forms a geometric 1-spread in the quotient geometry on V, and T_F is the set of lines that lie in a plane of F but do not contain V. Let T_V be the set of all lines through V in a fixed $(s+1)$-subspace on V. Then then every s-subspace on V contains a line of T_V. Hence, every s-space of $\mathrm{PG}(d,q)$ contains an element of $T = T_F \cup T_V$. The following result of Eisfeld and Metsch [17] shows that this example is optimal.

Result 2.4 *Suppose that T is a set of lines in $\mathrm{PG}(2s,q)$, such that every s-dimensional subspace contains a line of B. Then*

$$|B| \geq \frac{q^{2s+2} - q^2}{q^2 - 1} + \frac{q^{s+1} - 1}{q - 1}$$

and the above example is the only one in which equality holds.

Consider now the case $t = 1$ and $x = 2$, that is $d = 2s + 1$. Then V is a line. Let $P_0, \ldots, P_q$ be the points of V and let U_i be a subspace of dimension $s+2$ on V. For $s > 2$, let T_V consists of the lines l with $P_i \in l \subseteq U_i$ for some i. For $s = 2$, choose the subspaces U_i in such a way that they cover the whole space and let T_V consists of the lines with $l \not\subseteq V$ and $P_i \in l \subseteq U_i$ for some i. It seems likely that $T = T_F \cup T_V$ is a set of lines of smallest cardinality with the property that every s-space contains a line of T. For $s = 2$ and sufficiently large q, this was proved in [27].

Similar constructions of the set T_V probably give optimal sets $T = T_F \cup T_V$, at least for some values of s and t and for large q. For example, suppose that $s = t^2$ and $d = s + \frac{s}{t} = t^2 + t$. Then V is a point and $\mathrm{PG}(d,q)/V$ has dimension $t^2 + t - 1$. Thus, there exists a set T_V of t-subspaces of $\mathrm{PG}(d,q)$ that is a geometric $(t-1)$-spread of $\mathrm{PG}(d,q)/V$. It follows from Result 2.3 that all s-subspaces on V contain an element of T_V. Therefore all s-subspaces of $\mathrm{PG}(d,q)$ contain an element of $T = T_F \cup T_V$. Is it true that $T = T_F \cup T_V$ is a smallest example for a set of t-subspaces that blocks all s-subspaces?

Here is another nice construction by Eisfeld [15], which we give in a special case. It is easier to state when $t > s$. Consider a geometric spread T' of $\mathrm{PG}(7,q)$ and let T be the set consisting of the q^4+q^2+1 subspaces of dimension three spanned by two lines of T'. Then every line of $\mathrm{PG}(7,q)$ is incident with one element of T and moreover every line not in T' is incident with exactly one element of T. Is it true that this example is optimal, that is, is $q^4 + q^2 + 1$ the smallest number of 3-subspaces that cover all lines? This example can easily been generalised.

3 Variations of the Bose–Burton result in projective spaces

One can think of many problems that are similar to Problem 1. We give three examples.

3.1 Blocking subspaces by more than one point

One generalisation has been provided by Bose and Burton themselves. If T is a set of points of $\mathrm{PG}(d,q)$ such that every s-subspace has at least Θ_z points in T, then $|T| \geq \Theta_{d-s+z}$ with equality if and only if T consists of the points in a subspace of dimension $d - s + z$. Almost nothing is known when the required number of points in the s-subspaces does not have the special form Θ_z. Only some results on multiple blocking sets in the plane are known, see Ball [1]. This problem can be generalised to finding the smallest sets T of t-subspaces such that every s-subspace is incident with at least a certain number of elements of T.

3.2 Blocking as many s-spaces as possible by n t-subspaces

Another generalisation is the following question. Given a number n, what is the largest number of s-subspaces that can be blocked by n subspaces of dimension t? For example, a set of $n = q + 1 - k$ points in $\mathrm{PG}(d, q)$ can block at most all but kq^{d-1} hyperplanes. This is easy. The following special case is much more difficult and seems to be unknown. Is it true that a set of $n = q^2 + q + 1 - k$ points in $\mathrm{PG}(3, q)$ blocks at most all but kq^2 lines with equality if the points lie in a common plane?

3.3 Blocking sets

In the case $t = 0$, Problem 1 asked for the smallest sets of points that block all subspaces of dimension s in $\mathrm{PG}(d, q)$. The answer is, as we have seen, known. The optimal solution is the point set of a subspace of dimension $d - s$. Now we want to look for the second smallest minimal point set that blocks all s-subspaces.

A *blocking set* of $\mathrm{PG}(d, q)$ is a set T of points with the property that no line is contained in T and every hyperplane meets T. Asking for the smallest blocking set is merely asking for the second smallest minimal set of points that meets all hyperplanes. This has been quite extensively studied in the plane and many results are known. Bruen [9] showed that a blocking set T of $\mathrm{PG}(2, q)$ has at least $q + \sqrt{q} + 1$ points with equality if and only if T consists of the points of a Baer subplane, which implies that q is a square. If q is not a square, better bounds are known, see for example [35]. One particular case, which was open for quite some time was solved by Blokhuis [5], who showed that a blocking set of $\mathrm{PG}(2, q)$ with q a prime has at least $3(q+1)/2$ points. For further information on blocking sets, the reader is referred to the new edition of Hirschfeld's book [20]. One recent result that cannot be found there is due to Polverino [36]. It states the following: If $q = p^3$ with a prime $p \geq 7$, then the sizes of the smallest and of the second smallest blocking set of $\mathrm{PG}(2, q)$ are $q + p^2 + 1$ and $q + p^2 + p + 1$.

In $\mathrm{PG}(d, q)$, a *u-blocking set* is a set of points that meets every subspace of dimension $d - u$ and which contains no subspace of dimension u. The smallest u-blocking set in $\mathrm{PG}(d, q)$ is a cone with a vertex V of dimension $u - 2$ over a blocking set B of minimum cardinality in a plane π. Here V is a subspace of dimension $u - 2$ skew to π and the *cone* consists of the points on the lines PX with $P \in B$ and $X \in V$. If $u = 1$, then the cone reduces to the blocking set B. This was proved by Heim [19] who also studied related problems.

Thus, if $q = p^3$ with a prime $p \geq 7$, the smallest blocking set of $\mathrm{PG}(3, q)$ has size $q + p^2 + 1$. If $q = p^3$ with a prime $p \geq 11$, Storme [41] showed that a second smallest minimal blocking set of $\mathrm{PG}(3, q)$ is either a minimal blocking set in a plane or a subgeometry $\mathrm{PG}(3, q)$. Probably, a similar statement is also true for non-prime cubes q. Another result of [41] is that a set of at most $1.1q^2$

points of $\mathrm{PG}(n,p^2)$, p a prime, that meets every hyperplane either contains a line or a Baer subplane in a plane.

In $\mathrm{PG}(4,q)$, q a square, the smallest 2-blocking set is a cone with a point-vertex over a Baer subplane. These have size $q^2+q\sqrt{q}+q+1$. Storme and the author [31] showed that the second smallest minimal 2-blocking sets have size $q^2+q\sqrt{q}+q+\sqrt{q}+1$ and consist of the point set of a subgeometry $\mathrm{PG}(3,\sqrt{q})$. In particular, the subgeometry $\mathrm{PG}(3,\sqrt{q})$ is the smallest minimal point set of $\mathrm{PG}(4,q)$ that generates the space and meets every plane. Probably it is true in general that the smallest minimal point set of $\mathrm{PG}(2s,q)$, q a square, that generates the whole space and meets every s-subspace is a subgeometry $\mathrm{PG}(2s,\sqrt{q})$.

4 Spreads and partial spreads in $\mathrm{PG}(d,q)$

A *partial (s,t)-spread* is a set T of t-subspaces with the property that every s-subspace is incident with at most one element of T. Problem 2 asks for the largest partial (s,t)-spreads. If every s-subspace is incident with exactly one element of T, we call T an (s,t)-*spread*. A t-*spread* is a $(0,t)$-spread. Clearly, if an (s,t)-spread T exists, then $|T|$ is the common answer to Problem 1 and Problem 2.

We study affine and polar spaces later. Consider now only $\mathrm{PG}(d,q)$. A partial (s,t)-spread T is a partial $(d-s-1,d-t-1)$-spread of the dual space and therefore it suffices to handle the case $s < t$ (the case $s = t$ being trivial). Since $\dim(T_1 \cap T_2) \geq 2t-d$, we have $|T| \leq 1$ if $s \leq 2t-d$. If $2t-d < s < t$, then a partial (s,t)-spread is the same as a partial $(2t-s,t)$-spread and these are $(d-1+s-2t, d-1-t)$-spreads of the dual space where $d+1+s-2t < d-1-t$. Hence, it suffices to study partial (s,t)-spreads in the case when $s < t$ and $2t \leq d-1$.

4.1 Existence of (s,t)-spreads

We have already seen that $\mathrm{PG}(d,q)$ possesses a $(0,t)$-spread if and only if $t+1$ divides $d+1$. As was mentioned in [2], no (s,t)-spread for other parameters seems to be known. I state this as a conjecture.

Conjecture 4.1 $\mathrm{PG}(d,q)$ *has no* (s,t)-*spread with* $0 < s < t \leq d-1$.

In order to prove the conjecture in general, it suffices to show that it is true in the case $s = 1$. In fact, an (s,t)-spread of $\mathrm{PG}(d,q)$ gives rise to a $(1,t-s+1)$-spread in the quotient geometries $\mathrm{PG}(d,q)/U$ for the subspaces U of dimension $s-2$. If T is a $(1,t)$-spread of $\mathrm{PG}(d,q)$, then $2t \leq d+1$, since distinct members of T cannot contain a line, and also t must divide d, since T gives rise to a $(0,t-1)$-spread in the quotient geometries $\mathrm{PG}(d,q)/P$ for points P. A $(1,t)$-spread T has cardinality $|T| = \Theta_d\Theta_{d-1}/\Theta_t\Theta_{t-1}$, which gives a divisibility condition. For example $PG(d,q)$ can contain a $(1,2)$-spread only if $d \equiv 2 \pmod{6}$ or $d \equiv 4 \pmod{6}$.

4.2 Partial (s,t)-spreads

Since $PG(d,q)$ possesses an $(0,t)$-spread when $t+1$ divides $d+1$, Problem 2 is solved in this case. The answer is $|T| = (q^{d+1} - 1)/(q^{t+1} - 1)$. If $s = 0$ and if d is a multiple of $t+1$, then the solution to Problem 2 is also known. In this case it is quite easy to see that a partial (s,t)-spread has at most

$$1 + q^{t+2} \frac{q^{d-t-1} - 1}{q^{t+1} - 1}$$

elements. The following example due to Eisfeld [15] shows that the bound can be attained.

Example 4.2 Let $d \geq 2(t+1)$ be a multiple of $t+1 \geq 2$. Let F be a $(d-t-1)$-spread of $PG(2d-2t-1)$ and consider a subspace $PG(d,q)$ of $PG(2d-2t-1,q)$ that contains an element U of F. Then $PG(d,q)$ meets every other member of F in a subspace of dimension t. Using induction, it is possible to find $1 + q^{t+2}(q^{d-2t-2} - 1)/(q^{t+1} - 1)$ mutually skew t-subspaces in U. Together with the $|F| - 1 = q^{d-t}$ subspaces $PG(d,q) \cap V$ with $U \neq V \in F$, this gives the required number of mutually skew t-subspaces in $PG(d,q)$.

As far as I know, Problem 2 is unsolved for other values of s and t.

4.3 The completion problem

The question for the second largest (s,t)-spread is a completion problem. It asks for the smallest number $n = n(s,t,d)$ such that every partial (s,t)-spread of $PG(d,q)$ with more than n elements can be extended to a partial (s,t)-spread of maximal size. This problem has been studied extensively in the case when one knows that an (s,t)-spread exists, that is in the case $s = 0$ and $t+1$ divides $d+1$.

The case $(s,t) = (0,1)$ in $PG(3,q)$ has received a lot of attention. A (partial) $(0,1)$-spread of $PG(3,q)$ is simply called a (partial) spread. The spreads have $q^2 + 1$ lines. The largest known maximal partial spread of $PG(3,q)$ that is not a spread was constructed by Bruen [10] and has size $q^2 - q + 2$. It is not known whether all partial spreads with more than $q^2 - q + 2$ lines can be completed to a spread. This problem is related to blocking sets in the plane. In fact, if T is a maximal partial spread with $q^2 + 1 - \delta < q^2 + 1$ lines and if π is a plane that contains no line of T, then the points of π not covered by lines of T form a blocking set of π of size $q^2 + q + 1 - |T| = q + \delta$. Most known results rely on this connection with blocking sets. For example, if q is an odd prime, we have seen in Section 3.3 that every blocking set of $PG(2,q)$ has size at least $3(q+1)/2$ and therefore $|T| \leq q^2 - \frac{1}{2}(q+1)$ in this case. One result that does not depend on blocking sets was given in [28].

In 1998 Storme and the author [32] obtained new bounds on maximal partial t-spreads in $\mathrm{PG}(2t+1, q)$.

5 A result in affine spaces

Consider Problem 1 in affine spaces $\mathrm{AG}(d, q)$. There are some easy solutions. For example, there exist $(0, t)$-spreads for all $t \leq d$ so that q^{d-t} is the solution to Problem 1 (and Problem 2) in the case when $s = 0$.

However, in general, Problem 1 seems to be much harder in affine spaces $\mathrm{AG}(d, q)$ than in projective spaces. Even the case $t = 0$ and $s = d - 1$ is not easy. To see an example of a point set that meets every hyperplane, consider d lines on a point that span $\mathrm{AG}(d, q)$. If T consists of the points on these lines, then $|T| = d(q-1) + 1$ and every hyperplane meets T. There are many other examples of the same cardinality. Jamison [22] proved however that smaller ones do not exist.

Result 5.1 *If T is a set of points of $\mathrm{AG}(d, q)$ that meets every hyperplane, then $|T| \geq d(q-1) + 1$.*

The theorem of Jamison is more general; see [22]. All known proofs use algebraic techniques that can also be applied in other situation; see for example Bruen and Fisher [11]. A short proof for Result 5.1 was given by Brouwer and Schrijver [8]. In the case $d = 2$ a further proof was given by Szőnyi [35].

6 Ovoids and Spreads of finite classical polar spaces

Polar spaces form a class of geometries on which the classical orthogonal, symplectic and unitary groups act, analogous to the action of the linear groups on projective spaces. Many problems that have been studied for projective spaces have analogous for polar spaces. The rest of this paper considers problems of Bose–Burton type for finite polar spaces. Such a space is of symplectic type $W(m, q)$, m odd, of unitary type $U(m, q)$, or of orthogonal type.

Polar spaces of symplectic or unitary type are defined by a symplectic or unitary polarity $\perp$ of the *ambient* projective space $\mathrm{PG}(m, q)$. Such a polarity is induced by a non-singular sesquilinear form b on the underlying vector space V, by the rule that $U^\perp = \{v \in V \mid b(v, U) = 0\}$. The subspaces U of $\mathrm{PG}(m, q)$ that are contained in their *perp* $U^\perp$ are the elements of the polar space. These are also called the *totally isotropic* subspaces of $\mathrm{PG}(m, q)$.

A finite classical orthogonal polar space consists of the subspaces of its *ambient* projective space $\mathrm{PG}(m, q)$ that lie on a non-degenerate quadric, which is defined by a quadratic form f of the underlying vector space V. These subspaces are called the *totally singular* subspaces of $\mathrm{PG}(m, q)$. The related bilinear form b is defined by $b(v,w) = f(v+w) - f(v) - f(w)$. For each subspace U we define $U^\perp := \{v \in V \mid b(v,U) = 0\}$ and call $U^\perp$ the *perp* of U. The orthogonal polar spaces fall into three types, the hyperbolic type $Q^+(2n+1,q)$, the elliptic type $Q^-(2n+1,q)$, and the parabolic type $Q(2n,q)$. The map $U \to U^\perp$ is a polarity except in the parabolic case when q is even. In the elliptic and hyperbolic cases when q is even, the polarity is symplectic.

For even q the polar spaces $W(2n-1,q)$ and $Q(2n,q)$ are isomorphic, so there is an ambiguity concerning what to take as the ambient projective space. Since we consider this polar space only in its orthogonal representation $Q(2n,q)$, we consider $PG(2n,q)$ as its ambient projective space.

For all polar spaces we call the points of the ambient projective space $PG(m,q)$ that lie in the polar space the *singular points* of $PG(m,q)$. The maximal subspaces of the ambient projective space that belong to the polar space are called *generators*. Their common projective dimension d is the *dimension* of the polar space. For information on polar spaces, see for example [21] or [12].

The following table identifies the six types of finite polar spaces. The parameter d gives the dimension of the polar space, and m the dimension of its ambient projective space.

polar space	name	number of points	m	d
$Q^+(2n+1,q)$	hyperbolic	$\Theta_n(q^n+1)$	$2n+1$	n
$Q^-(2n+1,q)$	elliptic	$\Theta_{n-1}(q^{n+1}+1)$	$2n+1$	$n-1$
$Q(2n,q)$	parabolic	Θ_{2n-1}	$2n$	$n-1$
$W(2n+1,q)$	symplectic	Θ_{2n+1}	$2n+1$	n
$U(2n+1,q)$	unitary	$\Theta_n(q^n\sqrt{q}+1)$	$2n+1$	n
$U(2n,q)$	unitary	$\Theta_{n-1}(q^n\sqrt{q}+1)$	$2n$	$n-1$

Table 1: Finite classical polar spaces

An *ovoid* of a polar space is a set T of singular points such that every generator meets T in exactly one point. Clearly, if a polar space has an ovoid, then its ovoids are its smallest point sets that meet every generator and the cardinality of the ovoids is the answer to Problems 1 and 2 when $t = 0$ and $s = d$.

Table 2 gives the state of knowledge on the existence of ovoids in finite classical polar spaces. In the table, the integer p is the characteristic of the field $GF(q)$.

The existence of ovoids in $Q^+(5,q)$ follows from the existence of spreads in $PG(3,q)$ by applying the Klein Correspondence. The results for the existence of ovoids in $Q^+(7,q)$, q prime or $q \not\equiv 0$ or $2 \pmod{3}$, are obtained from the existence results of spreads in Table 3 by applying triality.

The Klein Correspondence is an incidence-preserving bijection from the points and the two families of generators of $Q^+(5,q)$ to the lines, points and planes respectively of $\mathrm{PG}(3,q)$. If we identify the two families of generators of $Q^+(7,q)$ as "first kind" and "second kind", the triality map is an incidence preserving bijection from the lines, points, first and second kinds of generators to lines, first and second kinds of generators, and points respectively [12].

polar space	ovoid exists	no ovoid
$Q^+(2n+1,q)$	$n=1,2$ $n=3$, q even [45, 46] $n=3$, q prime $n=3$, $q \not\equiv 1 \pmod 3$	$p^n > \binom{p+2n}{2n+1} - \binom{p+2n-2}{2n+1}$ where $q = p^e$, p prime [6]
$Q^-(2n+1,q)$		$n \geq 2$ [46]
$Q(2n,q)$	$n=2$ [34] $n=3$, $q=3^h$ [23]	$n \geq 3$, q even [46] $n=3$, $q=5,7$ [33] $n \geq 4$, q odd [18]
$W(2n+1,q)$	$n=1$, q even [43]	$n=1$, q odd [43] $n \geq 2$ [46]
$U(2n+1,q)$	$n=1$	$p^{2n+1} > \binom{p+2n}{2n+1}$ where $q = p^e$, p prime [6]
$U(2n,q)$		$n \geq 2$ [46]

Table 2: Existence of ovoids in finite classical polar spaces

The bound $p^n > \binom{p+2n}{2n+1} - \binom{p+2n-2}{2n+1}$ is satisfied if n is sufficiently large with respect to p. Thus for each fixed p there do not exist ovoids in the polar spaces $Q^+(2n+1, p^e)$ for all e and all sufficiently large n. For example, $Q^+(2n+1, 2^e)$ and $Q^+(2n+1, 3^e)$ have no ovoids for $n \geq 4$. These two special cases were proved earlier by Kantor [23] for $e=1$ and $q=2$ and Shult [38] for $e=1$ and $q=3$. Similar conclusions hold for $U(2n+1,q)$.

If a polar space has no ovoid or if it is not yet known whether or not a polar space has an ovoid, then the answer to Question 1 in the case when $t=0$ and $s=d$, that is, the smallest number of singular points needed to block all generators, is in general not known. It is known only in the elliptic case [29]. In fact, in the elliptic case much more is known, as we can see from the following theorem that was proved in [30].

Result 6.1 *Let Q be the elliptic polar space $Q^-(2n+1,q)$ and denote by $\perp$ the related polarity of the ambient space $\mathrm{PG}(2n+1,q)$. Let s be an integer with $0 \leq s \leq n-1$ and suppose that T is a set of points that meets every s-dimensional totally singular subspace.*

Then $|T| \geq (q^{n+1} + q^s)\Theta_{n-1-s}$ with equality if and only $T = (U^\perp \setminus U) \cap Q$ for a subspace U of dimension $s - 1$ of Q.

Some more answers to Problem 1 in polar spaces are given in the next two sections.

A *partial ovoid* is a set T of singular points such that every generator is incident with at most one element of T. When ovoids do not exist, then the maximal size of partial ovoids is not known. Thas [48] gave an upper bound in the elliptic case $Q^-(2n+1,q)$, which was recently improved to $q^{n+1} - q^{n-1} + 2$ by Klein [26]. A upper bound for partial ovoids for many polar spaces was given in [6]. Tallini [42] proved the upper bound $q^2 - q + 1$ for partial ovoids of $W(3, q)$.

A *spread* in a polar space is a set T of generators such that every point of the polar space lies in exactly one member of T. Clearly, if a polar space has a spread, the cardinality of the spreads is the answer to Problem 1 and Problem 2 in the case when $s = 0$ and $t = d$, the dimension of the polar space. Table 3 is obtained from [20] and gives the state of knowledge on the existence of spreads in finite classical polar spaces. The relevant papers for the results marked with (*) are [13, 14, 23–25, 37, 45, 46].

polar space	spread exists	no spread
$Q^+(2n+1,q)$	$n = 1$ n odd, q even [14, 45, 46] $n = 3$, q prime (*) $n = 3$, $q \not\equiv 1 \pmod 3$ (*)	n even
$Q^-(2n+1,q)$	$n = 2$, q odd [34] $n \geq 2$, q even [14, 45, 46]	
$Q(2n,q)$	$n \geq 2$, q even [14, 45, 46] $n = 3$, q prime (*) $n = 3$, $q \not\equiv 1 \pmod 3$ (*)	n even, q odd [43, 48]
$W(2n+1,q)$	$n \geq 1$ [44]	
$U(2n+1,q)$		$n \geq 1$ [46, 48]
$U(2n,q)$		$n = 2$, $q = 4$ [20]

Table 3: Existence of spreads of finite classical polar spaces

For more information on results on ovoids and spreads see [48] and, for the rank two case, [49]. For a common generalisation of ovoids and spreads in polar space we refer the reader to Shult and Thas [39, 40].

The existence of a spread provides a solution to Problems 1 and 2 in the case when $s = 0$ and $t = d$. But it also gives rise to more solutions. If $t + 1$ divides $d + 1$ and if a spread T' exists, then it is also possible to partition the singular points by subspaces of dimension t of the polar space, since then a $PG(d,q)$ possesses a t-spread. In this case $|T'|\Theta_d/\Theta_t$ is the common answer to Problems 1 and 2.

A *partial spread* of a polar space is a set consisting of mutually disjoint generators. If a spread does not exist, then it is usually not known how large a partial spread can be. There is one exception. If n is even, then the hyperbolic space $Q^+(2n+1,q)$ has two families of generators and members of the same family always have a non-trivial intersection. Thus for even n the largest partial spread of $Q^+(2n+1,q)$ has two elements. Tallini [42] proved the upper bound $q^2 - q + 1$ for partial spreads of $Q(4,q)$, q odd, and Thas [47, 48] showed that a partial spread of $U(2n+1,q^2)$, n odd, has size at most $q^{2n+1} - q^{n+1} + q^n + 1$.

7 Blocking lines by points in the polar spaces $Q^+(2n+1,q)$, $U(2n+1,q)$ and $Q(2n,q)$

In this section, we are interested in the smallest sets B of singular points that meet every line of a finite classical polar space. A lower bound for the size of such a set B can be obtained quite easily. Let d be the dimension of the finite classical polar space.

Since B meets every line of the polar space, the theorem of Bose and Burton shows that B meets every generator U in at least $(q^{d+1} - 1)/(q-1)$ points with equality if and only if $U \cap B$ is a hyperplane of U. Thus, the number of generators times $(q^{d+1} - 1)/(q-1)$ divided by the number of generators on a point is a lower bound for B. It is attained if and only if B meets every generator U in a hyperplane of U.

For some polar spaces, this bound can be attained. In fact, if the ambient projective space has a hyperplane that does not contain a generator, then the set B consisting of the singular points in such a hyperplane is a set of the required size that meets every line of the polar space.

For the hyperbolic polar spaces $Q^+(2n+1,q)$, the unitary polar spaces $U(2n+1,q)$ and the parabolic polar spaces $Q(2n,q)$ such a hyperplane exists. In the hyperbolic and unitary case, every non-tangent hyperplane can be taken. For the parabolic spaces $Q(2n,q)$ one has to take the so called *elliptic hyperplanes*, which meet $Q(2n,q)$ in $Q^-(2n-1,q)$. The singular points in such a hyperplane provide always a set of minimum cardinality that meets every line of the polar space. If $d \geq 2$, we show below that there are no other examples of that size. For $d = 1$, this is not true; the so-called non-classical ovoids provide counterexamples.

The elliptic polar spaces $Q^-(2n+1,q)$ and the unitary polar spaces $U(2n,q)$ are more difficult to handle because, for these, all hyperplanes contain generators. For these polar spaces, the above counting argument does not give the correct value for the minimum cardinality of B. For the elliptic case, we have seen in Result 6.1 that the smallest sets live in tangent hyperplanes and do not meet all generators in a subspace of codimension one. The unitary case is much harder. It is handled in Section 8.

For the results we prove below, we need the following lemma which can be found in a more general form in [21, Chapter 26.3].

Lemma 7.1 *Suppose the subspaces of a polar space $\mathcal{P}$ are represented in an incidence and dimension preserving way by subspaces of a projective space $\mathrm{PG}(d, q)$. Then the points of the polar space generate in $\mathrm{PG}(m, q)$ a subspace of dimension $2n+1$ if $\mathcal{P} = Q^+(2n+1, q)$, a subspace of dimension m if $\mathcal{P} = H(m, q)$, and a subspace of dimension $2n - 1$ or $2n$ if $\mathcal{P} = Q(2n, q)$.*

7.1 The hyperbolic case

In this section we prove Theorem 1.1 for the hyperbolic space $\mathcal{Q} := Q^+(2n+1, q)$. Let $\mathrm{PG}(2n+1, q)$ be the ambient projective space and let $\perp$ be the related polarity. Suppose that B is a set of points that meets every singular line. The above counting argument shows that $|B| \geq |Q(2n, q)|$ with equality if and only if B meets every generator M in a hyperplane of M. We suppose from now on that $|B| = |Q(2n, q)|$ and that $n \geq 2$ and show that B consists of the singular points in a non-tangent hyperplane. This is not true for $n = 1$ and $q \geq 4$; counterexamples are easy to construct.

Since $|B| = |Q(2n, q)|$, our counting argument shows that B meets every generator M in a hyperplane of M. Thus a totally singular subspace U either is contained in B or meets B in a subspace of codimension one of U.

Consider a point P of $\mathcal{Q}$ that is not in B. Each totally singular subspace U on P meets B in a subspace of codimension one of U. It follows that the subspaces contained in $B \cap P^\perp$ form a polar space isomorphic to the polar space $Q^+(2n - 1, q)$ that is induced by $\mathcal{Q} = Q^+(2n + 1, q)$ on the quotient geometry $P^\perp/P$.

Lemma 7.1 shows that the points in $B \cap P^\perp$ span a subspace S_P of dimension $2n - 1$ of $\mathrm{PG}(2n + 1, q)$. We have $P \notin S_P$ (otherwise every totally singular line on P would lie in S_P as it contains also a point of B; but, again by Lemma 7.1, the totally singular lines on P span the subspace $P^\perp$, which has dimension $2n$). Since every totally singular line on P meets B, it follows that $P^\perp \cap B$ consists of the singular points in S_P. Thus, we have shown the following:

For every point $P \in \mathcal{Q} \setminus B$, there exists a complement S_P of P in $P^\perp$ such that $P^\perp \cap B$ consists of the singular points of S_P.

Put $\mathcal{S} := \{S_P \mid P \in \mathcal{Q} \setminus B\}$. If $S \in \mathcal{S}$, then $S \cap \mathcal{Q}$ is a hyperbolic quadric $Q^+(2n - 1, q)$ and thus $S^\perp$ is a secant line of $\mathcal{Q}$. Hence $S \subseteq P^\perp$, that is $S = S_P$, for at most two points $P \in \mathcal{Q} \setminus B$. This shows that $|\mathcal{S}| \geq |\mathcal{Q} \setminus B|/2$. As $|B| = |Q(2n, q)|$, it follows that $|\mathcal{S}| \geq \frac{1}{2}q^n(q^n + 1)$.

Consider distinct points P and R of $\mathcal{Q} \setminus B$. Then either $S_P \subseteq R^\perp$ or else $S_P \cap R^\perp$ is a hyperplane of S_P. In both cases, the singular points of $S_P \cap R^\perp$ span $S_P \cap R^\perp$. Since the singular points of $S_P \cap R^\perp$ lie in S_R, it follows that

$S_P \cap R^\perp \subseteq S_R$. Thus either $S_P = S_R$ or $S_P \cap S_R$ is a subspace of codimension one in S_P and S_R.

Hence the subspaces of $\mathcal{S}$ have dimension $2n-1$ and distinct subspaces of $\mathcal{S}$ meet in a subspace of dimension $2n-2$. It follows that all spaces of $\mathcal{S}$ either pass through a common subspace of dimension $2n-2$ or lie in a common subspace of dimension $2n$, which is a hyperplane. The first case is not possible, since $|\mathcal{S}| > q^2 + q + 1$ and since a subspace of dimension $2n-2$ lies only in $q^2 + q + 1$ subspaces of dimension $2n-1$.

Hence there exists a hyperplane H containing all subspaces S of $\mathcal{S}$. If $X \in B$, then $X^\perp$ contains $1 + q|Q^+(2n-1,q)|$ singular points. Since $|B| = |Q(2n,q)| < q|Q^+(2n-1,q)|$, there exists a point $P \in \mathcal{Q} \cap X^\perp$ that is not in B; then $X \in S_P$. Hence the sets S of $\mathcal{S}$ cover B and therefore $B \subseteq H$.

We have to show that H is a non-tangent hyperplane. In view of $|B| = |Q(2n,q)|$, it follows then that B consists of all singular points of H.

Assume on the contrary that H is a tangent hyperplane, $H = P^\perp$ for some singular point P. Then every singular point $X \neq P$ of H lies on a totally singular line that meets H only in X. Since B meets all totally singular lines, it follows that all singular points of H except possibly P lie in B. But the tangent hyperplane $P^\perp$ has $q|Q^+(2n-1,q)|$ singular points different from P and that are too many for B, a contradiction.

This proves Theorem 1.1 in the hyperbolic case.

Remark: Via the Klein Correspondence, the problem in the case $n = 2$ can be translated to the following problem of $\mathrm{PG}(3,q)$: What is the smallest cardinality of a set B of lines in $\mathrm{PG}(3,q)$ such that for every incident point-plane pair (P, E), there exists a line $l \in B$ with $P \in l \in E$? It is easy to see that $|B| \geq q^3 + q^2 + q + 1$ with equality if and only if, for every point P, there exists a plane P^π on P such that the lines of B on P are the lines of P^π of P. In the case of equality, it is also not too difficult to see geometrically that the map $P \to P^\pi$ is the restriction of a symplectic polarity of the point set. Thus the sets B of minimal cardinality are the sets of the totally isotropic lines of symplectic polarities. This provides a geometric proof of the fact that the points of $Q^+(5,q)$ in a non-tangent hyperplane of $\mathrm{PG}(5,q)$ correspond via the Klein Correspondence to the totally isotropic lines of a symplectic polarity of $\mathrm{PG}(3,q)$ (see for example Cameron [12, Chapter 8.1]).

7.2 The unitary case $U(2n+1, q)$

In this section Theorem 1.1 is proved for the unitary polar space $U(2n+1, q)$. Let $\mathrm{PG}(2n+1, q)$ be the ambient projective space and $\perp$ the related polarity. Suppose that B is a set of points that meets every totally isotropic line. The above counting argument shows that $|B| \geq |U(2n, q)|$ with equality if and only if B meets every generator M in a hyperplane of M. We suppose from now on that $|B| = |U(2n, q)|$ and that $n \geq 2$. We show that B consists of the singular points in a non-tangent hyperplane. The techniques we use are the same as in the hyperbolic case.

As B meets every generator M in a hyperplane of M, every totally isotropic subspace T lies in B or meets B in a subspace of codimension one of T.

If P is a singular point that is not in B, then every totally isotropic subspace T on P meets B in a subspace of codimension one. Thus the subspaces of $B \cap P^\perp$ form a polar space isomorphic to the polar space $U(2n-1, q)$ that is induced by $U(2n+1, q)$ in the quotient geometry $P^\perp/P$. As in Section 7.1, Lemma 7.1 shows that P has a complement S_P in $P^\perp$ such that the points of $P^\perp \cap B$ are the singular points of S_P.

Let $\mathcal{S}$ be the set consisting of all subspaces S_P for singular points P outside B. The elements of $\mathcal{S}$ are subspaces of dimension $2n-1$ that meet $U(2n+1, q)$ in a $U(2n-1, q)$. Thus for $S \in \mathcal{S}$, the subspace $S^\perp$ is a secant line. Hence $S \subseteq P^\perp$, that is, $S = S_P$, for at most $\sqrt{q}+1$ singular points P. Since $|U(2n+1, q) \setminus B| = |U(2n+1, q)| - |U(2n, q)| = q^n(q^n\sqrt{q}+1)$, it follows that $|\mathcal{S}| \geq q^n(q^n\sqrt{q}+1)/(\sqrt{q}+1) > q^2 + q + 1$.

The elements of $\mathcal{S}$ are subspaces of dimension $2n-1$. As in the hyperbolic case, we see that distinct subspaces of $\mathcal{S}$ meet in a subspace of dimension $2n-2$. Thus, either there exists a subspace of dimension $2n-2$ contained in all elements of $\mathcal{S}$ or else there exists a hyperplane that contains all elements of $\mathcal{S}$. The first case however is not possible, since $\mathcal{S}$ has more than q^2+q+1 elements.

Hence there exists a hyperplane H containing all subspaces S of $\mathcal{S}$. Since these cover B, it follows that $B \subseteq H$. Since B meets every singular line, it follows that H is either a non-tangent hyperplane and then B consists of the singular points of H or else H is a tangent hyperplane $P^\perp$ for some singular point P and then B contains all singular points of $P^\perp \setminus \{P\}$. Since the second set is too large, the first case occurs.

This proves Theorem 1.1 for the unitary polar spaces $U(2n+1, q)$.

7.3 The parabolic case $Q(2n, q)$

Now we prove Theorem 1.1 for the parabolic polar space $\mathcal{Q} := Q(2n, q)$ defined by a quadratic form in $\mathrm{PG}(2n, q)$. Suppose that B is a set of points that meets every singular line. The above counting argument shows that $|B| \geq |Q^-(2n-1, q)|$ with equality if and only if B meets every generator M in a hyperplane of M. We suppose from now on that $|B| = |Q(2n, q)|$ and that $n \geq 3$. Then B meets every generator T of $\mathcal{Q}$ in a hyperplane of T. We show now that B consists of the singular points in an elliptic hyperplane, that is a hyperplane of $\mathrm{PG}(2n, q)$ that meets Q in a $Q^-(2n-1, q)$. We could proceed as in the above two cases, but there is another, probably more elegant, method.

Denote by $\mathcal{H}$ the set of the *hyperbolic hyperplanes* of $\mathrm{PG}(2n, q)$. These are the hyperplanes of $\mathrm{PG}(2n, q)$ that meet $\mathcal{Q}$ in a $Q^+(2n-1, q)$. Then $|\mathcal{H}| = \frac{1}{2}q^n(q^n+1)$. Consider a hyperplane $H \in \mathcal{H}$. Then the generators of $\mathcal{Q} \cap H$

are generators of $\mathcal{Q}$ and so they meet $B \cap H$ in a subspace of dimension $n-2$. This implies that $B \cap H$ is a point set of minimum cardinality that meets every totally singular line of H (see the counting argument at the beginning of Section 7). Since Theorem 1.1 is already proved for the hyperbolic case, it follows that $B \cap H$ consists of the singular points of a non-tangent hyperplane S_H of H, that is S_H is a subspace of codimension two of $\mathrm{PG}(2n, q)$ that meets $\mathcal{Q}$ in a $Q(2n-2, q)$.

Let $\mathcal{S}$ be the set consisting of all subspaces S_H with $H \in \mathcal{H}$. Since every subspace $S \in \mathcal{S}$ lies in $q+1$ hyperplanes, we have $|\mathcal{S}| \geq |\mathcal{H}|/(q+1) > q^2+q+1$.

If H and H' are different hyperbolic hyperplanes, then the singular points of $H' \cap S_H$ lie in $S_{H'}$ and this implies that $H' \cap S_H \subseteq S_{H'}$. Hence the subspaces of $\mathcal{S}$ have dimension $2n-2$ and distinct subspaces of $\mathcal{S}$ meet in a subspace of dimension $2n-3$. Since $|\mathcal{S}| > q^2+q+1$, it follows that there exists a hyperplane U that contains all elements S of $\mathcal{S}$. Since every singular point lies in a hyperbolic hyperplane, every point of B lies in a set S_H and thus the hyperplane U contains B.

Since B meets every line, either U is a non-tangent hyperplane and B consists of all singular points of U or otherwise U is the tangent hyperplane of a point $P \in \mathcal{Q}$ and then B contains all singular points of U different from P. Since tangent hyperplanes and hyperbolic hyperplanes have too many singular points for B, it follows that U is an elliptic hyperplane.

This proves Theorem 1.1 for the parabolic polar spaces $Q(2n, q)$.

7.4 Blocking planes by points in hyperbolic quadrics

In this section we prove Theorem 1.3. Consider $\mathcal{Q} := Q^+(2n+1, q)$. Let $\mathrm{PG}(2n+1, q)$ be the ambient projective space and $\perp$ the related polarity. A subspace of $\mathrm{PG}(2n+1, q)$ of codimension two is called *elliptic*, if it meets $\mathcal{Q}$ in a $Q^-(2n-1, q)$.

We suppose that B is a set of singular points that meets every totally singular plane. The result of Bose and Burton (see Theorem 2.1) shows that every generator M meets B in at least $(q^{n-1}-1)/(q-1)$ points with equality if and only if $M \cap B$ is a subspace of dimension $n-2$. Thus, the number of generators times $(q^{n-1}-1)/(q-1)$ divided by the number of generators on a point is a lower bound for B. This lower bound is attained if and only if B meets every generator M in a hyperplane of M. Since the set consisting of the singular points in an elliptic subspace of codimension two has the property that it meets every generator in a subspace of dimension $n-2$, it follows that $|B| \geq |Q^-(2n-1, q)|$ with equality if and only if B meets every generator in a subspace of dimension $n-2$.

Suppose from now on that $|B| = |Q^-(2n-1, q)|$ and $n \geq 3$. We show that B consists of the singular points in an elliptic subspace of codimension two. Since every generator meets B in a subspace of dimension $n-2$, every totally singular subspace T meets B in a subspace of codimension at most two in T.

Consider a point $P \in \mathcal{Q} \setminus B$. Each totally singular line on P meets B in no point or one. The quadric $\mathcal{Q}$ induces on the quotient geometry $P^\perp/P$ a quadric $\mathcal{Q}_P$ of type $Q^+(2n-1,q)$. Let B_P be the set of totally singular lines of $\mathcal{Q}$ on P that meet B in one point and consider B_P as a set of *points* of $\mathcal{Q}_P$. Then the *generators* of $\mathcal{Q}_P$, which have dimension $n-1$, meet B_P in subspaces of dimension $n-2$. This implies that B_P is a set of *points* of smallest cardinality that meets every *line* of $\mathcal{Q}_P$ (see the beginning of Section 7). Theorem 1.1 shows that B_P is the section of $\mathcal{Q}_P$ with a non-tangent *hyperplane*. Thus the elements of B_P are the *points* of a $Q(2n-2,q)$ in $\mathcal{Q}_P$.

For $\mathcal{Q}$ this means the following: There exists a subspace T_P of dimension $2n-1$ with $P \in T_P \subseteq P^\perp$ such that $\mathcal{Q} \cap T$ is a cone with vertex P over an $Q(2n-2,q)$ and such that the totally singular lines of T_P on P are the lines of B_P. In particular $P^\perp \cap B \subseteq T_P$. The points of $P^\perp \cap B$ span a subspace S_P of T_P. Since the lines of B_P span T_P, we have $\langle S_P, P\rangle = T_P$. Hence S_P has dimension $2n-2$ or $2n-1$. The set B meets each totally singular subspace U with $P \in U \subseteq T_P$ in a subspace of codimension one of U. Therefore the subspaces of $T_P \cap B$ form a polar space of type $Q(2n-2,q)$. Then Lemma 7.1 implies that S_P has dimension $2n-3$ or $2n-2$. It follows that S_P has dimension $2n-2$, so that $P \notin S_P$ and $S_P \cap \mathcal{Q}$ is a polar space of type $Q(2n-2,q)$. Moreover $P^\perp \cap B = S_P \cap \mathcal{Q}$. This holds for all points $P \in \mathcal{Q} \setminus B$.

Put $\mathcal{S} := \{S_P \mid P \in \mathcal{Q} \setminus B\}$. If $S \in \mathcal{S}$, then $S \cap \mathcal{Q}$ is a parabolic quadric $Q(2n-2,q)$ and thus $S^\perp$ is a plane that meets $\mathcal{Q}$ in a conic. Hence $S \subseteq P^\perp$ and thus $S = S_P$ for at most $q+1$ points $P \in \mathcal{Q} \setminus B$. This shows that $|\mathcal{S}| \geq |\mathcal{Q} \setminus B|/(q+1)$. As $|B| = |Q(2n,q)|$, it follows that $|\mathcal{S}| \geq q^n(q^n+1)$.

For different points P and R of $\mathcal{Q} \setminus B$, we have $\dim(S_R \cap P^\perp) \geq \dim(S_R) - 1 = 2n-3$. Since every point of $\mathcal{Q}$ in $S_R \cap P^\perp$ belongs to B, we have $S_R \cap P^\perp \subseteq S_P$. Hence either $S_P = S_R$ or $S_P \cap S_{P'}$ is a subspace of dimension $2n-3$.

Hence distinct subspaces of $\mathcal{S}$ meet in a subspace of dimension $2n-3$. Since the subspaces of $\mathcal{S}$ have dimension $2n-2$, it follows that either all spaces of $\mathcal{S}$ pass through a common subspace of dimension $2n-3$ or all spaces of $\mathcal{S}$ lie in a common subspace of dimension $2n-1$. The first case is not possible, since a subspace of dimension $2n-3$ lies only in $(q^2+1)(q^2+1+1)$ subspaces of dimension $2n-1$ and since $|\mathcal{S}| \geq (q^3+1)q^3$ (recall that $n \geq 3$).

Hence there exists a subspace U of dimension $2n-1$ containing all elements of $\mathcal{S}$ and thus all points of B. Since B meets all generators in a subspace of dimension $n-2$, it follows that U is an elliptic subspace of codimension two. This completes the proof of Theorem 1.3.

8 The unitary polar spaces $U(2n,q)$

This section considers the unitary polar space $U(2n,q)$. Its ambient projective space is $\mathrm{PG}(2n,q)$. The associated unitary polarity of $\mathrm{PG}(2n,q)$ is denoted by $\perp$. We note that q is a perfect square. A subspace S of $\mathrm{PG}(2n,q)$ is *anisotropic* if $S \cap S^\perp = \emptyset$. If S is anisotropic and $s = \dim(S)$, then S meets

$U(2n,q)$ in a $U(s,q)$; here $U(1,q)$ is a set of $\sqrt{q}+1$ collinear points, and $U(0,q)$ is empty.

Before we start the proof of Theorem 1.2 we prove the following lemma.

Lemma 8.1 *Let Z be a set of points of $U(2n,q)$, $n \geq 2$.*

(a) If $|Z| < q^{n-1}\sqrt{q}+1$, then for each point Y of $\mathrm{PG}(2n,q) \setminus Z$, there exists a generator U with $U \cap Z = \emptyset$ and $Y \in U^\perp$.

(b) If $|Z| < |U(2n,q)|/|U(3,q)|$, then $\mathrm{PG}(2n,q)$ has an anisotropic solid with $S \cap Z = \emptyset$.

Proof Let r_0 be the number of generators. Then each singular point lies on r_1 of these and each totally isotropic line lies on r_2 of these where $r_0 = r_1(q^n\sqrt{q}+1)$ and $r_1 = r_2(q^{n-1}\sqrt{q}+1)$.

(a) Assume $Y \in U(2n,q) \setminus Z$ and every generator on Y meets Z. Then $r_1 \leq |Z|r_2$, a contradiction.

Consider now the case that Y is not in $U(2n,q)$. Then $Y^\perp$ contains $r_1(q+1)$ generators and every singular point of $Y^\perp$ lies on $(q+1)r_2$ of these. Assume that every generator of $Y^\perp$ contains a point of Z. Then $r_1(q+1) \leq |Z|r_2(q+1)$, a contradiction.

(b) Let $\mathcal{S}$ be the set consisting of the anisotropic solids of $\mathrm{PG}(2n,q)$. Then each point of $U(2n,q)$ lies on r solids of $\mathcal{S}$ where $|U(2n,q)|r = |\mathcal{S}| \cdot |U(3,q)|$. Count pairs (S,P) of solids $S \in \mathcal{S}$ and points $P \in S \cap Z$ in two ways to obtain

$$\sum_{S \in \mathcal{S}} |S \cap Z| = |Z|r.$$

Since $|Z|r < |\mathcal{S}|$, it follows that $S \cap Z = \emptyset$ for some $S \in \mathcal{S}$. ∎

From now on, we consider a set B of singular points that meets every totally isotropic line and that satisfies

$$|B| \leq q|U(2n-2,q)| = \frac{q(q^{n-1}\sqrt{q}+1)(q^{n-1}-1)}{q-1}.$$

For $n \geq 3$ we show below in a series of lemmas that equality holds and that B consists of the singular points of $P_0^\perp \setminus \{P_0\}$ for some singular point P_0. However, we assume only that $n \geq 2$, since some of the lemmas hold also for $n = 2$ and these are needed to prove the case $n = 3$.

Lemma 8.2 *If T is a totally isotropic t-subspace, then $|T \cap B| \geq (q^t - 1)/(q-1)$. Equality holds if and only if $T \cap B$ is a hyperplane of T.*

Proof Since every line of T meets B, this follows from the Theorem of Bose and Burton; see Result 2.1. ∎

If S is an anisotropic subspace of dimension s and if P is a singular point with $S \subseteq P^\perp$, then $\langle S, P \rangle \cap U(2n,q)$ is the union of the lines PX for the singular points $X \in S$. We call this structure a *cone* with *vertex* P over a $U(s,q)$.

Lemma 8.3 *Suppose that T is a subspace of dimension $s + 1 \geq 4$ that meets $U(2n, q)$ in a cone with vertex P over a $U(s, q)$ where $P \notin B$. Suppose also that every totally isotropic line of T on P meets B in a unique point. Then T has a complement S in T such that the points of $T \cap B$ are the singular points of S.*

Proof Consider a totally isotropic subspace X of T on P and let x be its dimension. Since every totally isotropic line of T on P meets B in a unique point, the subspace X contains $(q^x - 1)/(q - 1)$ points of B. Lemma 8.2 shows that $X \cap B$ is a hyperplane of X. This holds for all totally isotropic subspaces X of T on P.

The map that maps each totally isotropic subspace X with $P \in X \subseteq T$ to the subspace $X \cap B$ is injective and its image consists of all totally isotropic subspaces of $T \cap B$. Since T meets $U(2n, q)$ in a cone with vertex P over a $U(s, q)$, it follows that the totally isotropic subspaces that are contained in $T \cap B$ form a polar space of type $U(s, q)$. Since $s \geq 3$, Lemma 7.1 shows that the points of $T \cap B$ span a subspace S of dimension s. We have $P \notin S$ (otherwise, T would contain every totally isotropic line of T on P; but these lines span T).

Hence S is a complement of P in T. The points of $T \cap B$ lie in S. If X is a singular point of S, then PX is a totally isotropic line, and, since it meets B, it follows that $X \in B$. Hence $T \cap B$ consists of the singular points in S. ∎

We call a generator T *tight* if $T \cap B$ is a hyperplane of T; then $|T \cap B| = (q^{n-1} - 1)/(q - 1)$, since $\dim(T) = n - 1$. We call a point P *tight* if there is a tight generator T with $T \subseteq P^\perp$; if P is singular this implies $P \in T$.

Lemma 8.4
    (a) *If P is a tight point of B, then $|P^\perp \cap B| \leq |B| - q^{2n-2}\sqrt{q}$.*
    (b) *If P is a non-singular tight point, then $|P^\perp \cap B| \leq |B| - q^{2n-2}\sqrt{q} + q^{n-1}\sqrt{q}$.*
    (c) *Suppose T is a tight generator. If the points P of $T \cap B$ with $|P^\perp \cap B| = |B| - q^{2n-2}\sqrt{q}$ span $T \cap B$, then $|X^\perp \cap B| = |U(2n - 2, q)|$ for every point $X \in T \setminus B$.*

Proof Let T be a tight generator, so that $T \cap B$ is a hyperplane of T. Denote by U a generator that always meets T in a hyperplane of T different from $T \cap B$. Then U meets $T \cap B$ in a subspace of dimension $n - 3$. Since U meets B in at least $(q^{n-1} - 1)/(q - 1)$ points, it follows that $U \setminus T$ contains at least q^{n-2} points of B. Every singular point that does not lie in $(T \cap B)^\perp$ occurs in exactly one of the sets $U \setminus T$.

(a) We may assume that $P \in T \cap B$. Then T has q^{n-1} hyperplanes that do not contain P and each of them lies in $q\sqrt{q}$ generators $U \neq T$. Thus for $q^n\sqrt{q}$ subspaces U we have $P \notin U$. Since all sets $U \setminus T$ meet B in at least q^{n-2} points, it follows that $|P^\perp \cap B| \leq |B| - q^n\sqrt{q} \cdot q^{n-2}$.

(b) We may assume that $P \in T^\perp \setminus T$. Each hyperplane of T lies in $q\sqrt{q}+1$ generators of which $(q-1)\sqrt{q}$ are not contained in $P^\perp$. Since T has $(q^n-q)/(q-1)$ hyperplanes different from $T \cap B$, exactly $t := (q^n-q)\sqrt{q}$ subspaces U are not contained in $P^\perp$. For these we have $U \cap P^\perp = U \cap T$. It follows that $B \setminus P^\perp$ contains at least tq^{n-2} points of B. Hence $|P^\perp \cap B| \leq |B| - tq^{n-2}$.

(c) Suppose that the points $P \in T \cap B$ with $|P^\perp \cap B| = |B| - q^{2n-2}\sqrt{q}$ span $T \cap B$. Since each subspace U does not contain one of these points, the proof of part (a) shows that all sets $U \setminus T$ meet B in exactly q^{n-2} points.

If $X \in T \setminus B$, then X lies in $s := (q^{n-1}-1)/(q-1)$ hyperplanes of T and each of them lies in $q\sqrt{q}$ subspaces U. Thus X lies in $sq\sqrt{q}$ subspaces U. Since each singular point of $X^\perp \setminus T$ lies in exactly one of them, it follows that $X^\perp \setminus T$ meets B in $sq^{n-1}\sqrt{q}$ points. Since $X^\perp \cap T = T$ meets B in $|T \cap B| = s$ points, we obtain $|X^\perp \cap B| = s(q^{n-1}\sqrt{q}+1)$. Hence $|X^\perp \cap B| = |U(2n-2,q)|$. ∎

Lemma 8.5 *Suppose that l is a secant line such that the points of l are tight and such that the singular points of l are in B. Suppose furthermore that $l^\perp$ meets B in at least $q|U(2n-4,q)|$ points. (Here $|U(0,q)|$ denotes the number zero.) Then*
 (a) $|B| = q|U(2n-2,q)|$ and $|l^\perp \cap B| = q|U(2n-4,q)|$.
 (b) $|P^\perp \cap B| = |B| - q^{2n-2}\sqrt{q}$ *for all singular points P of l.*

Proof The subspace $l^\perp$ has dimension $2n-2$. Put $\alpha := |l^\perp \cap B|$. The hyperplanes on $l^\perp$ are the hyperplanes $X^\perp$ with $X \in l$. Consequently, the sets $(X^\perp \setminus l^\perp) \cap B$ with $X \in l$ partition the set $B \setminus l^\perp$. If we apply Lemma 8.4 (a) to the $\sqrt{q}+1$ points X of $l \cap B$ and Lemma 8.4 (b) to the $q-\sqrt{q}$ points X of $l \setminus B$, we obtain the following upper bound for the size $|B| - \alpha$ of $B \setminus l^\perp$:

$$|B| - \alpha \leq (\sqrt{q}+1)\left(|B| - q^{2n-2}\sqrt{q} - \alpha\right)$$
$$+ (q-\sqrt{q})\left(|B| - q^{2n-2}\sqrt{q} + q^{n-1}\sqrt{q} - \alpha\right).$$

This gives

$$q(|B| - \alpha) \geq (q+1)q^{2n-2}\sqrt{q} - (q-\sqrt{q})q^{n-1}\sqrt{q}.$$

The number on the right hand side is equal to $q|U(2n-2,q)| - q|U(2n-4,q)|$. Using the bounds $|B| \leq q|U(2n-2,q)|$ and $\alpha \geq q|U(2n-4,q)|$, we obtain equality. This proves parts (a) and (b). ∎

Corollary 8.6 *Suppose that $n=2$. If there exists a secant line whose points are tight and whose singular points lie in B, then $|B| = q|U(2,q)|$.*

Proof Apply the previous lemma with $n=2$. ∎

Lemma 8.7 *There exists a point $P \in U(2n,q) \setminus B$ with*

$$|P^\perp \cap B| < |U(2n-2,q)| + q^{n-2}.$$

Proof Let r_2 be the number of generators on a line, let r_1 be the number of generators on a point and let r_0 be the number of generators. Then $r_1 = (q^{n-1}\sqrt{q}+1)r_2$ and $r_0 = (q^n\sqrt{q}+1)r_1$.

The generators meet B in at least $\Theta := (q^{n-1}-1)/(q-1)$ points. Put $a_S := |S \cap B| - \Theta$ for every generator S. Count pairs (P,S) of points $P \in B$ and generators S on P to obtain

$$\sum_S |S \cap B| = |B|r_1.$$

Since we use in this proof only that the points of B are singular and that $|B| \leq q|U(2n-2,q)|$ but no other property of B, we can assume in this proof that $|B| = q|U(2n-2,q)|$. Then we have

$$\sum_S a_S = |B|r_1 - r_0\Theta = r_1(|B| - \Theta(q^n\sqrt{q}+1)) = r_1(q^{n-1}-1).$$

Consider the triples (R,P,S) with generators S, points $R \in S \cap B$ and points $P \in S \setminus B$. Every generator S occurs in

$$|S \cap B| \cdot |S \setminus B| = |S \cap B|(q^{n-1} - a_S) \leq |S \cap B|q^{n-1} - (\Theta+1)a_S$$

such triples. Since r_2 is the number of generators on a totally isotropic line, each point $P \in U(2n,q) \setminus B$ occurs in $|P^\perp \cap B|r_2$ such triples (R,P,S). Since $|U(2n,q)\setminus B| = q^{2n-1}\sqrt{q}+1$, it follows that there exists a point $P \in U(2n,q)\setminus B$ for which

$$|P^\perp \cap B| \cdot (q^{2n-1}\sqrt{q}+1)r_2 \leq q^{n-1}\sum_S |S \cap B| - (\Theta+1)\sum_S a_S.$$

It follows that

$$|P^\perp \cap B| \cdot (q^{2n-1}\sqrt{q}+1)r_2 \leq q^{n-1}|B|r_1 - (\Theta+1)r_1(q^{n-1}-1).$$

Using $|B| = q|U(2n-2,q|$, the definition of Θ and $r_1 = (q^{n-1}\sqrt{q}+1)r_2$, now gives

$$|P^\perp \cap B| \leq |U(2n-2,q)|\frac{(q^n(q^{n-1}\sqrt{q}+1)-q^{n-1}-q+2)}{q^{2n-1}\sqrt{q}+1}.$$

This gives

$$|P^\perp \cap B| - |U(2n-2,q)| < \frac{|U(2n-2,q)|q^{n-1}(q-1)}{q^{2n-1}\sqrt{q}+1}.$$

Since $|U(2n-2,q)|(q-1) = (q^{n-1}\sqrt{q}+1)(q^{n-1}-1) < q^{2n-2}\sqrt{q}$, the right hand side is smaller than q^{n-2}. ∎

Lemma 8.8 *Suppose that $n \geq 3$ and consider a point P of $U(2n,q) \setminus B$ with $|P^\perp \cap B| < |U(2n-2,q)| + q^{n-2}$.*

(a) There exists an anisotropic solid S in $P^\perp$ such that the points of B in $\langle P, S \rangle$ are the singular points of S.

(b) If $X \neq P$ is a point of $P^\perp$ and if the line PX has at most one point in B, then X is a tight point.

Proof Let Z be the set consisting of the totally isotropic lines on P that meet B in more than one point. Since P lies on $|U(2n-2,q)|$ totally isotropic lines, we have $|Z| < q^{n-2}$.

The polar space $U(2n,q)$ induces in $P^\perp/P$ a polar space of type $U(2n-2,q)$. If we apply Lemma 8.1 (a) to this quotient space, considering Z as a set of some of its points, we obtain the following: If l is a line of $P^\perp$ on P that is not in Z, then $l \subseteq T^\perp$ for some generator T on P that contains no line Z. Thus every line of T on P has a unique point in B, so T is a tight generator. This shows that the points of l are tight, proving (b).

Since $n \geq 3$, we have $|Z| \leq q^{n-2} - 1 < |U(2n-2,q)|/|U(3,q)|$ and thus we can apply Lemma 8.1 (b) to the quotient space $P^\perp/P$. This shows that there exists an anisotropic solid S' in $P^\perp$ such that no line of Z lies in $\langle P, S' \rangle$. Therefore every totally isotropic line of $\langle P, S' \rangle$ on P meets B in a unique point. Lemma 8.3 shows that we can find a complement S of P in $\langle P, S' \rangle$ such that the points of B in $\langle P, S' \rangle$ are the singular points of S. This proves (a). ∎

Lemma 8.9 *Suppose that $n = 3$. Then we have*

(a) $|B| = q|U(4,q)|$.

(b) If P is a singular point that is not in B and if $|P^\perp \cap B| < |U(4,q)| + q$, then $|X^\perp \cap B| = |U(4,q)|$ for all singular points $X \in P^\perp \setminus B$.

Proof Let P be a point in $U(6,q) \setminus B$ with $|P^\perp \cap B| < |U(4,q)| + q$. Lemma 8.7 shows that such a point exists. The previous lemma says that $P^\perp$ contains an anisotropic solid S such that the points of B in $\langle P, S \rangle$ are the singular points of S. The singular points and totally isotropic lines of S form a $U(3,q)$.

Let Z be the set consisting of the totally isotropic lines on P that meet B in more than one point. Since P lies on $|U(4,q)|$ totally isotropic lines, we have $|Z| < q$. No line of Z meets S. If $g \in Z$, then $g^\perp$ meets S in a $U(2,q)$ so that $g^\perp$ contains $q^2 + q - q\sqrt{q} < q^2$ secant lines of S. Hence, fewer than $q^2|Z| < q^3$ secant lines of S lie in $g^\perp$ for a line g of Z. In other words, for at most $q^3 - 1$ secant lines l of S, the set $l^\perp$ contains a line of Z.

Let h be a totally isotropic line of S. Each point of h lies on q^2 secant lines of S. Thus there exist two secant lines l_1 and l_2 in S with the following properties: l_1 and l_2 meet h, the intersection points of h with l_1 and l_2 are distinct, and the sets $l_1^\perp$ and $l_2^\perp$ contain no line of Z.

The subspace $l_i^\perp$ has dimension four and meets S in a secant line l'_i. Since l_i and l'_i lie in S, all their singular points are in B. We want to apply Corollary 8.6 to the 4-space $l_i^\perp$ and its secant line l'_i in order to prove that $|l_i^\perp \cap B| \geq q|U(2,q)|$.

Before we can do that we have to show that the points of l'_i are *tight inside* $l_i^\perp$, which means that for each point X of l'_i there exists a totally isotropic line g in $l_i^\perp$ with $X \in g^\perp$ and $|g \cap B| = 1$. If $X \in l'_i \cap B$, then we can take $g := PX$. Now suppose that X is in $l'_i \setminus B$. Then $X^\perp \cap l_i^\perp$ is an anisotropic solid, which meets $U(6,q)$ in a $U(3,q)$ containing P. The point P lies on $\sqrt{q}+1$ totally isotropic lines of $X^\perp \cap l_i^\perp$. Since $l_i^\perp$ contains no line of Z, each totally isotropic line of $X^\perp \cap l_i^\perp$ on P meets B in a unique point and can therefore be taken for g. Thus $|l_i^\perp \cap B| \geq q|U(2,q)|$.

By Lemma 8.8 (b), the points of l_i are tight. Therefore Lemma 8.5 shows that $|B| = q|U(4,q)|$ and that $|X^\perp \cap B| = |B| - q^4\sqrt{q}$ for every point X of $l_i \cap B$. In particular, for $P_i := l_i \cap h$ we have $|P_i^\perp \cap B| = |B| - q^4\sqrt{q}$.

The plane $E := \langle P, h \rangle$ is totally isotropic and $E \cap B = h$. Since h is spanned by P_1 and P_2, Lemma 8.4 (c) shows that $|X^\perp \cap B| = |U(4,q)|$ for all points $X \in E \setminus B = E \setminus h$. In particular $|P^\perp \cap B| = |U(4,q)|$, which implies that $Z = \emptyset$.

Lemma 8.3 shows that P has a complement C in $P^\perp$ such that $P^\perp \cap B$ consists of the singular points of C. Then $C \cap H$ is a $U(4,q)$. Therefore in the above argument, the totally isotropic line h and the solid S can be replaced by any other totally isotropic line of C and every anisotropic solid of C on that line. This shows that every totally isotropic line h of C has the property that $|X^\perp \cap B| = |U(4,q)|$ for all points $X \in \langle P, h \rangle \setminus h$. Hence $|X^\perp \cap B| = |U(4,q)|$ for all totally isotropic points X of $P^\perp \setminus C$, that is for all singular points of $P^\perp \setminus B$. ∎

Lemma 8.10 *If $n = 3$, then there exists a singular point R such that B consists of the singular points of $R^\perp \setminus \{R\}$.*

Proof Let $\mathcal{P}$ be the set consisting of the points $P \in U(6,q) \setminus B$ for which $|P^\perp \cap B| = |U(4,q)|$. By Lemmas 8.7 and 8.9, this set is not empty. If $P \in \mathcal{P}$, then Lemma 8.3 shows that there exists a complement S_P of P in $P^\perp$ such that $P^\perp \cap B$ consists of the singular points of S_P.

Consider a point $P \in \mathcal{P}$. Then $S_P^\perp$ is a secant line. Suppose X is a singular point that is not in B and not on the secant line $S_P^\perp$. Then $X \in P'^\perp$ for some singular point P' of $P^\perp \setminus S_P$. Lemma 8.9 (b) shows that $P' \in \mathcal{P}$ and, since $X \in P'^\perp \setminus B$, the same lemma shows that $X \in \mathcal{P}$. Hence all points of $U(6,q) \setminus B$ except possibly the points of the secant line $S_P^\perp$ lie in $\mathcal{P}$. Since $|B| = q|U(4,q)|$, we have $|U(6,q)| - |B| = q^5\sqrt{q} + 1$ and thus $|\mathcal{P}| \geq q^5\sqrt{q} - \sqrt{q}$.

Let $\mathcal{S}$ be the set consisting of the subspaces S_P with $P \in \mathcal{P}$. If $S \in \mathcal{S}$, then $S^\perp$ is a secant line and thus we have $S = S_P$ for at most $\sqrt{q} + 1$ points P of $\mathcal{P}$. Therefore $|\mathcal{S}| \geq |\mathcal{P}|/(\sqrt{q}+1) > q^2 + q + 1$.

For points $P, R \in \mathcal{P}$ we have $\dim(R^\perp \cap S_P) \geq \dim(S_P) - 1 = 3$. Since $U(6,q) \cap S_P$ is a $U(4,q)$, the points of $R^\perp \cap S_P$ generate $R^\perp \cap S_P$. Since the singular points of S_P are in B, it follows that $R^\perp \cap S_P \subseteq S_R$. Hence $S_P = S_R$ or $S_P \cap S_R$ is a solid. Thus distinct subspaces of $\mathcal{S}$ meet in a solid. This implies that all subspaces of $\mathcal{S}$ are on a common solid or all subspaces of $\mathcal{S}$ lie in a hyperplane. The first case is not possible, since $|\mathcal{S}| > q^2 + q + 1$.

Hence there exists a hyperplane H containing all solids S of $\mathcal{S}$. We show now that H contains all points of B. Let X be a point of B. Suppose first that there exists a point of $P \in \mathcal{P}$ with $X \notin S_P^\perp$. Then there exists a singular point $Y \in S_P$ with $Y \notin X^\perp$. Then the totally isotropic line PY contains a point P' with $P' \in X^\perp$ and $P' \neq Y$. Thus $P' \in P^\perp \setminus S_P$ and therefore $P' \in \mathcal{P}$ and $X \in S_{P'} \subseteq H$. Suppose now that $X \in S_P^\perp$ for all points $P \in \mathcal{P}$, that is $X \in S^\perp$ for all $S \in \mathcal{S}$. Then all the secant lines $S^\perp$ must meet in X. Since the sets S span H, it follows that $X = H^\perp$. Since X is singular, it follows that $X \in H$. Hence, H contains all points of B.

Assume that H is a non-tangent hyperplane. Then H contains $|U(5,q)|$ singular points. Since $|B| = q|U(4,q)| < |U(5,q)|$, there exists a singular point $P \in H \setminus B$. Let l be a totally isotropic line on P with $l \cap H = P$. As $B \subseteq H$, the line l has no point in B, a contradiction.

Hence H is a tangent hyperplane, that is $H = R^\perp$ for a singular point R. Every singular point P of H with $P \neq R$ lies on a totally isotropic line that meets H only in P and therefore P must be in H. Hence all singular points of H except possibly R lie in B. Since $|B| = q|U(4,q)|$, it follows that B consists of the singular points of $R^\perp \setminus \{R\}$. ∎

Lemma 8.11 *If $n \geq 3$, then there exists a point $R \in U(2n, q)$ such that B consists of the singular points of $R^\perp \setminus \{R\}$.*

Proof We prove this by induction on n. The case $n = 3$ was handled in the previous lemma, so assume now that $n \geq 4$. By Lemma 8.7, there exists a point $P \in U(2n, q) \setminus B$ with $|P^\perp \cap B| \leq |U(2n-2, q)| + q^{n-2} - 1$ and Lemma 8.8 implies that $P^\perp$ contains a secant line l whose points are tight and whose singular points lie in B.

Since $U(2n, q) \cap l^\perp$ is a $U(2n-2, q)$, the induction hypothesis implies that $|l^\perp \cap B| \geq q|U(2n-4, q)|$. Lemma 8.5 shows therefore that $|l^\perp \cap B| = q|U(2n-4, q)|$ and $|B| = q|U(2n-2, q)|$. The induction hypothesis also gives the structure of $l^\perp \cap B$. Thus, there exists a singular point $R \in l^\perp$ such $l^\perp \cap B$ consists of the singular points of $(l^\perp \cap R^\perp) \setminus \{R\}$.

Since P lies on $|U(2n-2, q)|$ totally isotropic lines, it lies on at most $q^{n-2} - 1$ totally isotropic lines that meet B in more than one point. It follows that $R \neq P$. Hence PR is a secant line of $l^\perp$ and therefore $S := l^\perp \cap R^\perp \cap P^\perp$ is an anisotropic subspace of dimension $2n - 4$, which meets $U(2n, q)$ in a $U(2n-4, q)$. The singular points of S are points of B, since they lie in $l^\perp \cap R^\perp$. Thus the points of B in $\langle P, S \rangle$ are the singular points of S.

Consider any totally isotropic line h on R that is not in $l^\perp$. Then $\dim(h^\perp \cap S) \geq 2n - 5 \geq 3$ and therefore $h^\perp \cap S$ contains a secant line l'. The plane $\langle l', P \rangle$ lies in $P^\perp$ and the points of B in the plane $\langle P, l' \rangle$ are the singular points of l'. Lemma 8.8 (b) shows that the points of l' are tight. As for l, it follows that there exists a point $R' \in l'^{\perp}$ such that the points of B in $l'^{\perp}$ are the singular points of $(l'^{\perp} \cap R'^{\perp}) \setminus \{R'\}$.

The totally isotropic lines of $l'^{\perp}$ that meet B in exactly q points are therefore the totally isotropic lines of $l'^{\perp}$ on R'. Similarly the totally isotropic lines of $l^{\perp}$ that meet B in exactly q points are the totally isotropic lines of $l^{\perp}$ on R. But $R' \in l^{\perp}$, so $R = R'$. The line h lies in $l'^{\perp}$ and contains $R = R'$ and therefore all points of $h \setminus \{R\}$ lie in B.

This holds for every totally isotropic line h on R that is not in $l^{\perp}$. Above we have seen that the same holds for the totally isotropic lines h of $l^{\perp}$ on R. It follows that the singular points of $R^{\perp} \setminus \{R\}$ lie in B. Since there are $q|U(2n-2,q)|$ singular points in $R^{\perp} \setminus \{R\}$ and since $|B| \leq q|U(2n-2,q)|$, this proves the lemma. ∎

The preceding lemma completes the proof of Theorem 1.2.

9 Unsolved problems

1. Determine the smallest number of points that are needed to block all lines of $U(4,q)$. Probably, Corollary 8.6 can help.

2. Determine the smallest number of points that are needed to block all lines of $W(2n+1,q)$ for odd q.

References

[1] S. Ball, Multiple blocking sets and arcs in finite planes, *Journal of the London Mathematical Society, Series 2*, **54** (1996), 581–593.

[2] A. Beutelspacher, Parallelismen in unendlichen projektiven Räumen, *Geometriae Dedicata*, **7** (1978), 499–506.

[3] A. Beutelspacher, On t-covers in finite projective spaces, *Journal of Geometry*, **12** (1979), 10–16.

[4] A. Beutelspacher & J. Ueberberg, A characteristic property of geometric t-spreads in finite projective spaces, *European Journal of Combinatorics*, **12** (1991), 277–281.

[5] A. Blokhuis, On the size of blocking sets in PG$(2,p)$, *Combinatorica*, **14** (1994), 111–114.

[6] A. Blokhuis & G. E. Moorhouse, Some p-ranks related to orthogonal spaces, *Journal of Algebraic Combinatorics*, **4** (1995), 295–316.

[7] R. C. Bose & R. C. Burton, A characterization of flat spaces in a finite geometry and the uniqueness of the Hamming and the MacDonald codes, *Journal of Combinatorial Theory*, **1** (1966), 96–104.

[8] A. E. Brouwer & A. Schrijver, The blocking number of an affine space, *Journal of Combinatorial Theory, Series A*, **24** (1978), 251–253.

[9] A. A. Bruen, Blocking sets in finite projective planes, *SIAM Journal on Applied Mathematics*, **21** (1971), 380–392.

[10] A. A. Bruen, Partial spreads and replaceable nets, *Canadian Journal of Mathematics*, **23** (1971), 381–391.

[11] A. A. Bruen & J. C. Fisher, The Jamison method in galois geometries, *Designs, Codes and Cryptography*, **1** (1991), 199–205.

[12] P. J. Cameron, *Projective and Polar Spaces*, QMW Maths Notes, 13, Queen Mary and Westfield College, London (1991).

[13] J. H. Conway & P. B. Kleidman & R. A. Wilson, New families of ovoids in O_8^+, *Geometriae Dedicata*, **26** (1988), 157–170.

[14] R. H. Dye, Partitions and their stabilizers for line complexes and quadrics, *Designs, Codes and Cryptography*, **114** (1977), 173–179.

[15] J. Eisfeld, Private communication.

[16] J. Eisfeld, On smallest covers of finite projective spaces, *Archiv der Mathematik*, **68** (1997), 77–80.

[17] J. Eisfeld & K. Metsch, Blocking s-dimensional subspaces by lines in $PG(2s, q)$, *Combinatorica*, **17** (1997), 151–162.

[18] A. Gunawardena & G. E. Moorhouse, The non-existence of ovoids in $O_9(q)$, *European Journal of Combinatorics*, **18** (1997), 171–173.

[19] U. Heim, Proper blocking sets in projective spaces, *Discrete Mathematics*, **174** (1997), 167–176.

[20] J. W. P. Hirschfeld, *Projective Geometries over Finite Fields*, second edition, Clarendon Press, Oxford (1998).

[21] J. W. P. Hirschfeld & J. A. Thas, *General Galois Geometries*, Oxford Mathematical Monographs, Clarendon Press, Oxford (1991).

[22] R. Jamison, Covering finite fields with cosets of subspaces, *Journal of Combinatorial Theory, Series A*, **22** (1977), 253–266.

[23] W. M. Kantor, Spreads, translation planes and Kerdock sets I, *SIAM Journal on Algebraic Discrete Methods*, **3** (1982), 151–165.

[24] W. M. Kantor, Spreads, translation planes and Kerdock sets II, *SIAM Journal on Algebraic Discrete Methods*, **3** (1982), 308–318.

[25] W. M. Kantor, Ovoids and translation planes, *Canadian Journal of Mathematics*, **34** (1982), 1195–1203.

[26] A. Klein, Partielle Ovoide in klassischen endlichen Polarräumen, Preprint, Mathematisches Institut, Universität Gießen, 1998.

[27] K. Metsch, On the number of lines needed to block the planes of $PG(5,q)$, Preprint, Mathematisches Institut, Universität Gießen, 1998.

[28] K. Metsch, Improvement of Bruck's completion theorem, *Designs, Codes and Cryptography*, **1** (1991), 99–116.

[29] K. Metsch, The sets closest to ovoids in $Q^-(2n+1,q)$, *Bulletin de la Société mathématique de Belgique*, **5** (1998), 389–392.

[30] K. Metsch, The Bose-Burton theorem for elliptic polar spaces, *Designs, Codes and Cryptography*, submitted.

[31] K. Metsch & L. Storme, 2-blocking sets in PG(n,q), q square, *Journal of Statistical Planning and Inference*, submitted.

[32] K. Metsch & L. Storme, Partial t-spreads in PG$(2t+1,q)$, *Designs, Codes and Cryptography*, submitted.

[33] C. O'Keefe & J. A. Thas, Ovoids of the quadric $Q(2n,q)$, *European Journal of Combinatorics*, **16** (1995), 87–92.

[34] S. E. Payne & J. A. Thas, *Generalized Quadrangles*, Pitman, London (1984).

[35] T. Szőnyi, Blocking sets in desarguesian affine and projective planes, *Finite Fields and their Applications*, **3** (1997), 187–202.

[36] O. Polverino, Small blocking sets in PG$(2,p^3)$, *Designs, Codes and Cryptography*, submitted.

[37] E. E. Shult, A sporadic ovoid in $\Omega^+(8,7)$, *Algebras, Groups and Geometries*, **2** (1985), 495–513.

[38] E. E. Shult, Nonexistence of ovoids in $\Omega^+(10,3)$, *Journal of Combinatorial Theory, Series A*, **51** (1989), 250–257.

[39] E. E. Shult & J. A. Thas, m-systems of polar spaces, *Journal of Combinatorial Theory, Series A*, **168** (1994), 184–204.

[40] E. E. Shult & J. A. Thas, m-systems and partial m-systems of polar spaces, *Designs, Codes and Cryptography*, **8** (1996), 229–238.

[41] L. Storme, Minimal blocking sets in PG(n,q), *Journal of the London Mathematical Society*, submitted.

[42] G. Tallini, Blocking sets with respect to planes in $PG(3,q)$ and maximal spreads of a non singular quadric in $PG(4,q)$, *Mitteilungen aus dem Mathematischen Seminar Gießen*, **201** (1991), 141–147.

[43] J. A. Thas, Ovoidal translation planes, *Archiv der Mathematik*, **23** (1972), 110–112.

[44] J. A. Thas, Two infinite classes of perfect codes in metrically regular graphs, *Journal of Combinatorial Theory, Series B*, **23** (1977), 236–238.

[45] J. A. Thas, Polar spaces, generalized hexagons and perfect codes, *Journal of Combinatorial Theory, Series A*, **29** (1980), 87–93.

[46] J. A. Thas, Ovoids and spreads of finite classical polar spaces, *Geometriae Dedicata*, **10** (1981), 135–44.

[47] J. A. Thas, A note on spreads and partial spreads of hermitian varieties, *Simon Stevin*, **63** (1989), 101–105.

[48] J. A. Thas, Old and new results on spreads and ovoids of finite classical polar spaces, *Annals of Discrete Mathematics*, **52** (1992), 529–544.

[49] J. A. Thas & S. E. Payne, Spreads and ovoids in finite generalized quadrangles, *Geometriae Dedicata*, **52** (1994), 227–253.

Mathematisches Institut
Universität Gießen
Arndtstraße 2
D-35392 Gießen
Germany
klaus.metsch@math.uni-giessen.de

Geometric Graph Theory

János Pach

Summary A *geometric graph* is a graph drawn in the plane such that its vertices are points in general position and its edges are straight-line segments. The study of geometric graphs is a fairly new discipline abounding in open problems, but it has already yielded some striking results that have proved to be instrumental for the solution of various problems in combinatorial and computational geometry. These include the k-set problem, proximity questions, bounding the number of incidences between points and lines, designing various efficient graph drawing algorithms, etc. This paper surveys some Turán-type and Ramsey-type extremal problems for geometric graphs, and discusses their generalizations and applications.

1 Introduction, basic definitions

Let G be a finite graph with no loops or multiple edges, whose vertex set and edge set are denoted by $V(G)$ and $E(G)$, respectively. By a *drawing* of G we mean a representation of G in the plane such that each vertex is represented by a distinct point and each edge by a simple (non-self-intersecting) continuous arc connecting the corresponding two points. If it is clear that we are referring to a drawing, and not to the underlying "abstract" graph, these points and arcs will also be called *vertices* and *edges*, respectively.

Two edges (arcs) *cross* each other if they have an interior point in common. This point is called a *crossing*. A crossing p is called *proper* if in a small neighbourhood of p one edge passes from one side of the other edge to the other side. We assume throughout that in a drawing

1. no edge passes through any vertex other than its endpoints,

2. no two edges touch each other (i.e., if two edges cross, then they properly cross),

3. no three edges cross at the same point.

A drawing in which every edge is represented by a straight-line segment is called a *straight-line drawing* or a *geometric graph*. Two geometric graphs are *isomorphic* if and only if one of them can be carried into the other using a rigid motion of the plane. We always assume that the vertices of a geometric graph are in *general position*, i.e. no three of them are collinear. If the vertices of a geometric graph are in *convex position*, i.e. they form the vertex set of a convex polygon, then the graph is called a *convex geometric graph*.

We emphasize that two edges of a geometric graph G may cross each other. The *number of crossings* in G is denoted by $\mathrm{CR}(G)$. Two edges of G are *disjoint* if they do not cross and do not even share an endpoint.

A geometric graph G is said to be *complete* if its edge set consists of all $\binom{|V(G)|}{2}$ segments between its vertices. If $V(G) = V_1 \cup V_2$, and $E(G)$ consists of all segments between V_1 and V_2, then G is a *complete bipartite* geometric graph. A geometric graph G is said to be a *geometric tree* if the underlying abstract graph is a tree. A geometric graph G is a *geometric subgraph* of another geometric graph H if $V(G) \subseteq V(H)$ and $E(G) \subseteq E(H)$.

The systematic study of geometric graphs was initiated by S. Avital and H. Hanani [7], P. Erdős, Y. Kupitz [48], and M. Perles. They realized that many classical questions in extremal graph theory [10] have natural analogues for geometric graphs. Some of these questions turned out to be surprisingly difficult and required new techniques combining geometric and combinatorial tools. In this paper we survey some recent results of this type and some tantalizing open problems.

Section 2 focuses on crossing-free geometric graphs. Section 3 describes some conditions which guarantee that a geometric graph determines many crossings. Sections 4 and 5, and 6 concentrate on Turán-type and Ramsey-type problems for geometric graphs, respectively. In Section 7 we present a few applications of our results, while Section 8 is devoted to geometric hypergraphs.

As usual, for any positive integer-valued functions, $f(n)$ and $g(n)$, we use the notations $f(n) = o(g(n)), f(n) = O(g(n))$, and $f(n) = \Omega(g(n))$ to indicate that $\lim_{n \to \infty} f(n)/g(n) = 0$, $\sup f(n)/g(n) < \infty$, and $\inf f(n)/g(n) > 0$, respectively. We write $f(n) = \Theta(g(n))$ if $f(n) = O(g(n))$ and $f(n) = \Omega(g(n))$.

For many other problems and results in this area, see [56] and Chapter 14 of [57].

2 Crossing-free geometric graphs

Let $G = (V(G), E(G))$ be a geometric graph which has no crossing. Clearly, the underlying abstract graph, G_0, whose vertex set is $V(G)$ and whose edge set consists of those pairs of vertices which are connected in G by a segment, is *planar*. Of course, the reverse is not true: the fact that G_0 is planar does not imply that G must be crossing-free. However, by Fáry's theorem, for every abstract planar graph H, there is a crossing-free geometric graph G whose underlying graph G_0 is isomorphic to H. In simpler terms, we have the following:

Theorem 2.1 ([24], [78], [71]) *Every planar graph admits a crossing-free straight-line drawing.*

We can modify the problem by putting certain restrictions on the drawing. For instance, given an abstract planar graph H with n vertices and a set P of n points in general position in the plane, we can ask whether there exists a crossing-free geometric graph G with $V(G) = P$, whose underlying graph G_0 is isomorphic to H. To see that the answer is not necessarily in the affirmative, choose H to be a complete graph with 4 vertices, and let P consist of 4 points in convex position.

An abstract graph H is called *outerplanar* if it can be drawn in the plane without crossing so that all points representing the vertices of H lie on the boundary of the outer face of the resulting subdivision of the plane. Clearly, every tree is outerplanar. It follows from Theorem 2.1 that a graph is outerplanar if and only if it is the underlying graph of a crossing-free *convex* geometric graph (a triangulated cycle). However, these graphs also satisfy a stronger condition.

Theorem 2.2 ([30]) *For any outerplanar graph H with n vertices and for any set P of n points in the plane in general position, there is a crossing-free geometric graph G with $V(G) = P$, whose underlying graph is isomorphic to H.*

In other words, every outerplanar graph H admits a straight-line drawing without crossing such that its vertex set is mapped into an arbitrarily prespecified $|V(H)|$-element set. Often we do not have much freedom in how to select this mapping. But sometimes we do. Perles conjectured that if H is a tree, then we have at least one degree of freedom. More precisely, the following assertion is true.

Theorem 2.3 ([63], [37]) *Let T be a rooted tree, and let P be a set of $|V(T)|$ points in the plane in general position with a specified element $p \in P$.*

Then T admits a crossing-free straight-line drawing such that every vertex of T is represented by an element of P, and the root is represented by p.

It may be conjectured that for any forest F consisting of k rooted trees $T_1, T_2, \ldots, T_k$ with roots $r_1, r_2, \ldots, r_k$, and for any set $P = \{p_1, p_2, \ldots, p_K\}$ of $K = \sum_{i=1}^{k} |T_k|$ points in general position in the plane, F admits a crossing-free straight-line drawing such that every vertex of F is represented by a point of P, and r_i is represented by p_i ($i = 1, 2, \ldots, k$). A. Kaneko and M. Kano [39–41] verified this conjecture in several special cases, including when

1. $k = 2$;

2. every tree T_i is a *rooted star*, i.e. all of its vertices are adjacent to the root r_i;

3. all trees T_i are of the same size.

However, Kaneko and Kano believe that the above conjecture, in general, is false.

In some special cases, we may impose further restrictions on the drawings. For instance, colour the vertices of the tree T with red and blue such that no two vertices of the same colour are adjacent. Let n_1 and n_2 denote the number of red and blue vertices, respectively. Given an n_1-element set P_1 and an n_2-element set P_2 in general position in the plane, we wish to decide whether there is a crossing-free straight-line drawing of T, whose red and blue vertices are represented by the elements of P_1 and P_2, respectively. In the case when P_1 and P_2 are separated by a straight-line, $n_1 = n_2$, and T is a path, the answer is in the affirmative [35].

Abellanas, García, Hernández, Noy and Ramos studied the following relaxation of this question: what happens if, instead of embedding a fixed tree T, we want to find a drawing of at least one member of a given *class* of trees?

Problem 2.4 ([1]) Let $P_1 \cup P_2$ be a point set in general position in the plane, $|P_1| \geq |P_2| > 0$. Is it true that there always exists a crossing-free geometric tree on the vertex set $P_1 \cup P_2$, whose every edge runs between P_1 and P_2, and in which the degree of every vertex is at most an absolute constant times $|P_1|/|P_2|$?

Abellanas et al. showed that the answer is in the affirmative if

1. $P_1 \cup P_2$ is in convex position, or

2. P_1 and P_2 are separated by a straight line.

We close this section with a related question.

Problem 2.5 ([25]) What is the size of the smallest point set P in the plane with the property that for every abstract planar graph H of n vertices, there is a crossing-free geometric graph G, whose vertex set is a subset of P and whose underlying graph is isomorphic to H?

The next theorem shows that this minimum is $O(n^2)$ (see also [68]).

Theorem 2.6 ([25]) *For every planar graph H, there is a crossing-free geometric graph whose vertex set is a subset of an n by $2n$ piece of the integer grid, and whose underlying graph is isomorphic to H.*

3 Unavoidable crossings

The number of crossings in a geometric graph G is denoted by $\text{CR}(G)$. It follows from Euler's Polyhedral Formula that a geometric graph with n vertices and more than $3n - 6$ edges cannot be crossing-free ($n \geq 3$). In fact, this easily implies the following quantitative statement.

Lemma 3.1 *Any geometric graph G with $n \geq 3$ vertices and e edges has at least $e - 3n + 6$ crossings.*

Geometric Graph Theory

Proof By induction on e. If $e \leq 3n-6$ then the statement is void. Assume $e > 3n - 6$. Then G has at least one crossing. Applying the induction hypothesis to the graph G' obtained from G by deleting an edge which participates in at least one crossing, we obtain that

$$\text{CR}(G) = 1 + \text{CR}(G') \leq 1 + [(e-1) - 3n + 6)] = e - 3n + 6. \quad \blacksquare$$

It is easy to see that Lemma 3.1 cannot be improved if $3n-6 \leq e \leq 4n-8$. However, as was shown by R. Eggleton and R. Guy [20], if $e > 4n - 8$ then we can also find an edge which participates in at least *two* crossings. Using this fact, the above inductional argument yields that G has at least

$$[e - (3n-6)] + [e - (4n-8)] = 2e - 7n + 14$$

crossings.

This argument can be pushed a little further. Pach and G. Tóth proved the following.

Theorem 3.2 ([65]) *Let G be a geometric graph with $n \geq 3$ vertices and e edges such that every edge crosses at most k others. Then, for $0 \leq k \leq 4$, we have*

$$e \leq (k+3)(n-2).$$

For $0 \leq k \leq 2$, this bound cannot be improved.

As before, this leads to a further improvement of Lemma 3.1.

Corollary 3.3 ([65]) *Any geometric graph G with $n \geq 3$ vertices and e edges has at least*

$$\sum_{k=0}^{4} [e - (k+3)(n-2)] = 5e - 25n + 50$$

crossings. That is, we have $\text{CR}(G) \geq 5e - 25 + 50$.

Ajtai, Chvátal, Newborn and Szemerédi [3] and independently T. Leighton [51] discovered that from Lemma 3.1 (or from the slightly stronger Corollary 3.3), by a simple trick, one can deduce a much better lower bound for the number of crossings in 'dense' graphs.

Theorem 3.4 ([3], [51], [65]) *Let G be a geometric graph with n vertices and e edges. Then $\text{CR}(G)$, the number of crossings in G, satisfies*

$$\text{CR}(G) \geq \frac{1}{33.75} \frac{e^3}{n^2} - 0.9n > 0.029 \frac{e^3}{n^2} - 0.9n.$$

Proof Suppose first that $e \geq 7.5n$. Construct a *random* geometric subgraph $G' \subseteq G$ by selecting each vertex of G independently with probability $p = 7.5n/e \leq 1$, and letting G' be the geometric subgraph induced by the selected vertices. The expected number of vertices of G', $E[n'] = pn$. Similarly, the expected number of edges and the expected number of crossings in G' satisfy $E[e'] = p^2 e$ and $E[\text{CR}(G')] = p^4 \text{CR}(G)$, respectively.

Applying Corollary 3.3 to G' and taking expectations, we obtain

$$p^4 \text{CR}(G) = E[\text{CR}(G')] > 5E[e'] - 25E[n'] = 5p^2 e - 25pn.$$

This implies that

$$\text{CR}(G) \geq \frac{1}{33.75} \frac{e^3}{n^2},$$

whenever $e \geq 7.5n$. In fact, using Corollary 3.3 in the range $e < 7.5n$, it is easy to check that the slightly weaker inequality in Theorem 3.4 is valid for every geometric graph G. ∎

Theorem 3.4 does not remain true if we replace the constant $1/33.75 \approx 0.029$ by 0.06. To see this, let $n \ll e \ll n^2$, and consider a geometric graph whose vertex set can be obtained by a slight perturbation of a $\lfloor \sqrt{n} \rfloor$ by $\lfloor \sqrt{n} \rfloor$ piece of the integer grid, and connect two gridpoints, p and q, by an edge if and only if their distance satisfies

$$|p - q| \leq \sqrt{\frac{2}{\pi} \cdot \frac{e}{n}}.$$

It is not hard to show that the number of crossings in this geometric graph is at most $0.06 e^3/n^2$.

One may believe that extending Theorem 3.2 to larger values of k would result in a further improvement of the constant factor in Theorem 3.4. However, this is not so. Although the estimate in Theorem 3.2 is tight for small values of k, it becomes weak as k increases. It follows from Theorem 3.4 that, if every edge of a geometric graph with n vertices and e edges participates in at most $k \geq 1$ crossings, then

$$e \leq \sqrt{16.875 kn} \approx 4.108 \sqrt{kn}, \tag{1}$$

and this bound is tight apart from the value of the constant.

Although Theorem 3.4 is tight up to a constant factor, somewhat stronger estimates can be obtained for some special classes of geometric graphs. Let $\mathcal{K}_{r,s}$ denote the class of geometric graphs consisting of all rs segments between some r-element set and some s-element set in the plane. In other words, $\mathcal{K}_{r,s}$ is the family of all *complete bipartite* geometric graphs with r and s vertices in their vertex classes.

Geometric Graph Theory

Theorem 3.5 ([61]) *Let G be a geometric graph with n vertices and $e \geq 4n$ edges, which does not contain a complete bipartite geometric subgraph belonging to $\mathcal{K}_{r,s}$.*
Then the number of crossings in G satisfies

$$\mathrm{CR}(G) \geq c_{r,s} \frac{e^{3+1/(r-1)}}{n^{2+1/(r-1)}},$$

where $c_{r,s} > 0$ is a suitable constant. These bounds are tight up to a constant factor, whenever $s > (r-1)!$.

We close this section with another useful lower bound on the number of crossings in a geometric graphs. To state this result, we need a definition.

Definition Let G_0 be an *abstract* graph with vertex set $V(G_0)$ and edge set $E(G_0)$. The *bisection width*, $b(G_0)$, of G_0 is defined as the minimum number of edges, whose removal splits the graph into two roughly equal subgraphs. More precisely, $b(G_0)$ is the minimum number of edges running between V_1 and V_2, over all partitions of the vertex set of G_0 into two disjoint parts $V_1 \cup V_2$ such that $|V_1|, |V_2| \geq |V(G_0)|/3$.

The *bisection width* of a *geometric* graph G is defined as the bisection width of the underlying abstract graph G_0, and is denoted by $b(G)$.

Leighton observed that there is an intimate relationship between the bisection width and the number of crossings in a geometric graph [51], which is based on the Lipton–Tarjan separator theorem for planar graphs [52]. The following version of this relationship was obtained by Pach, F. Shahrokhi, and M. Szegedy.

Theorem 3.6 ([58]) *Let G be a geometric graph of n vertices with degrees $d_1, d_2, \ldots, d_n$. Then*

$$\mathrm{CR}(G) \geq \frac{1}{40} b^2(G) - \frac{1}{16} \sum_{i=1}^{n} d_i^2.$$

4 Forbidden geometric subgraphs—Multiple crossings

In the spirit of Turán's classical theorem for abstract graphs [77], one can raise the following general question. Given a class $\mathcal{H}$ of so-called *forbidden geometric subgraphs*, what is the maximum number of edges that a geometric graph (respectively a convex geometric graph) of n vertices can have without containing a geometric subgraph which belongs to $\mathcal{H}$? Denote this maximum by $\mathrm{ex}(\mathcal{H}, n)$ (respectively by $\mathrm{ex}_c(\mathcal{H}, n)$).

For any $k \geq 2$, let $\mathcal{C}_k$ denote the class of all geometric graphs consisting of k pairwise *crossing* edges.

The simple fact that a crossing-free geometric graph of n vertices has at most $3n - 6$ edges, can now be expressed by

$$\text{ex}(\mathcal{C}_2, n) = 3n - 6,$$

for every $n \geq 3$. Similarly, inequality (1) can be rewritten as

$$\text{ex}(\mathcal{X}_{1,k}, n) \leq 4.108\sqrt{k}n,$$

where $\mathcal{X}_{1,k}$ denotes the class of all geometric graphs consisting of $k + 1$ edges such that one of them crosses the other k. (Note that $\mathcal{X}_{1,1} = \mathcal{C}_2$.)

The order of magnitude of the function $\text{ex}(\mathcal{C}_3, n)$ has been determined by Agarwal, Aronov, Pach, Pollack and Sharir.

Theorem 4.1 ([2]) *The maximum number of edges that a geometric graph of n vertices can have without containing 3 pairwise crossing edges satisfies*

$$\text{ex}(\mathcal{C}_3, n) = O(n).$$

For $k \geq 4$, we do not know whether $\text{ex}(\mathcal{C}_k, n)$ is linear in n. However, as was shown by Pach, Shahrokhi and Szegedy, one can apply Theorems 3.6 and 4.1 to deduce that $\text{ex}(\mathcal{C}_k, n)$ is bounded from above by n times a polylogarithmic factor.

Corollary 4.2 ([58]) *For every $k \geq 3$, we have*

$$\text{ex}(\mathcal{C}_k, n) = O(n \log^{2k-6} n).$$

Proof The assertion is true for $k = 3$. We sketch the proof only for $k = 4$. The other cases can be treated similarly.

Let G be a geometric graph of n vertices with degrees $d_1, d_2, \ldots, d_n$. Suppose that G has no 4 pairwise crossing edges and $|E(G)| = \text{ex}(\mathcal{C}_4, n)$. For every $e \in E(G)$, the edges crossed by e form a geometric graph with no 3 pairwise crossing edges. Therefore, by Theorem 4.1, every edge crosses at most $O(n)$ other edges, and $\text{CR}(G) = O(|E(G)|n)$. Since

$$\sum_{i=1}^{n} d_i^2 \leq \left(\sum_{i=1}^{n} d_i\right) \left(\max_{1 \leq i \leq n} d_i\right) \leq 2|E(G)|n,$$

it follows from Theorem 3.6 that $b(G) = O\left(\sqrt{|E(G)|n}\right)$.

Consider a partition of $V(G)$ into two parts, V_1 and V_2, such that $|V_1| = n_1 \geq n/3$, $|V_2| = n_2 \geq n/3$, and the number of edges connecting them is $b(G)$. Let G_1 and G_2 denote the geometric subgraphs of G induced by V_1 and V_2, respectively. Since neither G_1 nor G_2 contains 4 pairwise crossing edges, we obtain

$$|E(G)| = |E(G_1)| + |E(G_2)| + b(G) \leq \text{ex}(\mathcal{C}_4, n_1) + \text{ex}(\mathcal{C}_4, n_2) + O(|E(G)|n),$$

which yields by an easy induction argument that

$$|E(G)| = \text{ex}(\mathcal{C}_4, n) = O(n \log^2 n). \blacksquare$$

Using the theory of generalized Davenport–Schinzel sequences [45], P. Valtr proved the following.

Theorem 4.3 ([79]) *Let G be a geometric graph with n vertices, all of whose edges can be intersected by a line. If G has no k pairwise crossing edges, then its number of edges is at most $c_k n$, where c_k is a suitable constant depending only on k.*

This result immediately implies that, for every $k \geq 4$, the exponent $2k - 6$ of the logarithmic factor in Corollary 4.2 can be reduced to 1.

Corollary 4.4 ([79]) *For every $k \geq 4$, we have*
$$\mathrm{ex}\,(\mathcal{C}_k, n) = O(n \log n).$$

Proof Let G be a geometric graph with n vertices, with no k pairwise crossing edges, and suppose that $|E(G)| = \mathrm{ex}\,(\mathcal{C}_k, n)$. Take a vertical line l which does not pass through any of the vertices and which partitions the vertex set of G into two roughly equal parts, V_1 and V_2, with $|V_1| = \lfloor n/2 \rfloor$ and $|V_2| = \lceil n/2 \rceil$. Let G_1 and G_2 denote the geometric subgraphs of G induced by G_1 and G_2, respectively, and let G' be the geometric graph formed by the edges of G intersecting l. Clearly, the number of edges of G is equal to

$$\mathrm{ex}\,(\mathcal{C}_k, n) = |E(G_1)| + |E(G_2)| + |E(G')| \leq \mathrm{ex}\,(\mathcal{C}_k, \lfloor n/2 \rfloor) + \mathrm{ex}\,(\mathcal{C}_k, \lceil n/2 \rceil) + c_k n,$$

where we used Theorem 4.3 to estimate $|E(G')|$. The result follows. ∎

It is conjectured that, for every fixed k, $\mathrm{ex}\,(\mathcal{C}_k, n)$ is linear in n. One can also make the following stronger conjecture.

Conjecture 4.5 *For any $k \geq 3$, there exists an integer C_k with the property that the edges of every geometric graph which has no k pairwise crossing edges can be coloured by C_k colours so that no two edges of the same colour cross each other.*

A somewhat weaker conjecture which would also imply the linearity of $\mathrm{ex}\,(\mathcal{C}_k, n)$ is the following. For any $k \geq 3$, there exists a positive constant γ_k such that every geometric graph without k pairwise crossing edges has a crossing-free geometric subgraph which has at least $\gamma_k |E(G)|$ edges.

Unfortunately, we do not even know if Conjecture 4.5 is true for $k = 3$. S. McGuinness [54] settled the case $k = 3$ for every geometric graph whose edges can be intersected by a straight line. Conjecture 4.5 is also known to be true for convex geometric graphs. The following theorem of A. Kostochka and J. Kratochvíl slightly improves an old bound of A. Gyárfás [31, 32] (see also [33]) and Kostochka [46].

Theorem 4.6 ([47]) *Let G be a convex geometric graph which has no k pairwise crossing edges. Then one can colour the edges of G with 2^{k+5} colours in such a way that no two edges of the same colour cross each other.*

Kostochka [46] constructed a series of convex geometric graphs with no k pairwise crossing edges, which cannot be coloured with fewer than a constant times $k \log k$ colours.

Theorem 4.6 plays an important role in the proof of Theorem 4.1. It also implies that $\text{ex}_c(\mathcal{C}_k, n) \leq 2^{k+5}(2n-3) = O(n)$ for every fixed k. However, V. Capoyleas and Pach determined the exact value of $\text{ex}_c(\mathcal{C}_k, n)$, which is one of the few non-trivial precise results in the area.

Theorem 4.7 ([13]) *Let $n \geq 2k$. The maximum number of edges that a convex geometric graph of n vertices can have without containing k pairwise crossing edges satisfies*

$$\text{ex}_c(\mathcal{C}_k, n) = 2(k-1)n - \binom{2k-1}{2}.$$

All of the results presented in this section can be rephrased in the following way. If a geometric graph G with n vertices has sufficiently many edges, then it must have a k-tuple of pairwise crossing edges. In fact, if $|E(G)|$ is very large, there are many such k-tuples. In the previous section we described this phenomenon for $k = 2$ by a fairly tight quantitative formula. The proof of Theorem 3.4 may serve as a prototype of how to obtain many similar results.

Theorem 4.8 ([55]) *For any $k \geq 2$ there exists a constant $c(k) > 0$ such that every convex geometric graph with n vertices and $e \geq (2k-1)n$ edges has at least $c(k)e^{2k-1}/n^{2k-2}$ k-tuples of pairwise crossing edges.*

For $k = 2$ and 3, the last statement can be extended to all (not necessarily convex) geometric graphs. The same is true for every fixed k such that $\text{ex}(\mathcal{C}_k, n) = O(n)$. By Theorem 4.8, if G has constant times n^2 edges, then the total number of k-tuples of edges in G is at least $c'n^{2k}$, for some $c' > 0$. A positive per cent of them are pairwise crossing.

5 Forbidden geometric subgraphs—Non-crossing configurations

In the previous section we discussed how many edges guarantee the existence of k pairwise crossing edges in a geometric graph. Here we address the 'dual' question where, in place of pairwise crossing edges, we look for pairwise *disjoint* edges. Recall that two disjoint edges cannot share even an endpoint.

For any $k \geq 2$, let $\mathcal{D}_k$ denote the class of all geometric graphs consisting of k pairwise *disjoint* edges.

The following observation of Kupitz settles the question for convex geometric graphs.

Theorem 5.1 ([49]) *Let* $\text{ex}_c(\mathcal{D}_k, n)$ *denote the maximum number of edges that a convex geometric graph of n vertices can have without containing k pairwise disjoint edges. Then for every k and $n \geq 2k$, we have*

$$\text{ex}_c(\mathcal{D}_k, n) = (k-1)n.$$

Proof Let G be a convex geometric graph with n vertices. Assume without loss of generality that $V(G)$ is the vertex set of a regular n-gon. Partition $E(G)$ into n classes so that two segments belong to the same class if and only if they are parallel. If G has no k pairwise crossing edges then each class contains at most $k - 1$ elements. Hence, $|E(G)| \leq (k-1)n$, and this bound can be attained. ∎

The case of non-convex geometric graphs is more difficult. Hopf and Pannwitz [36] and (independently) Erdős proved a long time ago the following.

Theorem 5.2 ([36]) *Let* $\text{ex}(\mathcal{D}_2, n)$ *denote the maximum number of edges that a geometric graph with n vertices can have without containing two disjoint edges. Then, for every $n \geq 3$,*

$$\text{ex}(\mathcal{D}_2, n) = n.$$

Proof (M. Perles) We say that an edge xy is to the *left of* xz if the ray xz can be obtained from the ray xy by a clockwise turn of less than π. A vertex x is called *pointed* if all edges incident to it lie in a half-plane whose boundary contains x.

For every pointed vertex of G, delete the leftmost edge incident with it. If G had no two disjoint edges then the resulting graph is empty. Indeed, if there were at least one remaining edge, xz, then we could find an edge xy to the left of xz and another edge zv to the left of $zx = xz$, and these two edges would be disjoint. Since we deleted at most one edge for each vertex, $|E(G)| \leq |V(G)|$, and this bound can be attained. ∎

Alon and Erdős [5] were the first to show the linearity of $\text{ex}(\mathcal{D}_3, n)$. The best known lower and upper bounds,

$$2.5n - 4 \leq \text{ex}(\mathcal{D}_3, n) \leq 3n$$

are due to Perles (unpublished) and Goddard, Katchalski and Kleitman [28], respectively. Recently, Tóth and Valtr [76] showed that

$$4n - 9 \leq \text{ex}(\mathcal{D}_4, n) \leq 8.5n.$$

Pach and Törőcsik proved that $\text{ex}(\mathcal{D}_k, n) = O(n)$, for every fixed k.

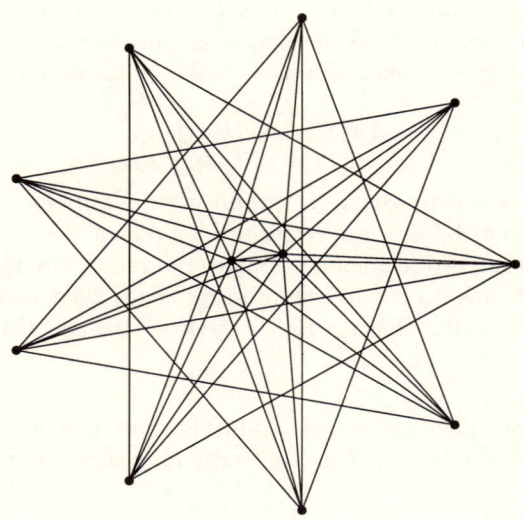

Figure 1: A geometric graph with no 4 pairwise disjoint edges.

Theorem 5.3 ([64]) *Let* $\mathrm{ex}\,(\mathcal{D}_k, n)$ *denote the maximum number of edges that a geometric graph of n vertices can have without containing k pairwise disjoint edges. Then, for every k and n, we have*

$$\mathrm{ex}\,(\mathcal{D}_k, n) \leq (k-1)^4 n.$$

Proof Let G be a geometric graph with n vertices, containing no k pairwise disjoint edges. For any vertex v, let $x(v)$ and $y(v)$ denote the x-coordinate and the y-coordinate of v, respectively. Assume without loss of generality that no two vertices have the same x-coordinate.

An edge $e \in E(G)$ is said to lie *below* another edge $e' \in E(G)$ if there is no vertical line which intersects both e and e', and whose intersection with e' is strictly below its intersection with e. (According to this definition, if the projections of e and e' onto the x-axis do not overlap, then we say that e lies below e' and also e' lies below e.)

Define four binary relations $\prec_1$, $\prec_2$, $\prec_3$, and $\prec_4$ on $E(G)$, as follows. One necessary condition for two edges to be in any of these relations is that they must be disjoint. Let $e = v_1 v_2$ and $e' = v_1' v_2'$ be two disjoint edges of G with $x(v_1) < x(v_2)$ and $x(v_1') < x(v_2')$. Then

- $e \prec_1 e'$ if $x(v_1) < x(v_1')$, $x(v_2) < x(v_2')$ and e lies below e';

- $e \prec_2 e'$ if $x(v_1) > x(v_1')$, $x(v_2) > x(v_2')$ and e lies below e';

- $e \prec_3 e'$ if $x(v_1) < x(v_1')$, $x(v_2) > x(v_2')$ and e lies below e';

Geometric Graph Theory

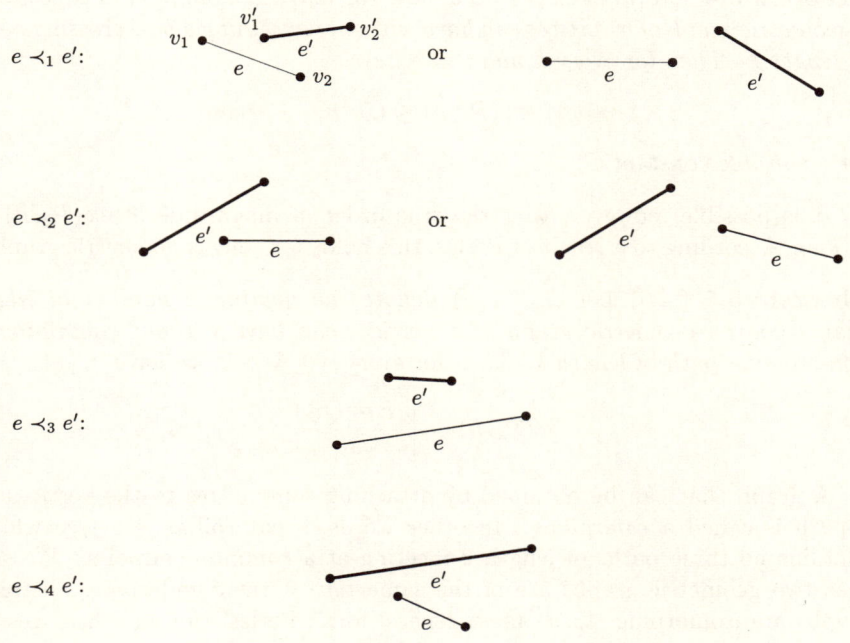

Figure 2: The relations $\prec_i$.

- $e \prec_4 e'$ if $x(v_1) > x(v'_1)$, $x(v_2) < x(v'_2)$ and e lies below e'.

Obviously, each of the relations $\prec_i$ defines a partial order on $E(G)$, and any pair of disjoint edges is comparable by at least one of them. Since G has no k pairwise disjoint edges, there is no chain of length k in the partially ordered set $(E(G), \prec_i)$ $(1 \leq i \leq 4)$. This implies that, for each i, $E(G)$ can be partitioned into at most $k-1$ classes (*antichains*) such that no two edges belonging to the same class are comparable by $\prec_i$. Overlaying these four partitions, we obtain a decomposition of $E(G)$ into at most $(k-1)^4$ classes E_j $(1 \leq j \leq (k-1)^4)$ such that no two elements of E_j are comparable by any of the relations $\prec_i$. Thus, none of the graphs $G_j := (V(G), E_j)$ has two disjoint edges. In view of Theorem 5.2, $|E(G_j)| = |E_j| \leq n$ $(1 \leq j \leq (k-1)^4)$, and the result follows. ∎

Recently, Tóth improved the above result by showing that the dependence of $\text{ex}(\mathcal{D}_k, n)$ on k is at most quadratic. Indeed, he proved a somewhat stronger result.

The *length* of a path is the number of its edges. Let $\mathcal{P}_k$ denote the class of all non-crossing paths of length k. Clearly, every non-crossing path of length $2k-1$ has k pairwise disjoint edges.

Theorem 5.4 ([75]) *Let* $\text{ex}(\mathcal{P}_k, n)$ *denote the maximum number of edges that a geometric graph of n vertices can have without containing a non-crossing path of length k. Then for every k and n, we have*

$$\text{ex}(\mathcal{P}_k, n) \leq Ck^2 n,$$

for a suitable constant C.

It is possible, however, that the true order of magnitude of $\text{ex}(\mathcal{P}_k, n)$ is $O(kn)$. According to a result of Perles, this holds for convex geometric graphs.

Theorem 5.5 ([44]) *Let* $\text{ex}_c(\mathcal{P}_k, n)$ *denote the maximum number of edges that a convex geometric graph of n vertices can have without containing a non-crossing path of length k. Then for every $n \geq k+1$, we have*

$$\text{ex}_c(\mathcal{P}_k, n) = \left\lfloor \frac{(k-1)n}{2} \right\rfloor.$$

A graph that can be obtained by attaching some edges to the vertices of a path is called a *caterpillar*. In other words, a caterpillar is a tree which contains no three paths of length 2 meeting at a common endpoint. We say that two geometric graphs are of the same *type* if their underlying abstract graphs are isomorphic. In its most general form, Perles' (unpublished) result is the following.

Theorem 5.6 *Let $\mathcal{CP}_k$ denote the class of all non-crossing caterpillars of a fixed type, consisting of k edges. Then, for every $n \geq k+1$, we have*

$$\text{ex}_c(\mathcal{CP}_k, n) = \left\lfloor \frac{(k-1)n}{2} \right\rfloor.$$

We mention a related result for (not necessarily convex) geometric graphs, which can easily be deduced from Theorem 2.2 and from Turán's theorem quoted at the beginning of this section.

Corollary 5.7 *Let $\mathcal{OP}_k$ denote the class of all non-crossing outerplanar graphs of a fixed type, which can be obtained from a cycle of length k by adding some (or no) diagonals. Then, for every $n \geq k+1$, we have*

$$\text{ex}(\mathcal{OP}_k, n) = \text{ex}_c(\mathcal{OP}_k, n) = \frac{k-2}{2(k-1)}(n^2 - r^2) + \binom{r}{2},$$

where r is the remainder of n upon division by $k-1$.

As we pointed out at the end of the preceding section, whenever $\text{ex}(\mathcal{H}, n) = O(n)$ holds for a class of forbidden geometric subgraphs $\mathcal{H}$, we can use the idea of the proof of Theorem 3.4 to show that, if $|E(G)| = e \gg \text{ex}(\mathcal{H}, n)$, then G has many subgraphs belonging to $\mathcal{H}$. In particular, Theorem 5.3 can be used to establish the following.

Geometric Graph Theory

Theorem 5.8 ([55]) *For any $k \geq 2$ there exists a constant $c(k) > 0$ such that every geometric graph with n vertices and $e \geq k^4 n$ edges has at least $c(k)e^{2k-1}/n^{2k-2}$ k-tuples of pairwise disjoint edges.*

Theorem 5.4 implies that the last statement also holds with the weaker assumption $e \geq Ck^2 n$.

6 Ramsey-type results

In classical Ramsey-theory, one wants to find large monochromatic subgraphs in a complete graph whose edges are coloured with several colours [12,29]. Most questions of this type can be generalized to complete geometric graphs, where the monochromatic subgraphs are required to satisfy certain geometric conditions.

The next two statements were conjectured by Bialostocki and Dierker and proved by Károlyi, Pach and Tóth.

Theorem 6.1 ([42], [43]) *If the edges of a finite complete geometric graph are coloured by two colours, there exists a non-crossing spanning tree, all of whose edges are of the same colour.*

Proof Let $V = \{v_1, v_2, \ldots, v_n\}$ denote the vertex set of a complete geometric graph K_n whose edges are coloured with red and blue. Suppose without loss of generality that no two vertices have the same x-coordinate and that the vertices are listed in increasing order of their x-coordinates. The assertion is trivial for $n \leq 2$. Thus, we can assume that $n \geq 3$ and the theorem has already been proved for all complete geometric graphs having fewer than n vertices. We can also assume that all edges along the boundary of the convex hull of V are of the same colour (say, red). Indeed, if two consecutive edges of the convex hull have different colours, then remove their common endpoint from K_n. By the induction hypothesis, the remaining graph has a monochromatic non-crossing spanning tree. Depending on its colour, this spanning tree can be completed to a monochromatic non-crossing spanning tree of K_n, by putting back one of the two previously deleted edges of the convex hull of V.

For every i, $1 < i < n$, let K_i^l and K_i^r denote the subgraphs of K_n induced by the points $\{v_1, \ldots, v_i\}$ and $\{v_i, \ldots, v_n\}$, respectively. By the induction hypothesis, both K_i^l and K_i^r have a monochromatic non-crossing spanning tree, T_i^l respectively T_i^r. We can assume that these two trees have different colours; otherwise their union will meet the requirements of the theorem. We can also assume that T_2^l is red and T_2^r is blue. Otherwise, T_2^r would be red, and it could be completed to a non-crossing red spanning tree of K_n by the addition of any edge of the convex hull of V incident to v_1. Similarly, we can suppose that T_{n-1}^l is blue and T_{n-1}^r is red. Hence, there exists an i, $1 < i < n-1$, such that

- T_i^l is red and T_i^r is blue,

- T_{i+1}^l is blue and T_{i+1}^r is red.

Connecting T_i^l and T_{i+1}^r by any edge of the convex hull of V which intersects a vertical line separating v_i and v_{i+1}, we obtain a non-crossing red spanning tree of K_n, as required. ∎

Theorem 6.2 ([42], [43]) *If the edges of a complete geometric graph with $3n - 1$ vertices are coloured by two colours, there exist n pairwise disjoint edges of the same colour. This result is best possible.*

The analogues of Theorems 6.1 and 6.2 for *abstract* graphs, i.e. when the geometric constraints are ignored, were found by Erdős and Rado (see [22]) and Gerencsér and Gyárfás [27], respectively. In fact, Gerencsér and Gyárfás proved the stronger result that, for any 2-colouring of the edges of a complete graph with $3n-1$ vertices, there exists a monochromatic path of length $2n-1$. This statement, as well as Theorem 6.2, is best possible, as is shown by the following example. Take the disjoint union of a complete graph of $n-1$ vertices and a complete graph of $2n - 1$ vertices, all of whose edges are red and blue, respectively, and colour all edges between the two parts red.

Theorem 6.2 also has an "asymmetric" version.

Theorem 6.3 ([42], [43]) *Let k and l be positive integers, $n = k + l + \max\{k, l\} - 1$. If the edges of a complete geometric graph with n vertices are coloured by red and blue, one can find either k disjoint red edges or l disjoint blue edges. This result cannot be improved.*

To formulate some further results of this type, it is convenient to introduce some notation.

Let $\mathcal{G}_1$ and $\mathcal{G}_2$ be not necessarily different classes of geometric graphs. Let $R(\mathcal{G}_1, \mathcal{G}_2)$ denote the smallest positive number R with the property that any complete geometric graph of R vertices, whose edges are coloured with red and blue, contains a geometric subgraph which is either totally red and belongs to $\mathcal{G}_1$ or totally blue and belongs to $\mathcal{G}_2$. If $\mathcal{G}_1 = \mathcal{G}_2 = \mathcal{G}$, we write $R(\mathcal{G})$ instead of $R(\mathcal{G}_1, \mathcal{G}_2)$. If we restrict our attention to *convex* geometric graphs, then the corresponding function is denoted by $R_c(\mathcal{G})$. Clearly, $R_c(\mathcal{G}) \leq R(\mathcal{G})$.

There is an obvious relationship between $R(\mathcal{G})$ and the function $\mathrm{ex}(\mathcal{G}, n)$ discussed in the preceding sections:

$$\mathrm{ex}(\mathcal{G}, n) < \frac{1}{2}\binom{n}{2} \implies R(\mathcal{G}) \leq n.$$

Indeed, if we colour the edges of a complete geometric graph with two colours, one of the colour classes will contain at least $\frac{1}{2}\binom{n}{2}$ elements. If $\mathrm{ex}(\mathcal{G}, n) < \frac{1}{2}\binom{n}{2}$, then there is a geometric graph belonging to $\mathcal{G}$, all of whose edges are in this colour class. Similarly, we have

$$\text{ex}_c(\mathcal{G}, n) < \frac{1}{2}\binom{n}{2} \implies R_c(\mathcal{G}) \leq n. \tag{2}$$

In the special case when $\mathcal{G}$ is the class $\mathcal{T}_n$ of all non-crossing geometric trees of n vertices and the class $\mathcal{D}_n$ of all geometric graphs consisting of n disjoint edges, respectively, the values of $R(\mathcal{G})$ are given in Theorems 6.1 and 6.2. Theorem 6.3 provides the exact value of $R(\mathcal{G}_1, \mathcal{G}_2)$, when $\mathcal{G}_1 = \mathcal{D}_k$ and $\mathcal{G}_2 = \mathcal{D}_l$.

$$R(\mathcal{T}_n) = R_c(\mathcal{T}_n) = n, \quad R(\mathcal{D}_n) = R_c(\mathcal{D}_n) = 3n - 1,$$

$$R(\mathcal{D}_k, \mathcal{D}_l) = R_c(\mathcal{D}_k, \mathcal{D}_l) = k + l + \max\{k, l\} - 1.$$

Next we show

Theorem 6.4 ([44]) *Let $\mathcal{CC}_n$ and $\mathcal{CC}'_n$ denote, respectively, the class of all non-crossing cycles of length n and the class of all non-crossing cycles of length n triangulated from a vertex. Then we have*

$$(n-1)^2 < R_c(\mathcal{CC}_n) \leq R(\mathcal{CC}'_n) \leq 2(n-1)(n-2) + 2.$$

Proof First we prove that $R_c(\mathcal{CC}_n) > (n-1)^2$. Take $(n-1)^2$ points on a circle and partition them into $n-1$ groups, each containing $n-1$ consecutive points. Colour red all edges between points in different groups, and colour blue all edges between points belonging to the same group.

Any non-crossing red cycle contains at most one point from each group; hence it cannot have more than $n-1$ points. On the other hand, all vertices of a blue cycle are from the same group, so there is no blue cycle with more than $n-1$ points.

Next we show that $R(\mathcal{CC}'_n) \leq 2(n-1)(n-2) + 2$. Let V denote the vertex set of a complete geometric graph K of $2(n-1)(n-2)+2$ vertices, whose edges are coloured with red and blue. Let v be a vertex of the convex hull of V. Since there are $2(n-1)(n-2)+1$ edges incident to v, at least $(n-1)(n-2)+1$ of them are of the same colour, say, red. Let $v_1, v_2, \ldots, v_{(n-1)(n-2)+1}$ be the vertices of K, listed in clockwise order of visibility from v, such that each edge vv_i is red. A path $v_{i_1} v_{i_2} \ldots v_{i_j}$ is said to be *monotone* if $i_1 < i_2 < \ldots < i_j$.

Define a partial ordering of the vertices $v_1, v_2, \ldots, v_{(n-1)(n-2)+1}$, as follows. Let $v_i < v_j$ if $i < j$ and there is a monotone red path connecting v_i to v_j. As in the proof of Theorem 5.3, there are either $n-1$ elements that form a chain (totally ordered subset), or n elements that form an antichain (i.e. they are pairwise incomparable). In the former case, there is a monotone red path $u_1, u_2, \ldots, u_{n-1}$, and we can complete it to a non-crossing red cycle $v, u_1, u_2, \ldots, u_{n-1}$ of size n, together with the corresponding diagonals from v.

In the latter case, there is a complete blue subgraph of n vertices, $u_1, u_2, \ldots, u_n$, because any two incomparable elements are connected by a blue edge. By Theorem 2.2, this complete blue subgraph contains a non-crossing cycle of length n, triangulated from a vertex. ∎

Corollary 6.5 ([44]) *For any 2-colouring of the edges of a complete geometric graph with n vertices, there exist monochromatic non-crossing cycles of length $3, 4, \ldots, \lfloor\sqrt{n/2}\rfloor$, having the same colour. The order of magnitude of this bound cannot be improved.*

Let $\mathcal{P}_n$ denote the class of all non-crossing paths of length n. It is an immediate consequence of the above results that

$$R(\mathcal{P}_n) \leq R(\mathcal{CC}_{n+1}) \leq 2n(n-1) + 2.$$

However, we have a better upper bound on $R(\mathcal{P}_n)$.

Theorem 6.6 ([44]) *For any 2-colouring of the edges of a complete geometric graph with n vertices, there exists a monochromatic non-crossing path whose length is at least a constant times $n^{2/3}$. Equivalently,*

$$R(\mathcal{P}_n) = O(n^{3/2}).$$

It is very likely that $R(\mathcal{P}_n)$ is linear in n. For convex geometric graphs we can verify this conjecture.

Theorem 6.7 ([44]) *For any 2-colouring of the edges of a convex complete geometric graph with $n \geq 3$ vertices, there exists a monochromatic non-crossing path of length $\lfloor\frac{n+1}{2}\rfloor$. This bound cannot be improved. Consequently, for every n, we have*

$$R_c(\mathcal{P}_n) = 2n - 1. \tag{3}$$

Proof We prove only the easy part of (3): if n is even then $R_c(\mathcal{P}_n) \leq 2n - 1$. In view of (2) and Theorem 5.6, it is enough to verify that

$$\left\lfloor \frac{(n-1)(2n-1)}{2} \right\rfloor < \frac{1}{2}\binom{2n-1}{2},$$

which is trivially true. ∎

Some other Ramsey numbers for convex geometric graphs were studied by H. Harborth and H. Lefmann [34].

For any positive integer n, let $n\mathcal{G}$ denote the class of all geometric graphs that can be obtained by taking the union of n pairwise disjoint members of $\mathcal{G}$, any two of which can be separated by a straight line.

Theorem 6.8 ([44]) *Let $\mathcal{G}$ be any class of geometric graphs, each of which has at least two vertices.*

1. If n is a power of 2 then
$$R(n\mathcal{G}) \leq (R(\mathcal{G})+1)n - 1.$$

2. For any $n > 0$,
$$R(n\mathcal{G}) \leq \left\lceil \frac{3(R(\mathcal{G})+1)}{2} \right\rceil n - \left\lceil \frac{R(\mathcal{G})+1}{2} \right\rceil.$$

3. For any $n > 0$,
$$R_c(n\mathcal{G}) \leq (R_c(\mathcal{G})+1)n - 1.$$

In particular, if $\mathcal{G} = \mathcal{D}_1$ is the class of all geometric graphs consisting of a single edge, then $R(\mathcal{D}_1) = 2$, and part 1 of Theorem 6.8 implies that
$$R(n\mathcal{D}_1) = R(\mathcal{D}_n) \leq 3n-1,$$
provided that n is a power of 2. Theorem 6.2 shows that this inequality holds for every n, and it is best possible.

Suppose next that $\mathcal{G} = \Delta (= \mathcal{CC}_3)$ is the class of all triangles. Then we have $R(\Delta) = 6$, so, by part 1 of Theorem 6.8, we now obtain
$$R(n\Delta) \leq 7n-1,$$
provided that n is a power of 2. The following theorem shows that this result cannot be improved.

Theorem 6.9 ([44]) *Let Δ denote the class of triangles and let n be a positive integer. Then*
$$R(n\Delta) \geq (R(\Delta)+1)n - 1 = 7n-1.$$

7 Applications

This section outlines a few fairly direct applications of the results presented before. L. Székely [72] discovered that the following celebrated result of Szemerédi and Trotter readily follows from Theorem 3.4.

Corollary 7.1 ([73], [74]) *The maximum number of incidences between m distinct lines and n distinct points in the plane is at most*
$$O\left(m^{2/3}n^{2/3} + m + n\right).$$

This bound is tight up to a constant factor.

Proof Assume without loss of generality that every line is incident to at least one point. Let I denote the number of incidences between our points and lines. Define a geometric graph G whose vertex set is the given set of n points, and whose two vertices are connected by a segment if they are consecutive vertices along one of the lines. Clearly, $|E(G)| = I - m$.

On the other hand, by Theorem 3.4, either $|E(G)| < 4n$ or

$$\text{CR}(G) \geq \frac{1}{33.75} \frac{(I-m)^3}{n^2}.$$

The result follows by comparing this bound with the trivial relation $\text{CR}(G) \leq \binom{m}{2}$. ∎

The same idea was explored by T. Dey [16] to achieve a breakthrough in the so-called k-set problem mentioned in the abstract, which plays an important role in the analysis of many algorithms in computational geometry.

Given a set P of n points in the plane in general position (no 3 on a line), a k-element subset $P' \subseteq P$ is said to be a *k-set*, if there is an open half-plane H such that $H \cap P = P'$. It was shown by Erdős, Lovász, Simmons and Straus [23, 53] that the number of k-sets of an n-element set is $O(n^{3/2})$, and this was slightly improved by Pach, Steiger and Szemerédi [62]. The order of magnitude of the best known lower bound is $n \log n$.

Theorem 7.2 ([16]) *For every $0 < k < n$, the number of k-sets of an n-element point set in general position in the plane is at most $6.5(k+1)^{1/3}n$.*

Proof Let P be a fixed set of n points in the plane in general position. Let H_k be a geometric graph defined on the vertex set P such that two points $u, v \in P$ are connected by a segment if and only if there are precisely $k - 1$ points of P strictly on one side of the line uv. It is easy to see that the number of edges of H_k is equal to the number of k-sets of P. Orient every edge of H_k from left to right, and assume without loss of generality that at least $|E(H_k)|/2$ edges have precisely $k - 1$ points on their *left-hand* side. Let G_k denote the graph formed by these edges.

Let the *successor* of an edge $\vec{uv} \in E(G_k)$ be the edge $\vec{vw}$ with minimum slope among all edges $\vec{vw'}$ whose slopes exceed that of $\vec{uv}$. This relation arranges the edges of G_k into pairwise disjoint convex chains. It is not hard to verify that there is precisely one chain starting (ending) at each of the k leftmost (rightmost) points of P, and no chain starts (ends) anywhere else.

Assign to each crossing $p \notin V(G)$ of two chains, γ_1 and γ_2, the uniquely determined segment $u_p v_p$ below p, whose supporting line is a common tangent to γ_1 and γ_2. (We say that a segment is *below* p if it intersects the half-line starting at p and pointing to the direction of the negative y-axis.) Clearly, $u_p, v_p \in P$, and no segment is assigned to two different crossings. The number of tangent segments assigned to crossings is at most kn, because each vertex

occurs at most once as the left endpoint of a tangent segment, whose other endpoint lies on a chain not passing through p. Thus, the total number of crossings,
$$\mathrm{CR}(G_k) \leq nk.$$
On the other hand, by Theorem 3.4, either $|E(G_k)| < 4n$ or
$$\mathrm{CR}(G_k) \geq \frac{1}{33.75} \frac{|E(G_k)|^3}{n^2}.$$
Comparing these two bounds, we obtain
$$|E(H_k)| \leq 2|E(G_k)| \leq 6.5(k+1)^{1/3}n. \quad \blacksquare$$

The proofs of Theorems 3.4 and 3.6 easily generalize to arbitrary graph drawings, where the arcs representing the edges are not necessarily straight-line segments. Using these generalizations, the next two theorems, due to Clarkson, Edelsbrunner, Guibas, Sharir and Welzl [15] and Spencer, Szemerédi and Trotter [70], can be established by minor modifications of the proof of Corollary 7.1.

Corollary 7.3 ([15], [18]) *The total number of sides of m distinct cells in an arrangement of n lines in the plane is at most $O(m^{2/3}n^{2/3} + m)$.*

Corollary 7.4 ([70]) *The number of unit distances determined by n points in the plane is $O(n^{4/3})$.*

A family Γ of curves in the plane is said to have d *degrees of freedom* if there exists an integer s such that

- no two curves in Γ have more than s points in common, and

- for any d points, there are at most s curves in Γ passing through all of them.

Pach and Sharir applied the above mentioned (straightforward) extension of Theorem 3.4 to obtain the following common generalization of Corollaries 7.1 and 7.4.

Theorem 7.5 ([60], [59]) *Let Γ be a family of curves in the plane with d degrees of freedom. Then the maximum number of incidences between n points in the plane and m elements of Γ is*
$$O(n^{d/(2d-1)}m^{(2d-2)/(2d-1)} + n + m).$$

Using the same technique, Székely slightly improved a well-known result of Chung, Szemerédi and Trotter [14].

Theorem 7.6 ([72]) *Any set of n points in the plane has an element, from which there are at least $cn^{4/5}$ distinct distances to the remaining points. (Here $c > 0$ is an absolute constant.)*

Next we present some easy consequences of Theorem 3.6 and its extension to arbitrary graph drawings. The proof of Corollary 4.2 was also based on this result.

Corollary 7.7 *The number of crossings in an arbitrary drawing of a random graph with n vertices and $e \geq 4n$ edges is at least Ce^2, where $C > 0$ is a suitable constant. This bound is evidently tight, apart from the constant.*

Proof It is sufficient to verify that the bisection width of such a random graph is almost surely at least constant times e. ∎

A drawing of a graph is called *polygonal* if every edge is represented by a non-crossing polygonal path. In Section 2, we saw that every outerplanar graph of n vertices admits a straight-line drawing without crossing such that its vertex set is mapped into an arbitrarily prespecified n-element set (Theorem 2.2). If we also want to specify the location of each vertex, then usually we cannot insist on straight-line edges. Of course, even in this case we can always find a crossing-free *polygonal* drawing. We want to minimize the total length (i.e. the total number of segments) of the polygonal paths representing the edges.

Corollary 7.8 ([66]) *For every n, there exist an abstract outerplanar graph G_n of n vertices and an assignment of locations (distinct points) for the vertices such that, in any planar drawing of G_n with the property that each vertex is mapped into the point assigned to it and each edge is represented by a polygonal path, there are at least $n/100$ edges consisting of at least $n/100$ segments.*

Proof We show only the weaker assertion that there exist a G_n and an assignment of locations such that the total number of segments forming the edges of G_n in any polygonal drawing of G_n with the given point locations, is at least constant times n^2.

Let G_n be a matching of size $\lfloor n/2 \rfloor$, and let the set of locations be the vertex set P of a regular n-gon. Randomly assign to each vertex of G_n a distinct element of P. Assume that with this assignment the matching can be realized with pairwise disjoint polygonal paths of total length L. Add to this drawing the edges of the convex hull of P. In this way, we produce a polygonal drawing of a graph G'_n obtained from a cycle by adding a random set of $\lfloor n/2 \rfloor$ vertex-disjoint diagonals. As every edge of a polygonal path crosses at most two convex hull edges, the number of crossings in this drawing is at most $2L$. On the other hand, it is easy to see that, almost surely, the bisection width of G'_n is at least Cn, for a suitable positive constant C. By the obvious

Geometric Graph Theory

generalization of Theorem 3.6 to polygonal drawings, we have that the number of crossings is at least

$$\frac{1}{40}C^2n^2 - \frac{1}{16}\sum_{i=1}^{n} 3^2 > \frac{1}{50}C^2n^2,$$

provided that n is large enough. Thus, with probability tending to 1 as $n \to \infty$, we have $L > nC^2/100$. ∎

Pach and Wenger [66] also designed a drawing algorithm for arbitrary planar graphs, with the property that every edge is represented by a polygonal path of length $O(n)$.

The proof of the following result is almost identical to the last argument.

Corollary 7.9 ([58]) *There exist two point sets $\{p_1, \ldots, p_n\}, \{q_1, \ldots, q_n\}$ in the plane with the property that any piecewise linear homeomorphism from the plane to the itself, for which $f(p_i) = q_i$ $(i = 1, \ldots, n)$, consists of at least constant times n^2 linear pieces.*

D. Souvaine and R. Wenger [69] showed that there always exists a homeomorphism meeting the requirements of the last corollary, which consists of $O(n^2)$ linear pieces.

Finally, we mention an old application of Theorem 5.2.

Corollary 7.10 ([36]) *The maximum number of point pairs realizing the diameter of a n-element set in the plane is equal to n.*

Proof It is enough to notice that connecting two elements of a set P by a segment if their distance is equal to the diameter of P, we obtain a geometric graph with no two disjoint edges. The result follows from Theorem 5.2. ∎

In view of this statement, there is always a point whose degree in the graph of diameters is at most 2. Thus, the chromatic number of this graph is at most 3, and the diameter of each colour class is smaller than 1. This is a very special case of Borsuk's famous conjecture [11], which turned out to be false in high dimensions [38].

Corollary 7.11 *Every finite point set in the plane can be partitioned into three subsets of smaller diameter.*

Z. Füredi [26] applied a Turán-type result with a class of forbidden 5-point geometric subgraphs to prove that the maximum number of times that the same distance can occur among n vertices of a convex polygon is $O(n \log n)$. The best known lower bound, $2n - 7$, is due to Edelsbrunner and Hajnal [19]. Erdős and L. Moser conjectured that the true order of magnitude of this function is $\Theta(n)$.

8 Geometric hypergraphs

It seems plausible that to extend the incidence results to higher dimensions, to improve the upper bound for the number of times the unit distance can occur among n points in 3-space, or to make progress concerning the higher dimensional analogue of the k-set problem, one has to find the right generalizations of the results discussed in the preceding sections to systems of surfaces or surface patches in d-space. For simplicity, we discuss only the case when these surface patches are flat (simplices).

Definition A *d-dimensional geometric r-hypergraph* H_r^d is a pair (V, E), where V is a set of points in general position in $\mathbb{R}^d$, and E is a set of closed $(r-1)$-dimensional simplices induced by some r-tuples of V. The sets V and E are called the *vertex set* and *edge set* of H_r^d, respectively.

Clearly, a *geometric graph* is a 2-dimensional geometric 2-hypergraph.

Given any class $\mathcal{F}$ of *forbidden* geometric hypergraphs, let $\text{ex}_r^d(\mathcal{F}, n)$ denote the maximum number of edges that a d-dimensional geometric r-hypergraph H_r^d of n vertices can have without containing a geometric subhypergraph which belongs to $\mathcal{F}$. Most bounds in this section are asymptotic: d and r are thought to be fixed, while n tends to infinity.

Akiyama and Alon proved the following theorem.

Theorem 8.1 ([4]) *Let $V = V_1 \cup \ldots \cup V_d$ ($|V_1| = \ldots = |V_d| = n$) be a dn-element set in general position in $\mathbb{R}^d$, and let E consist of all $(d-1)$-dimensional simplices having exactly one vertex in each V_i. Then E contains n disjoint simplices.*

According to a well-known result of Erdős [21], any abstract d-uniform hypergraph with n vertices and at least $n^{d-(1/k)^{d-1}}$ hyperedges contains a complete d-partite subhypergraph with k elements in each of its classes. Combining this with the previous result, we obtain

Theorem 8.2 ([4]) *Let $\mathcal{D}_k^d$ denote the class of all d-dimensional geometric d-hypergraphs consisting of k pairwise disjoint edges. Then the maximum number of hyperedges that a d-dimensional geometric d-hypergraph of n vertices can have without containing k pairwise disjoint hyperedges, satisfies*

$$\text{ex}\left(\mathcal{D}_k^d, n\right) < n^{d-(1/k)^{d-1}}.$$

However, it is conjectured that the true order of magnitude of $\text{ex}\left(\mathcal{D}_k^d, n\right)$ is $O(n^{d-1})$. Theorems 5.3 and 5.4 show that this is the case when $d = 2$.

If we want to exclude *crossings* rather than disjoint edges, or want to generalize Theorem 3.4 to geometric hypergraphs, we face the following problem. Even if we restrict our attention to systems of triangles induced by

3-dimensional point sets in general position, it is not completely clear how a "crossing" should be defined. If two segments cross, they do not share an endpoint. Should this remain true for triangles? We have to clarify the terminology.

Definition Two simplices are said to have a *non-trivial intersection*, if their relative interiors have a point in common. If, in addition, the two simplices are vertex disjoint, then they are said to *cross*.

More generally, k simplices are said to have a *non-trivial intersection*, if their relative interiors have a point in common. If, in addition, all simplices are vertex disjoint, then they are said to *cross*.

Consider k simplices. The fact that *every pair* of them has a non-trivial intersection does not imply that *all* of them do. To emphasize that this stronger condition is satisfied, we often say that the simplices have a *non-trivial intersection in the strong sense*, or simply that they *strongly intersect*. Similarly, a set of *pairwise* crossing simplices is not necessarily *crossing*. If we want to emphasize that they *all* cross, we say that they *cross in the strong sense*, or, in brief, that they *strongly cross*.

Definition Let $\mathcal{I}_k^r$ (respectively, $\mathcal{SI}_k^r$) denote the class of all geometric hypergraphs consisting of k $(r-1)$-dimensional simplices, any two of which have a non-trivial intersection (respectively, all of which are strongly intersecting). Similarly, let $\mathcal{C}_k^r$ (respectively, $\mathcal{SC}_k^r$) denote the class of all geometric hypergraphs consisting of k pairwise crossing (respectively, strongly crossing) $(r-1)$-simplices.

Theorem 8.3 ([8], [18]) *For any fixed $k > 1$, one can select at most $O(n^{\lceil d/2 \rceil})$ d-dimensional simplices induced by n points in d-space with the property that no k of them share a common interior point. This bound cannot be improved. That is,*

$$\mathrm{ex}_{d+1}^d(\mathcal{I}_k^{d+1}, n) = \Theta(n^{\lceil d/2 \rceil}), \qquad \mathrm{ex}_{d+1}^d(\mathcal{SI}_k^{d+1}, n) = \Theta(n^{\lceil d/2 \rceil}).$$

Proof Clearly, we have

$$\Omega(n^{\lceil d/2 \rceil}) \leq \mathrm{ex}_{d+1}^d(\mathcal{I}_k^{d+1}, n) \leq \mathrm{ex}_{d+1}^d(\mathcal{SI}_k^{d+1}, n),$$

where the first inequality follows from the fact that there are triangulations of size $\Omega(n^{\lceil d/2 \rceil})$ with n vertices in $\mathbb{R}^d$. Consider, for example, the vertical projection of the lower part of any cyclic polytope of n vertices in $\mathbb{R}^{d+1}$.

To see that $\mathrm{ex}_{d+1}^d(\mathcal{SI}_k^{d+1}, n) \leq O(n^{\lceil d/2 \rceil})$, we set up a charging scheme. Let us regard $\mathbb{R}^{d-1}$ as the coordinate hyperplane in $\mathbb{R}^d$ spanned by the first $d-1$ axes, and let X_d denote the last coordinate axis. Suppose that X_d is vertical. Fix a geometric hypergraph $H_{d+1}^d = (V, E)$ which has no k edges with a common interior point and whose n vertices are in general position. For any l-dimensional simplex Δ induced by V, where $l \le \lfloor (d-1)/2 \rfloor$, let $E_\Delta \subseteq E$ denote the set of all edges of H_{d+1}^d that contain Δ on their boundaries. It follows from the condition on H_{d+1}^d that the infinite vertical cylinder $\Delta + X_d$ based on Δ intersects the interior of at most $2(k-1)$ elements of E_Δ. Let us charge Δ one unit for each of these edges. Since the total number of l-simplices with $l \le \lfloor (d-1)/2 \rfloor$ is at most $\lceil d/2 \rceil n^{\lceil d/2 \rceil}$, it remains to show that every edge $e \in E$ has been charged for. Indeed, by Radon's theorem [67], the vertex set of the orthogonal projection of e into $\mathbb{R}^{d-1}$ can be partitioned into two parts, S_1 and S_2, such that their convex hulls cross each other and $|S_1| + |S_2| = d+1$. Suppose without loss of generality that $|S_1| \le \lfloor (d+1)/2 \rfloor$. Then the convex hull of S_1 is an l-dimensional simplex Δ_1 for some $l \le \lfloor (d-1)/2 \rfloor$, and we had to charge Δ_1 for e.

Theorem 8.4 ([18]) *Let E be any set of d-dimensional simplices induced by an n-element point set $V \subseteq \mathbb{R}^d$. If E has no two crossing elements, then $|E| = O(n^d)$, and this bound is asymptotically tight. In notation,*

$$\mathrm{ex}_{d+1}^d(\mathcal{C}_2^{d+1}, n) = \Theta(n^d).$$

The last two theorems were about d-dimensional geometric $d+1$-hypergraphs, i.e. systems of full-dimensional simplices induced by a set of points.

Next we present a few statements on d-dimensional geometric d-hypergraphs. They can be viewed as natural extensions of the corresponding results for geometric graphs (the case $d = 2$) discussed in the preceding sections.

Theorem 8.5 ([18], [79]) *Let E be a family of $(d-1)$-dimensional simplices induced by an n-element point set $V \subseteq \mathbb{R}^d$ such that E has no k members with pairwise non-trivial intersections $(d, k > 1)$. Then, for $k = 2$ and 3, we have $|E| = O(n^{d-1})$. Otherwise, $|E| = O(n^{d-1} \log n)$. In notation,*

$$\mathrm{ex}_d^d(\mathcal{I}_k^d, n) = O(n^{d-1}) \quad \text{if } k = 2, 3;$$

$$\mathrm{ex}_d^d(\mathcal{I}_k^d, n) = O(n^{d-1} \log n) \quad \text{otherwise.}$$

This result is asymptotically tight if $d, k \le 3$.

Proof For $d = 2$, the assertion is true, by Theorem 4.1 and by Corollary 4.4. Assume that $d \ge 3$. For any $(d-3)$-simplex Δ induced by V, let E_Δ denote the family of all members of E that contain Δ as a face. Pick any point p_Δ in the relative interior of Δ, and let F_Δ denote the 3-dimensional flat orthogonal to Δ and passing through p_Δ.

Every $e \in E_\Delta$ meets F_Δ in a polygon, whose two sides incident to p_Δ are the intersections of F_Δ with the two $(d-2)$-faces of e containing Δ. Thus, the total number of sides incident to p_Δ that occur in some $e \cap F_\Delta$ ($e \in E_\Delta$) is

Geometric Graph Theory

at most $n - d + 2 < n$. Take a small 2-dimensional sphere $S^2 \subseteq F_\Delta$ centered at p_Δ. The intersections of S^2 with the elements of E_Δ form the edge set of a graph with at most n vertices. It follows from the properties of E that this graph has no k pairwise crossing edges, so, by the planar results, its number of edges, $|E_\Delta|$, satisfies

$$|E_\Delta| = O(n) \quad \text{if } k = 2, 3;$$

$$|E_\Delta| = O(n \log n) \quad \text{otherwise.}$$

Summing over all $(d-3)$-simplices Δ induced by V, we obtain $\binom{d}{2}|E| = \sum_\Delta |E_\Delta|$, and hence the upper bound.

To show that the result is tight for $d = 3, k = 2$, consider a nested sequence of $n/2$ pyramids based on the same 2-dimensional convex $n/2$-gon. These pyramids have a total of $n^2/4$ triangular faces, no two of which have a nontrivial intersection. ∎

It is an outstanding open problem to decide whether the order of magnitude of the above bound can be improved, e.g., for $d = 4, k = 2$. However, we have the following related result.

Theorem 8.6 ([18]) *Let E be a family of $(d-1)$-dimensional simplices induced by an n-element point set $V \subseteq \mathbb{R}^d$. If E has no two crossing members, then $|E| = O(n^{d-1})$, and this bound cannot be improved. In notation,*

$$\mathrm{ex}_d^d(\mathcal{C}_2^d, n) = \Theta(n^{d-1}).$$

This can be generalized as follows.

Theorem 8.7 *Let E be a family of $(d-1)$-dimensional simplices induced by an n-element point set $V \subseteq \mathbb{R}^d$, where $d, k > 1$. If E has no k pairwise crossing members, then $|E| = O(n^{d-(1/d)^{k-2}})$. In notation,*

$$\mathrm{ex}_d^d(\mathcal{C}_k^d, n) = O(n^{d-(1/d)^{k-2}}).$$

The proof of this result is based on a natural extension of Theorem 4.8, which provides us with a recipe for giving a lower bound on the *number* of crossing k-tuples of edges, if we know how many edges are necessary to guarantee the existence of *one* such k-tuple.

Theorem 8.8 ([18]) *Assume that, for suitable constants c_1 and $0 \leq \delta \leq 1$, we have $\mathrm{ex}_r^d(\mathcal{SC}_k^r, n) < c_1 \binom{n}{r}/n^\delta$ and $e \geq (c_1 + 1)\binom{n}{r}/n^\delta$.*

Then there exists $c_2 > 0$ such that the minimum number of strongly crossing k-tuples of edges in a d-dimensional r-hypergraph with n vertices and e edges is at least

$$c_2 \binom{n}{kr} e^\gamma / \binom{n}{r}^\gamma,$$

where $\gamma = 1 + \frac{(k-1)r}{\delta}$.

As Theorem 3.4 can be applied to the planar k-set problem (compare Theorem 7.2), the last result, which is a far-reaching generalization of Theorem 3.4, is relevant to its higher dimensional analogues.

Given a set P of n points in 3-space, in general position, a k-element subset $P' \subseteq P$ is called a k-set, if there is a half-space H such that $H \cap P = P'$. Bárány, Füredi and Lovász [9] gave the first non-trivial (subcubic) upper bound for the number of $\lfloor n/2 \rfloor$-sets of an n-element set in 3-space. The best known bound is given by the following theorem.

Theorem 8.9 ([17], [6]) *The number of $\lfloor n/2 \rfloor$-sets of an n-element set in general position in 3-space is $O(n^{8/3})$.*

Proof Let P be a set of n points in general position in 3-space. Similarly to the 2-dimensional case, it is sufficient to bound the number of edges in the 3-dimensional geometric 3-hypergraph $H = (V, E)$, where $V = P$ and E consists of all triangles whose supporting planes pass through 3 elements of P, and which have precisely $\lfloor n/2 \rfloor$ points on one of their sides. Let $e := |E|$.

In view of Theorem 8.6, we can apply Theorem 8.8 with $d = r = 3$, $k = 2$, and $\delta = 1$ to obtain that the number of (strongly) crossing pairs of edges of H is at least $\Omega(e^4/n^6)$, provided that $e > cn^2$ for a suitable constant $c > 0$. Notice that if two triangles cross each other, then one of them must have an edge that crosses the other triangle. Thus, there is an edge crossing at least $\Omega(e^3/n^6)$ triangles. On the other hand, by an old (and easy) observation of Lovász, every straight line crosses at most n^2 edges of H. We have

$$n^2 = \Omega(e^3/n^6),$$

and the result follows. ∎

We close this section with a generalization of Theorem 6.2.

Theorem 8.10 ([18]) *Let us colour with two colours all $(d-1)$-dimensional simplices induced by $(d+1)n - 1$ points in general position in $\mathbb{R}^d$. Then one can always find n disjoint simplices of the same colour. This result cannot be improved.*

Acknowledgements

I thank Professor Micha Perles, who gave a course on geometric graph theory at Rutgers University in 1989, during the Special Year in Discrete and Computational Geometry, the first "special year" of DIMACS (Center for Discrete Mathematics and Theoretical Computer Science). Many of the results surveyed here grew out of discussions following his exciting lectures ten years ago.

This research was supported by National Science Foundation grant CCR-9732101, by PSC–City University of New York Research Award 667339, and by a grant from the Hungarian National Fund for Scientific Research (T-020914).

References

[1] M. Abellanas, J. García, G. Hernández, M. Noy & P. Ramos, Bipartite embeddings of trees in the plane, in *Graph Drawing '96* (ed. S. North), *Lecture Notes in Computer Science*, 1190, Springer-Verlag, Berlin (1997), pp. 1–10.

[2] P. K. Agarwal, B. Aronov, J. Pach, R. Pollack & M. Sharir, Quasi-planar graphs have a linear number of edges, *Combinatorica*, **17** (1997), 1–9.

[3] M. Ajtai, V. Chvátal, M. Newborn & E. Szemerédi, Crossing-free subgraphs, *Annals of Discrete Mathematics*, **12** (1982), 9–12.

[4] J. Akiyama & N. Alon, Disjoint simplices and geometric hypergraphs, in *Combinatorial Mathematics* (eds. G. S. Bloom et al.), *Annals of New York Academy of Sciences*, 555 (1989), pp. 1–3.

[5] N. Alon & P. Erdős, Disjoint edges in geometric graphs, *Discrete and Computational Geometry*, **4** (1989), 287–290.

[6] B. Aronov, B. Chazelle, H. Edelsbrunner, L. Guibas, M. Sharir & R. Wenger, Points and triangles in the plane and halving planes in space, *Discrete and Computational Geometry*, **6** (1991), 435–442.

[7] S. Avital & H. Hanani, Graphs (in Hebrew), *Gilyonot Lematematika*, **3** (1966), 2–8.

[8] I. Bárány & Z. Füredi, Empty simplices in Euclidean space, *Canadian Mathematical Bulletin*, **30** (1987), 436–445.

[9] I. Bárány, Z. Füredi & L. Lovász, On the number of halving planes, *Combinatorica*, **10** (1990), 175–183.

[10] B. Bollobás, *Extremal Graph Theory*, Academic Press, New York (1978).

[11] K. Borsuk, Drei Sätze über die n-dimensionale euklidische Sphäre, *Fundamenta Mathematicae*, **20** (1933), 177–190.

[12] S. Burr, Generalized Ramsey-theory for graphs – a survey, in *Graphs and Combinatorics* (eds. R. Bari & F. Harary), *Lecture Notes in Mathematics*, 406, Springer-Verlag, Berlin (1974), pp. 52–75.

[13] V. Capoyleas & J. Pach, A Turán-type theorem on chords of a convex polygon, *Journal of Combinatorial Theory, Series B*, **56** (1992), 9–15.

[14] F. Chung, E. Szemerédi & W. T. Trotter, The number of distinct distances determined by a set of points in the Euclidean plane, *Discrete and Computational Geometry*, **7** (1992), 1–11.

[15] K. Clarkson, H. Edelsbrunner, L. Guibas, M. Sharir & E. Welzl, Combinatorial complexity bounds for arrangements of curves and surfaces, *Discrete and Computational Geometry*, **5** (1990), 99–160.

[16] T. K. Dey, Improved bounds for planar k sets and related problems, *Discrete and Computational Geometry*, **19** (1998), 373–382.

[17] T. K. Dey & H. Edelsbrunner, Counting triangle crossings and halving planes, *Discrete and Computational Geometry*, **12** (1994), 281–289.

[18] T. K. Dey & J. Pach, Extremal problems for geometric hypergraphs, *Discrete and Computational Geometry*, **19** (1998), 473–484.

[19] H. Edelsbrunner & P. Hajnal, A lower bound on the number of unit distances between the points of a convex polygon, *Journal of Combinatorial Theory, Series A*, **56** (1991), 312–316.

[20] R. B. Eggleton & R. K. Guy, The crossing number of the n-cube, *Notices of the American Mathematical Society*, **17** (1970), 757.

[21] P. Erdős, On extremal problems on graphs and generalized graphs, *Israel Journal of Mathematics*, **2** (1964), 183–190.

[22] P. Erdős, A. Gyárfás & L. Pyber, Vertex coverings by monochromatic cycles and trees, *Journal of Combinatorial Theory, Series B*, **51** (1991), 90–95.

[23] P. Erdős, L. Lovász, A. Simmons & E. G. Straus, Dissection graphs of planar point sets, in *A Survey of Combinatorial Theory* (eds. J. N. Srivastava et al.), North-Holland, Amsterdam (1973), pp. 139–149.

[24] I. Fáry, On straight line representation of planar graphs, *Acta Scientiarum Mathematicarum (Szeged)*, **11** (1948), 229–233.

[25] H. deFraysseix, J. Pach & R. Pollack, How to draw a planar graph on a grid, *Combinatorica*, **10** (1990), 41–51.

[26] Z. Füredi, The maximum number of unit distances in a convex n-gon, *Journal of Combinatorial Theory, Series A*, **55** (1990), 316–320.

[27] G. Gerencsér & A. Gyárfás, On Ramsey-type problems, *Annales Universitatis Scientarium Budapestinensis Roland Eötvös, Sectio Mathematica*, **X** (1967), 167–170.

[28] W. Goddard, M. Katchalski & D. J. Kleitman, Forcing disjoint segments in the plane, *European Journal of Combinatorics*, **17** (1996), 391–395.

[29] R. L. Graham, B. L. Rothschild & J. H. Spencer, *Ramsey Theory*, second edition, Wiley, New York (1990).

[30] P. Gritzmann, B. Mohar, J. Pach & R. Pollack, Embedding a planar triangulation with vertices at specified points (Solution to Problem E3341), *American Mathematical Monthly*, **98** (1991), 165–166.

[31] A. Gyárfás, On the chromatic number of multiple interval graphs and overlap graphs, *Discrete Mathematics*, **55** (1985), 161–166.

[32] A. Gyárfás, Corrigendum, *Discrete Mathematics*, **62** (1986), 333.

[33] A. Gyárfás & J. Lehel, Covering and coloring problems for relatives of intervals, *Discrete Mathematics*, **55** (1985), 167–180.

[34] H. Harborth & H. Lefmann, Coloring arcs of convex sets, preprint, Technische Universität Braunschweig, 1998.

[35] J. Hershberger & S. Suri, Applications of a semi-dynamic convex hull algorithm, *BIT*, **32** (1992), 249–267.

[36] H. Hopf & E. Pannwitz, Aufgabe No. 167, *Jahresbericht der deutschen Mathematiker-Vereinigung*, **43** (1934), 114.

[37] Y. Ikebe, M. Perles, A. Tamura & S. Tokunaga, The rooted tree embedding problem into points in the plane, *Discrete and Computational Geometry*, **11** (1994), 51–63.

[38] J. Kahn & G. Kalai, A counterexample to Borsuk's conjecture, *Bulletin of the American Mathematical Society*, **29** (1993), 60–62.

[39] A. Kaneko & M. Kano, Straight-line embeddings of two rooted trees in the plane, preprint, Kogakuin University, Tokyo and Ibaraki University, Hitachi, 1998.

[40] A. Kaneko & M. Kano, Straight-line embeddings of rooted star forests in the plane, preprint, Kogakuin University, Tokyo and Ibaraki University, Hitachi, 1998.

[41] A. Kaneko & M. Kano, A balanced partition of points in the plane and tree embedding problems, preprint, Kogakuin University, Tokyo and Ibaraki University, Hitachi, 1998.

[42] G. Károlyi, J. Pach & G. Tóth, Ramsey-type results for geometric graphs, I, *Discrete and Computational Geometry*, **18** (1997), 247–255.

[43] G. Károlyi, J. Pach, G. Tardos & G. Tóth, An algorithm for finding many disjoint monochromatic edges in a complete 2-colored geometric graph, in *Intuitive Geometry* (eds. I. Bárány & K. Böröczky), *Bolyai Society Mathematical Studies*, 6, Bolyai Society, Budapest (1997), pp. 367–372.

[44] G. Károlyi, J. Pach, G. Tóth & P. Valtr, Ramsey-type results for geometric graphs, II, *Discrete and Computational Geometry*, **20** (1998), 375–388.

[45] M. Klazar & P. Valtr, Generalized Davenport–Schinzel sequences, *Combinatorica*, **14** (1994), 463–476.

[46] A. Kostochka, On upper bounds on the chromatic numbers of graphs (in Russian), *Transactions of the Institute of Mathematics (Siberian Branch of the Academy of Sciences of the USSR)*, **10** (1988), 204–226.

[47] A. Kostochka & J. Kratochvíl, Covering and coloring polygon-circle graphs, *Discrete Mathematics*, **163** (1997), 299–305.

[48] Y. Kupitz, *Extremal Problems in Combinatorial Geometry, Aarhus University Lecture Notes Series*, 53, Aarhus University, Denmark (1979).

[49] Y. Kupitz, On pairs of disjoint segments in convex position in the plane, *Annals of Discrete Mathematics*, **20** (1984), 203–208.

[50] D. Larman, J. Matoušek, J. Pach & J. Törőcsik, A Ramsey-type result for planar convex sets, *Bulletin of the London Mathematical Society*, **26** (1994), 132–136.

[51] T. Leighton, *Complexity Issues in VLSI, Foundations of Computing Series*, MIT Press, Cambridge, MA (1983).

[52] R. Lipton & R. Tarjan, A separator theorem for planar graphs, *SIAM Journal of Applied Mathematics*, **36** (1979), 177–189.

[53] L. Lovász, On the number of halving lines, *Annales Universitatis Scientarium Budapestinensis Roland Eötvös, Sectio Mathematica*, **XIV** (1971), 107–108.

[54] S. McGuinness, Colouring arcwise connected sets in the plane, I, preprint, University of Umeå, 1997.

[55] J. Pach, Notes on geometric graph theory, in *Discrete and Computational Geometry: Papers from the DIMACS Special Year* (eds. J. E. Goodman, R. Pollack & W. Steiger), *DIMACS Series in Discrete Mathematics and Theoretical Computer Science*, 6, American Mathematical Society, Providence, RI (1991), pp. 273–285.

[56] J. Pach, Geometric graphs and hypergraphs, *Graph Theory Notes of New York*, **XXXI** (1996), 39–43.

[57] J. Pach & P. K. Agarwal, *Combinatorial Geometry*, Wiley, New York (1995).

[58] J. Pach, F. Shahrokhi & M. Szegedy, Applications of the crossing number, *Algorithmica*, **16** (1996), 111–117.

[59] J. Pach & M. Sharir, Repeated angles in the plane and related problems, *Journal of Combinatorial Theory, Series A*, **59** (1990), 12–22.

[60] J. Pach & M. Sharir, On the number of incidences between points and curves, *Combinatorics, Probability and Computing*, **7** (1998), 121–127.

[61] J. Pach, J. Spencer & G. Tóth, New bounds on crossing numbers, preprint, Mathematical Institute of the Hungarian Academy of Sciences, 1999.

[62] J. Pach, W. Steiger & E. Szemerédi, An upper bound on the number of planar k-sets, *Discrete and Computational Geometry*, **7** (1992), 109–123.

[63] J. Pach & J. Törőcsik, Layout of rooted trees, in *Planar Graphs* (ed. W. T. Trotter), *DIMACS Series in Discrete Mathematics and Theoretical Computer Science*, 9, American Mathematical Society, Providence, RI (1993), pp. 131–137.

[64] J. Pach & J. Törőcsik, Some geometric applications of Dilworth's theorem, *Discrete and Computational Geometry*, **12** (1994), 1–7.

[65] J. Pach & G. Tóth, Graphs drawn with few crossings per edge, *Combinatorica*, **17** (1997), 427–439.

[66] J. Pach & R. Wenger, Embedding planar graphs with fixed vertex locations, in *Graph Drawing '98*, *Lecture Notes in Computer Science*, Springer-Verlag, Berlin (1999), in press.

[67] J. Radon, Mengen konvexer Körper, die einen gemeinsamen Punkt enthalten, *Mathematische Annalen*, **63** (1921), 113–115.

[68] W. Schnyder, Embedding planar graphs on the grid, in *First Annual ACM–SIAM Symposium on Discrete Algorithms* SIAM, Philadelphia, PA (1990), pp. 138–147.

[69] D. Souvaine & R. Wenger, Constructing piecewise linear homeomorphisms, Technical Report 94-52, DIMACS, New Brunswick, NJ, 1994.

[70] J. Spencer, E. Szemerédi & W. T. Trotter, Unit distances in the Euclidean plane, in *Graph Theory and Combinatorics* (ed. B. Bollobás), Academic Press, New York (1984), pp. 293–303.

[71] E. Steinitz, Polyeder- und Raumeinteilungen, in *Geometrie, Part 3AB12*, *Enzyklopädie der mathematischen Wissenschaften*, 3, (1922), pp. 1–139.

[72] L. A. Székely, Crossing numbers and hard Erdős problems in discrete geometry, *Combinatorics, Probability and Computing*, **6** (1997), 353–358.

[73] E. Szemerédi & W. T. Trotter, Extremal problems in discrete geometry, *Combinatorica*, **3** (1983), 381–392.

[74] E. Szemerédi & W. T. Trotter, A combinatorial distinction between the Euclidean and projective planes, *European Journal of Combinatorics*, **4** (1983), 385–394.

[75] G. Tóth, Note on geometric graphs, preprint, Massachusetts Institute of Technology, 1998.

[76] G. Tóth & P. Valtr, Geometric graphs with few disjoint edges, in *Proceedings of the 14th Annual Symposium on Computational Geometry*, ACM Press, New York (1998), pp. 184–191.

[77] P. Turán, On the theory of graphs, *Colloquium Mathematicum*, **3** (1954), 19–30.

[78] W. T. Tutte, How to draw a planar graph, *Proceedings of the London Mathematical Society*, **13** (1963), 743–768.

[79] P. Valtr, On geometric graphs with no k pairwise parallel edges, *Discrete and Computational Geometry*, **19** (1998), 461–469.

Mathematical Institute of the
Hungarian Academy of Sciences
H-1364 Budapest, P.O.B. 127
and
Department of Computer Science
City College, City University of New York
pach@cims.nyu.edu

Recent Excluded Minor Theorems for Graphs

Robin Thomas

Summary A graph is a *minor* of another if the first can be obtained from a subgraph of the second by contracting edges. An excluded minor theorem describes the structure of graphs with no minor isomorphic to a prescribed set of graphs. Splitter theorems are tools for proving excluded minor theorems. We discuss splitter theorems for internally 4-connected graphs and for cyclically 5-connected cubic graphs, the graph minor theorem of Robertson and Seymour, linkless embeddings of graphs in 3-space, Hadwiger's conjecture on t-colourability of graphs with no K_{t+1} minor, Tutte's edge 3-colouring conjecture on edge 3-colourability of 2-connected cubic graphs with no Petersen minor, and Pfaffian orientations of bipartite graphs. The latter are related to the even directed circuit problem, a problem of Pólya about permanents, the 2-colourability of hypergraphs, and sign-nonsingular matrices.

1 Introduction

All *graphs* in this paper are finite, and may have loops and parallel edges. A graph is a *minor* of another if the first can be obtained from a subgraph of the second by contracting edges. An H *minor* is a minor isomorphic to H. The following is Wagner's reformulation [75] of Kuratowski's theorem [27].

Theorem 1.1 *A graph is planar if and only if it has no minor isomorphic to K_5 or $K_{3,3}$.*

Kuratowski's theorem is important, because it gives a good characterization (in the sense of J. Edmonds) of planarity, but we can also think of it as a structural theorem characterizing graphs with no K_5 or $K_{3,3}$ minor. What about excluding only one of these graphs? Wagner [75] characterized those classes. To state his theorems we need one definition.

Let G_1 and G_2 be graphs with disjoint vertex-sets, let $k \geq 0$ be an integer, and for $i = 1, 2$ let $X_i \subseteq V(G_i)$ be a set of cardinality k of pairwise adjacent vertices. For $i = 1, 2$ let G'_i be obtained from G_i by deleting a (possibly empty) set of edges with both ends in X_i. Let $f \colon X_1 \to X_2$ be a bijection, and let G be the graph obtained from the union of G'_1 and G'_2 by identifying x with $f(x)$ for all $x \in X_1$. In those circumstances we say that G is a k-*sum* of G_1 and G_2.

Theorem 1.2 *A graph has no minor isomorphic to $K_{3,3}$ if and only if it can be obtained from planar graphs and K_5 by means of 0-, 1-, and 2-sums.*

By V_8 we mean the graph obtained from a circuit of length eight by joining each pair of diagonally opposite vertices by an edge.

Theorem 1.3 *A graph has no minor isomorphic to K_5 if and only if it can be obtained from planar graphs and V_8 by means of 0-, 1-, 2- and 3-sums.*

There are many similar results in Graph Theory, known as excluded minor theorems (see for example [5, 6, 16, 19, 27, 75, 76]). Such characterizations can be useful: we often need to exclude certain minors when they are obvious obstructions to some desired property, but knowledge of the structure which their exclusion forces may enable us to establish that property for the remaining graphs. Surveys of excluded minor theorems are given in [12] (for finite minors) and [45] (for infinite minors). We show that Theorem 1.1 is not an isolated result, but rather a beginning of a rich theory. We do not attempt to give a complete survey, but instead concentrate on the developments of this decade.

2 Seymour's splitter theorem

Seymour's splitter theorem is a tool for proving excluded minor theorems. We say that a simple graph G is obtained from a simple graph H by *splitting a vertex* if H is obtained from G by contracting an edge e, where both ends of e have degree at least three in G. Since H is simple, it follows that e belongs to no triangle of G. A graph is a *wheel* if it is obtained from a circuit on at least three vertices by adding a vertex joined to every vertex on the circuit. (*Paths* and *circuits* have no "repeated" vertices.) A graph G is *k-connected* if it has at least $k+1$ vertices, and $G \setminus X$ is connected for every set $X \subseteq V(G)$ with $|X| < k$. (We use $\setminus$ for deletion.) The following is a classical result of Tutte [71].

Theorem 2.1 *Every simple 3-connected graph can be obtained from some wheel by repeatedly applying the operations of adding an edge between two nonadjacent vertices and splitting a vertex.*

The converse also holds: if a graph can be obtained from a wheel (in fact, any simple 3-connected graph) by means of the above two operations, then it is simple and 3-connected. Seymour [61] proved the following strengthening.

Theorem 2.2 *Let H be a simple 3-connected minor of a simple 3-connected graph G such that if H is a wheel, then H is the largest wheel minor of G. Then a graph isomorphic to G can be obtained from H by repeatedly applying the operations of adding an edge between two nonadjacent vertices and splitting a vertex.*

Thus a simple 3-connected graph can be built starting from almost any simple 3-connected minor of itself, not necessarily a wheel. To illustrate the use of Seymour's theorem, let us deduce Theorem 1.2 from it. A *separation* in a graph G is a pair of subgraphs (G_1, G_2) such that $G_1 \cup G_2 = G$ and $E(G_1) \cap E(G_2) = \emptyset$. The order of (G_1, G_2) is $|V(G_1) \cap V(G_2)|$.

Proof of Theorem 1.2 The "if" part is easy. For the "only if" part let G be a graph with no minor isomorphic to $K_{3,3}$, and assume that the theorem holds for all graphs H with $|V(H)| + |E(H)| < |V(G)| + E(G)|$. Assume first that G is not a simple 3-connected graph. If v is an isolated vertex of G, then the theorem follows by considering $G \setminus v$. Otherwise, G has a separation (G_1, G_2) of order k, where $k \leq 2$, such that $|E(G_1)|, |E(G_2)| \leq |E(G)| - 1$, and the inequality is strict if $k = 2$. Moreover, we may assume that (G_1, G_2) is chosen with k minimum. For $i = 1, 2$ let G'_i be G_i if the order of (G_1, G_2) is less than two, and otherwise let G'_i be obtained from G_i by adding an edge joining the two vertices of $V(G_1) \cap V(G_2)$. It follows from the minimality of k that G'_1 and G'_2 are minors of G. Thus, both G'_1 and G'_2 can be obtained from planar graphs and K_5 by means of 0-, 1-, and 2-sums. However, G is a k-sum of G'_1 and G'_2, as desired.

Thus we may assume that G is simple and 3-connected. If G is planar, then the theorem holds, and so we may assume that G is not planar. By Theorem 1.1, G has a minor isomorphic to K_5. We claim that G is isomorphic to K_5. Indeed, if it is not, then, by Theorem 2.2 applied to $H = K_5$ and to G, a graph isomorphic to G can be obtained from K_5 as stated in Theorem 2.2. Since K_5 is a complete graph, the next graph in the sequence is obtained from it by splitting a vertex. There is, up to isomorphism, only one way to split a vertex of K_5. It is easy to check that the resulting graph has a minor isomorphic to $K_{3,3}$, and hence so does G, a contradiction. Thus G is isomorphic to K_5, as desired. ∎

To prove Theorem 1.3 we need the following lemma. We say that a graph G is *internally* 4-*connected* if G is simple, 3-connected and for every separation (G_1, G_2) of G of order three, either $|E(G_1)| \leq 3$ or $|E(G_2)| \leq 3$.

Lemma 2.3 *Let G be an internally* 4-*connected nonplanar graph. Then either G is isomorphic to $K_{3,3}$, or it has a minor isomorphic to K_5 or V_8.*

Proof Let G be an internally 4-connected nonplanar graph. By Theorem 1.1 the graph G has a minor isomorphic to K_5 or $K_{3,3}$. In the former case we are done, and so we may assume that G has a $K_{3,3}$ minor, and that it is not isomorphic to $K_{3,3}$. Thus G has six distinct vertices $v_1, v_2, \ldots, v_6$ and nine paths P_{ij} ($i = 1, 2, 3; j = 4, 5, 6$) such that P_{ij} has ends v_i and v_j, and the paths are disjoint, except possibly for their ends. Let H denote the union of the nine paths. We claim that we may assume the following.

(∗) For distinct integers $i, k = 1, 2, 3$ and $j, l = 4, 5, 6$ the graph G has no path with one end in $V(P_{ij}) - \{v_i, v_j\}$, the other end in $V(P_{kl}) - \{v_k, v_l\}$, and otherwise disjoint from H.

Indeed, otherwise the union of H and the path give a V_8 minor, as desired. Thus we may assume that (∗) holds.

Since G is internally 4-connected and is not isomorphic to $K_{3,3}$, we deduce that at least two of the graphs $(P_{14} \cup P_{15} \cup P_{16}) \setminus \{v_4, v_5, v_6\}$, $(P_{24} \cup P_{25} \cup P_{26}) \setminus \{v_4, v_5, v_6\}$, $(P_{34} \cup P_{35} \cup P_{36}) \setminus \{v_4, v_5, v_6\}$ belong to the same component of $G \setminus \{v_4, v_5, v_6\}$. By symmetry and (*) we may assume that G has a path P with one end in $V(P_{14}) - \{v_4\}$ and the other end in $V(P_{24}) - \{v_4\}$.

Similarly, there exist an integer $i \in \{1, 2, 3\}$, distinct integers $k, l \in \{4, 5, 6\}$ and a path Q in $G \setminus \{v_1, v_2, v_3\}$ with one end in $V(P_{ik}) - \{v_i\}$ and the other end in $V(P_{il}) - \{v_i\}$. By considering the graph $H \cup P \cup Q$ we deduce that G has a K_5 minor, as desired. ∎

Proof of Theorem 1.3 Again, the "if" part is easy. For the "only if" part let G be a graph with no minor isomorphic to K_5, and assume that the theorem holds for all graphs with fewer edges. If G is not internally 4-connected, then we conclude the proof in a similar way as in the proof of Theorem 1.2. Thus we may assume that G is internally 4-connected. If G is planar, then it satisfies the conclusion of the theorem, and so we may assume that G is not planar. By Theorem 1.1, G has a minor isomorphic to $K_{3,3}$. By Lemma 2.3 either G is isomorphic to $K_{3,3}$, or it has a V_8 minor. In the former case the theorem holds, because $K_{3,3}$ is a 3-sum of two planar graphs. Thus we may assume that G has a V_8 minor. Now it follows from Theorem 2.2 as in the proof of Theorem 1.2 that G is isomorphic to V_8, as desired. ∎

3 A splitter theorem for internally 4-connected graphs

Many excluded minor theorems (e.g. the results of [17, 18, 75, 77]) can be deduced using Theorem 2.2 as in the above proofs of Theorems 1.2 and 1.3. For others, however, it is desirable to have versions of Theorem 2.2 for different kinds of connectivity. Robertson [38] and Kelmans [25] obtained one such version. This section discusses a splitter theorem for internally 4-connected graphs, and its applications. We consider yet another splitter theorem in the next section.

The straightforward analogue of Theorem 2.2 does not hold for internally 4-connected graphs for various reasons. Let us consider the following example. Let H be a graph, and let C be a circuit in H with vertices $v_1, v_2, \ldots, v_t$ (in order). Assume that each v_i has degree three, and let u_i be the neighbour of v_i other than its two neighbours on C. Let G be obtained from H by adding, for $i = 1, 2, \ldots, t$, an edge e_i joining v_i and u_{i+1} (where u_{t+1} means u_1). Then, in general, there is no sequence $J_0, J_1, \ldots, J_k$ of internally 4-connected graphs such that $J_0 = H$, $J_k = G$, and for $i = 1, 2, \ldots, k$, J_{i-1} is isomorphic to a minor of J_i and differs from J_i only "a little". (Notice that if H' is obtained from H by adding a nonempty proper subset of $\{e_1, e_2, \ldots, e_t\}$, then H' is not internally 4-connected, because it has a vertex of degree three that belongs to a circuit of length three.) Thus in the theorem to follow we allow the intermediate graphs to fail the requirement of internal 4-connectivity, but only in one area, and we insist that the next operation to be performed repairs this connectivity violation, possibly at the expense of creating another violation elsewhere.

Let us make this precise now. Let e be an edge of a graph G, and let v be a vertex of degree three adjacent to both ends of e. We say that e is a *violating edge*, and that (v, e) is a *violating pair*. We say that a graph G is *almost 4-connected* if G is simple, 3-connected and, for every separation (G_1, G_2) of order three, either $|E(G_1)| \leq 4$ or $|E(G_2)| \leq 4$. Thus if a graph G is obtained from an internally 4-connected graph H by applying one of the two operations of Theorem 2.2, then G is almost 4-connected, and has at most two violating edges. It turns out that we need two additional operations, which we now introduce.

Let H be a graph, let e be a violating edge in H, let v be a vertex of H such that v is not incident with or adjacent to either end of e, and let H have no violating pair (w, e) such that v is adjacent to w in H. Let G be a graph obtained from H by deleting e, and adding a new vertex and three edges joining the new vertex to v and the two ends of e. We say that G was obtained from H by a *special addition*.

Let H be a simple graph, let (v, e) be a violating pair in H, let u be the neighbour of v that is not incident with e, let u have degree at least five, and let G be obtained from H by splitting u, and then adding an edge between v and the new vertex not adjacent to v in such a way that both new vertices have degree at least four in G. We say that G was obtained from H by a *special split*.

Finally, we need several exceptional families that will play the same roles that the wheels played in Theorem 2.2. We say that an internally 4-connected graph G is a *biwheel* if G has two vertices u, v such that $G \setminus \{u, v\}$ is a circuit, and we say that it is a *ladder* if it belongs to one of the four infinite families indicated in Figure 1. The following is a result of [21].

Theorem 3.1 *Let H be an internally 4-connected minor of an internally 4-connected graph G such that H has at least seven vertices and, if H is a ladder or a biwheel, then it has at least nine vertices and it is the largest ladder or biwheel minor of G. Then a graph isomorphic to G can be obtained from H by repeatedly applying the operations of adding an edge between two nonadjacent vertices, splitting a vertex, special addition and special split in such a way that each intermediate graph is almost 4-connected, with at most one violating edge, and no edge is a violating edge of two consecutive graphs in the sequence.*

For an application, let us consider the following unpublished theorem of N. Robertson.

Theorem 3.2 *An internally 4-connected graph G has no V_8 minor if and only if one of the following holds.*

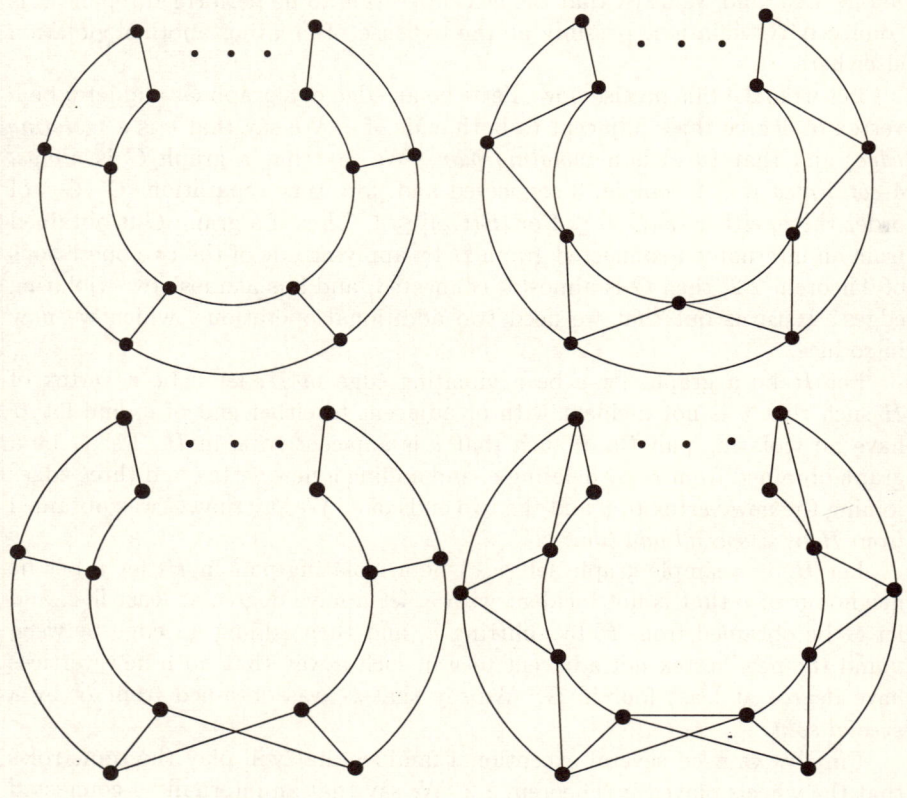

Figure 1: Ladders

(a) G is planar, or

(b) G has two vertices u, v such that $G \setminus \{u, v\}$ is a circuit, or

(c) there is a set $X \subseteq V(G)$ of cardinality four such that every edge of G has at least one end in X, or

(d) G is isomorphic to the line graph of $K_{3,3}$, or

(e) G has at most seven vertices.

It is easy to see that in order to characterize graphs with no V_8 minor it suffices to restrict oneself to internally 4-connected graphs. Thus Theorem 3.2 can be turned into a characterization of all graphs with no V_8 minor. One step in the proof of the theorem is to show the following.

Recent Excluded Minor Theorems for Graphs

Lemma 3.3 *If an internally 4-connected graph has a minor isomomorphic to the line graph of $K_{3,3}$, and has no V_8-minor, then it is isomorphic to the line graph of $K_{3,3}$.*

Lemma 3.3 can be proved using Theorem 3.1 just as we used Theorem 2.2 to prove Theorems 1.2 and 1.3. In fact, all we need to do is to verify that two graphs have V_8 minors. This time we do need the stronger Theorem 3.1, because the operations used in Theorem 2.2 produce graphs which are not internally 4-connected and have no V_8 minors.

4 A splitter theorem for cyclically 5-connected cubic graphs

A graph is *cubic* if every vertex has degree three. To motivate the next splitter theorem let us mention a special case of a theorem of Tutte [70] (the proof is easy).

Theorem 4.1 *Let G, H be 3-connected cubic graphs, and let H be a minor of G. Then a graph isomorphic to G can be obtained from H by repeatedly subdividing two distinct edges and joining the new vertices by an edge.*

A cubic graph G is *cyclically 5-connected* if it is simple, 3-connected, and for every set $F \subseteq E(G)$ of cardinality at most four, at most one component of $G \setminus F$ has circuits. For the results discussed in Section 10 below we need a similar theorem for cyclically 5-connected cubic graphs. An ideal analogue of Theorem 4.1 for cyclically 5-connected cubic graphs would assert that G can be obtained as in Theorem 4.1 in such a way that all the intermediate graphs are cyclically 5-connected. That is, unfortunately, not true, but the exceptions can be conveniently described. We will do so now.

Let G be a cyclically 5-connected cubic graph. Let e, f be distinct edges of G with no common end and such that no edge of G is adjacent to both e and f, and let G' be obtained from G by subdividing e and f and joining the new vertices by an edge. We say that G' is a *handle expansion* of G. It can be shown that G' is cyclically 5-connected. Let e_1, e_2, e_3, e_4, e_5 (in order) be the edges of a circuit of G of length five. Let us subdivide e_i by a new vertex v_i, add a circuit (disjoint from G) with vertices u_1, u_2, u_3, u_4, u_5 (in order), and for $i = 1, 2, \ldots, 5$ let us add an edge joining u_i and v_i to form a graph G''. In these circumstances we say that G'' is a *circuit expansion* of G.

Let p be an integer such that $p \geq 5$ if p is odd and $p \geq 10$ if p is even. Let G be a cubic graph with vertex-set $\{u_0, u_1, \ldots, u_{p-1}, v_0, v_1, \ldots, v_{p-1}\}$ such that for $i = 0, 1, \ldots, p-1$, u_i has neighbours u_{i-1}, u_{i+1} and v_i, and v_i has neighbours u_i, v_{i-2} and v_{i+2}, where the index arithmetic is taken modulo p (see Figure 4). We say that G is a *biladder* on $2p$ vertices. We remark that the Petersen graph is a biladder on 10 vertices, and that the dodecahedron is a biladder on 20 vertices. The following theorem [51] generalizes [1, 7, 10, 33, 34].

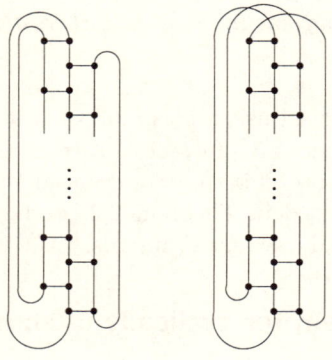

Figure 2: Biladders

Theorem 4.2 *Let G, H be cyclically 5-connected cubic graphs, let H be a minor of G, and assume that if H is a biladder, then it is the largest biladder minor of G. Then a graph isomorphic to G can be obtained from H by repeatedly applying the operations of handle expansion and circuit expansion.*

5 Excluding a general graph

We have mentioned several excluded minor theorems for specific excluded minors. Robertson and Seymour [43] found a general structure theorem for graphs with no minor isomorphic to an arbitrary fixed graph H. The theorem gives only a necessary condition for excluding H, but the condition is necessary and sufficient in the sense that no graph that possesses the structure has a minor isomorphic to some other graph H', where H' is much larger than H.

We say that a graph G is a *clique-sum* of two graphs G_1 and G_2 if G is an i-sum of G_1 and G_2 for some integer $i \geq 0$. Roughly speaking, the theorem of Robertson and Seymour says that for every graph H there exists an integer k such that every graph with no H minor can be obtained by means of clique-sums from the class of graphs that are obtained by adding at most k vertices (and any number of edges incident with these vertices) to graphs that can be "almost" drawn on a surface the graph H cannot be drawn on. The word almost means that the graph can be drawn in the surface, except for k disks, where crossings are permitted, but that the number of edges crossing is controlled in a certain way depending on k. Let us be more precise now.

Let G be a graph, and let U be a cyclic ordering of a subset of its vertices. We say that $(X_u)_{u \in U}$ is a *vortex decomposition* of the pair (G, U) if

(V1) $u \in X_u$ for every $u \in U$,

(V2) $\bigcup_{u \in U} X_u = V(G)$, and every edge of G has both ends in some X_u, and

(V3) if u_1, u_2, u_3, u_4 occur in U in the order listed, then $X_{u_1} \cap X_{u_3} \subseteq X_{u_2} \cup X_{u_4}$.
Let us remark that axiom (V3) is equivalent to saying that, for every vertex $v \in V(G)$, the set of all $u \in U$ with $v \in X_u$ is empty, or a contiguous interval, or the whole of U. We say that $(X_u)_{u \in U}$ has width less than k if $|X_u| \leq k$ for every $u \in U$.

A *surface* is a compact connected 2-manifold with (possibly empty) boundary; the surface is *closed* if its boundary is empty. The unique surface obtained from a closed surface Σ by removing the interiors of k disjoint closed discs will be denoted by $\Sigma - k$. The components of the boundary of a surface Σ are the *cuffs* of Σ. Thus, each cuff of a surface is homeomorphic to the unit circle.

Let G be a graph, and Σ a surface with cuffs $C_1, \ldots, C_k$. We say that G *can be nearly drawn* in Σ if G has a set X of at most k vertices (where k is the number of cuffs of Σ) such that $G \setminus X$ can be written as $G_0 \cup G_1 \cup \ldots \cup G_k$, where

(N1) G_0 is embedded in Σ;
(N2) the graphs G_i ($i = 1, \ldots, k$) are pairwise disjoint, and $U_i := V(G_0) \cap V(G_i) = V(G_0) \cap C_i$ for each $i = 1, 2, \ldots, k$;
(N3) for each $i = 1, \ldots, k$, the pair (G_i, U_i) has a vortex decomposition $(X_u)_{u \in U_i}$ of width less than k, where the ordering of U_i is determined by the cyclic ordering of points on C_i.

We can now state the excluded minor theorem of Robertson and Seymour [43].

Theorem 5.1 *For every graph H there exists an integer $k \geq 0$ such that every finite graph with no H minor can be obtained by means of clique-sums from graphs that can be nearly drawn in $\Sigma - k$ for some closed surface Σ such that H cannot be drawn in Σ.*

6 The graph minor theorem

Is there an analogue of Theorem 1.1 for other surfaces? The following is a result of Archdeacon [5], and Glover, Huneke and Wang [14].

Theorem 6.1 *A graph G admits an embedding in the projective plane if and only if G has no minor isomorphic to a member of an explicit list of 35 graphs.*

For other surfaces no such theorem is known, and there is some evidence that the list of graphs is too large to be useful. On the other hand, the following landmark result of Robertson and Seymour [44] guarantees that the lists are finite.

Theorem 6.2 *Every infinite set of graphs includes two distinct elements such that one is isomorphic to a minor of the other.*

The proof is based on Theorem 5.1. Let $\mathcal{F}$ be an infinite set of graphs, and let $F \in \mathcal{F}$. We may assume that no other member of $\mathcal{F}$ has an F minor, and hence every member of $\mathcal{F}$ has a structure as described in Theorem 5.1. That structure can be exploited to conclude the proof, but the argument is lengthy and depends on the results of several other papers.

The following is another deep result of Robertson and Seymour [42].

Theorem 6.3 *For every graph H there exists an $O(n^3)$ algorithm to decide whether an input graph on n vertices has a minor isomorphic to H.*

Theorems 6.2 and 6.3 have some surprising consequences.

Corollary 6.4 *For every class of graphs closed under isomorphisms and taking minors there exists an $O(n^3)$ algorithm to decide if an input graph on n vertices belongs to the class.*

Proof Let $\mathcal{L}'$ be the class of all graphs G such that $G \notin \mathcal{F}$, but every proper minor of G belongs to $\mathcal{F}$, and let $\mathcal{L}$ contain one graph from each isomorphism class of graphs in $\mathcal{L}'$. Then no member of $\mathcal{L}$ is isomorphic to a minor of another, and hence $\mathcal{L}$ is finite by Theorem 6.2. Thus membership to $\mathcal{F}$ can be tested using Theorem 6.3 by testing the absence of minors isomorphic to a member of $\mathcal{L}$. ∎

The above proof guarantees the existence of an algorithm, but gives no clue as to how to construct one. Let us look at a special case. We say that a piecewise-linear embedding of a graph G in 3-space is *knotless* if every circuit of G forms a trivial knot. It is easy to see that contracting an edge in a knotless embedding results in a knotless embedding. Thus, by Corollary 6.4 there *exists* a polynomial-time algorithm to test whether an input graph has a knotless embedding. Curiously, at the moment we know of *no* explicit algorithm (let alone a polynomial-time one) to decide whether a given graph has a knotless embedding.

7 Linklessly embeddable graphs

Related to knotless embeddings are the following two concepts, introduced by Sachs [56, 57] and Böhme [8], respectively. We say that a (piecewise-linear) embedding of a graph in 3-space is *linkless* if every two disjoint circuits of the graph have zero linking number. We say that an embedding is *flat* if every circuit of the graph bounds a (topological) disk disjoint from the rest of the graph. By the *Petersen family* we mean the set of seven graphs depicted in Figure 7. Those are precisely the graphs that can be obtained from K_6 by means of Y–Δ and Δ–Y exchanges. The Petersen graph belongs to this set, and hence the name.

Sachs showed that no member of the Petersen family has a linkless embedding, and conjectured that conversely every graph has a linkless embedding unless it has a minor isomorphic to a member of the Petersen family. It turns out that the related notion of flat embeddings has an interesting theory. The following three results are proved in [50].

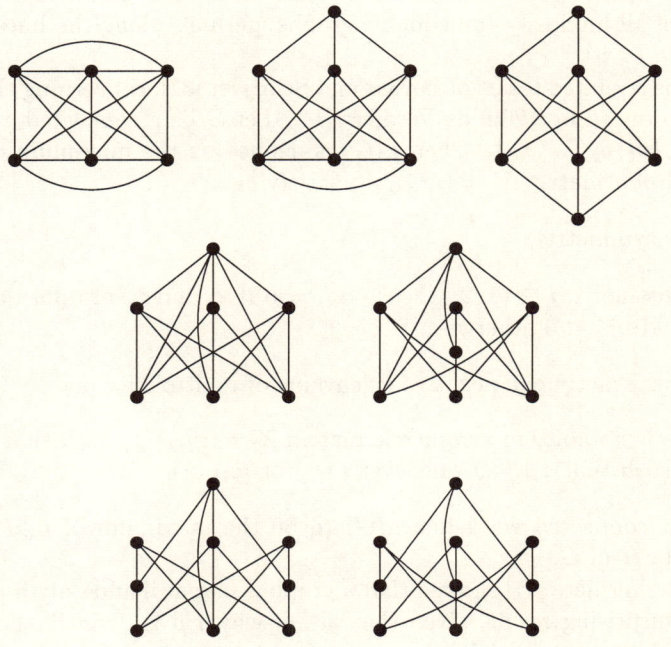

Figure 3: The Petersen family

Theorem 7.1 *A piecewise-linear embedding of a graph G in 3-space is flat if and only if the fundamental group of the complement in 3-space of every subgraph of G is free.*

Theorem 7.2 *Every two flat embeddings of a 4-connected graph in 3-space are related by a homeomorphism of the 3-space.*

Theorem 7.2 can be regarded as an analogue of Whitney's classical result [78] which states that every 3-connected planar graph has a unique planar embedding. The following implies Sach's conjecture.

Theorem 7.3 *For a graph G, the following conditions are equivalent:*

(i) *G has a flat embedding,*

(ii) *G has a linkless embedding,*

(iii) *G has no minor isomorphic to a member of the Petersen family.*

As a structure theorem, Theorem 7.3 is not completely satisfactory, because we do not know how to test in nondeterministic polynomial time whether a given embedding is flat. It would be nice to have a graph-theoretical description of all linklessly embeddable graphs, perhaps along the lines of Theorem 10.3.

There is a related result of Lovász and Schrijver [31], concerning the parameter μ introduced by Colin de Verdière [11]. Let G be a connected graph with vertex-set $\{v_1, v_2, \ldots, v_n\}$. Then $\mu(G)$ is defined as the maximum dimension of a kernel of a matrix $M = (m_{ij})_{i,j=1}^n$ satisfying

(i) M is symmetric,

(ii) for distinct $i, j \in \{1, 2, \ldots, n\}$, $m_{ij} = 0$ if v_i and v_j are not adjacent in G, and $m_{ij} < 0$ otherwise,

(iii) M has exactly one negative eigenvalue of multiplicity one,

(iv) there is no nonzero symmetric matrix $X = (x_{ij})_{i,j=1}^n$ such that $MX = 0$ and such that $x_{ij} = 0$ whenever $i = j$ or $m_{ij} \neq 0$.

If G is not connected we define $\mu(G)$ to be the maximum of $\mu(H)$ over all components H of G.

Colin de Verdière [11] showed that a graph is planar if and only if $\mu(G) \leq 3$. This is a surprising result, given the way in which μ is defined. Lovász and Schrijver [31] proved the following generalization, conjectured in [46].

Theorem 7.4 *A graph G has a linkless embedding if and only if $\mu(G) \leq 4$.*

It follows from [11] that this is indeed a generalization of Colin de Verdiére's result. It is tempting to ask whether there is any relationship between knotlessly embeddable graphs and $\mu(G) \leq 5$. As far as I am aware, it is not even known whether $K_{1,1,3,3}$, the complete 4-partite graph with parts of sizes $1, 1, 3, 3$, respectively, has a knotless embedding.

8 The four colour theorem

Our work on linkless embeddings was partly motivated by the fact that the conjectured answer involved the Petersen family, which was of interest to us because it includes both K_6 and the Petersen graph—two graphs whose exclusion is important for the $p = 5$ case of Hadwiger's conjecture and Tutte's conjectures (see Sections 9 and 10 below). The latter problems generalize the Four Colour Theorem (4CT), whose history dates back to 1852 when Francis Guthrie, while trying to colour the map of the counties of England, noticed that four colours sufficed, and asked whether the same could be true for any map. Since then the conjecture has attracted a lot of attention and motivated many new developments. A proof was finally found by Appel and Haken [2,3], reprinted in [4], formally as follows.

Theorem 8.1 *Every loopless planar graph is 4-colourable.*

However, the history seems not to end here. The proof by Appel and Haken is not completely satisfactory, because it relies on the use of computers, and even the computer-free part is so complicated that no one has been able to check it. This was partly remedied in a new proof recently found by Robertson, Sanders, Seymour and the author [41], but their proof is still computer-assisted. See [39, 40, 65] for recent surveys.

Another aspect of the 4CT is that there are several conjectures that, if true, would generalize the 4CT. It might be possible to reduce some of them to the 4CT, while others may require a strengthening of the proof of the Four Colour Theorem. We will discuss two such generalizations in the next two sections.

9 Hadwiger's conjecture

Hadwiger [15] made the following conjecture.

Conjecture 9.1 *For every integer $p \geq 1$, every loopless graph with no K_{p+1} minor is p-colourable.*

Conjecture 9.1 is trivial for $p \leq 2$, for $p = 3$ it was shown by Hadwiger [15] and Dirac [13] (the proof is not very difficult), but for $p \geq 4$ it seems very difficult, because it implies the Four Colour Theorem. To see this let $p \geq 4$, and let G be a planar graph. Let H be obtained from G by adding $p-4$ vertices adjacent to each other and to every vertex of G. Then H has no K_{p+1} minor (because no planar graph has a K_5 minor by the "easy" half of Theorem 1.1), and hence H has a p-colouring by the assumed truth of Conjecture 9.1. In this p-colouring vertices of G receive at most four colours, and so G is 4-colourable, as desired.

Theorem 1.3 implies that Hadwiger's conjecture for $p = 4$ is, in fact, equivalent to the 4CT. Robertson, Seymour and the author managed to prove that the next case (that is, $p = 5$) is also equivalent to the 4CT. More specifically, in [47] they proved the following (without using the 4CT), which immediately implies (assuming the 4CT) Hadwiger's conjecture for $p = 5$. We say that a graph G is *apex* if $G \setminus v$ is planar for some $v \in V(G)$.

Theorem 9.2 *Let G be a loopless graph with no K_6 minor such that G is not 5-colourable, and, subject to that, $|V(G)|$ is minimum. Then G is apex.*

While Theorem 1.3 gives a structural description of graphs with no K_5 minor, Theorem 9.2 does not do the same for graphs with no K_6 minor. Jorgensen [22] made the following beautiful conjecture, which implies Theorem 9.2 by a result of Mader [32].

Conjecture 9.3 *Every 6-connected graph with no minor isomorphic to K_6 is apex.*

At present, Hadwiger's conjecture is open for all $p \geq 6$.

10 Tutte's edge 3-colouring conjecture

Tait [64] showed that the Four Colour Theorem is equivalent to the following statement.

Theorem 10.1 *Every 2-connected cubic planar graph is edge 3-colourable.*

The smallest 2-connected cubic graph that is not edge 3-colourable is the Petersen graph. Tutte [72] conjectured that Theorem 10.1 holds with "planar" replaced by "no Petersen minor". Robertson, Sanders, Seymour and the author were recently able to settle Tutte's conjecture, as follows.

Theorem 10.2 *Every 2-connected cubic graph with no minor isomorphic to the Petersen graph is edge 3-colourable.*

The proof proceeds in two steps. First we showed in [53] that Theorem 10.2 holds in general as long as it holds for two classes of graphs: apex (defined above) and *doublecross* graphs (graphs that can be drawn in the plane with two crossings on the same region). Then we adapted our proof of the Four Colour Theorem [41] to show the edge 3-colourability of 2-connected apex [59] and doublecross graphs [58]. For the first part we used Theorem 4.2 to prove the following in [52]. (*Starfish* is the graph depicted in Figure 4.)

Theorem 10.3 *Let G be a cyclically 5-connected cubic graph with no Petersen minor, and assume that for every set $A \subseteq V(G)$ with $|A|, |V(G)-A| \geq 6$ there are at least six edges of G incident with both A and $V(G) - A$. Then G is apex, or it is doublecross, or it is isomorphic to Starfish.*

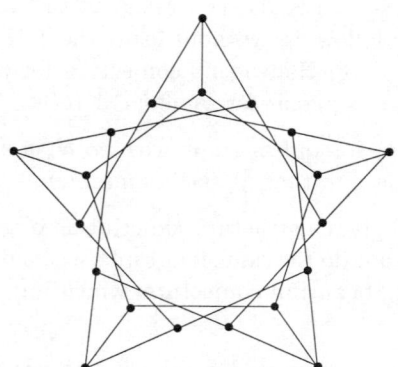

Figure 4: Starfish

Another consequence of Theorem 10.3 is the result [55] that every cubic graph of girth at least six has a subgraph isomorphic to a minor of the Petersen graph. Huck [20] used this to show that the 5-cycle double cover conjecture holds for cubic graphs with no Petersen minor.

We say that a graph G has a *nowhere-zero* 4-*flow* if there exists a function f mapping $E(G)$ into the nonzero elements of the Abelian group $\mathbf{Z}_2 \times \mathbf{Z}_2$ in such a way that, for every vertex v of G, the sum of $f(e)$, over all edges e incident with v, is zero. It follows that a cubic graph has a nowhere-zero 4-flow if and only if it is edge 3-colourable. Tutte [72] also made the following more general conjecture, known as the 4-flow conjecture.

Conjecture 10.4 *Every 2-connected graph with no Petersen minor has a nowhere-zero 4-flow.*

It may be possible to extend the proof of Theorem 10.2 to prove Conjecture 10.4, but no work in that direction has yet been done. Tutte made two other conjectures about nowhere-zero flows, known as the 3-flow [72] and 5-flow conjectures [69]. Both of them are still open. We refer to [62] for a survey on nowhere-zero flows.

11 Pfaffian orientations

Finally, I discuss a structural result pertaining to matching theory. An orientation D of a graph G is *Pfaffian* [23,24,30] if every even circuit C of G such that $G \setminus V(C)$ has a perfect matching has an odd number of edges directed in D in the direction of each orientation of C. The significance of Pfaffian orientations is that if a graph G has one, then the number of perfect matchings of G can be computed in polynomial time. Furthermore, the problem of deciding whether a bipartite graph has a Pfaffian orientation is equivalent to several other problems of interest—we mention these later. Little [28] obtained the following "excluded minor" characterization. We say that a graph H is a *matching minor* of a graph G if G has a subgraph K such that $G \setminus V(K)$ has a perfect matching, and H is obtained from K by repeatedly contracting pairs of edges incident with a common vertex of degree two.

Theorem 11.1 *A bipartite graph has a Pfaffian orientation if and only if it has no matching minor isomorphic to $K_{3,3}$.*

Theorem 11.1 is a beautiful result, but unfortunately it seems not to imply a polynomial-time algorithm to test if a given bipartite graph has a Pfaffian orientation. The next theorem, proven independently by McCuaig [35, 36] and by Robertson, Seymour and Thomas [54], can be used to design such an algorithm. We say that a bipartite graph is a *brace* if every matching of size at most two can be extended to a perfect matching. An argument similar to the

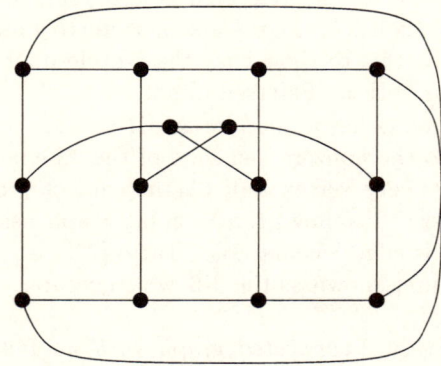

Figure 5: The Heawood graph

one in the proof of Theorem 1.2 shows that it suffices to characterize braces that have a Pfaffian orientation. The *Heawood* graph is depicted in Figure 11.

Let G_0 be a graph, and let C be a circuit of G_0 of length four such that $G_0 \setminus V(C)$ has a perfect matching. Let G_1, G_2 be two subgraphs of G_0 such that $G_1 \cup G_2 = G_0$, $G_1 \cap G_2 = C$, $V(G_1) - V(G_2) \neq \emptyset$ and $V(G_2) - V(G_1) \neq \emptyset$, and let G be obtained from G_0 by deleting a (possibly empty) subset of $E(C)$. In these circumstances we say that G is a C_4-*sum* of G_1 and G_2. The following result gives the desired characterization.

Theorem 11.2 *A brace has a Pfaffian orientation if and only if either it is isomorphic to the Heawood graph, or it can be obtained from planar braces by repeated applications of the C_4-sum operation.*

Using Theorem 11.2 we were able to design a polynomial-time algorithm [54] to decide if an input graph has a Pfaffian orientation:

Theorem 11.3 *There exists an $O(n^3)$ algorithm that, given an input graph G on n vertices, either outputs a Pfaffian orientation of G, or a valid statement that G has no Pfaffian orientation.*

I now describe some consequences of Theorem 11.2. Pólya [37] asked whether given a square 0,1-matrix A there is a matrix B obtained from A by changing some of the 1's into -1's in such a way that the determinant of B equals the permanent of A. This cannot be done for all matrices. However, given that the computing of permanents is #P-complete [73] it would seem desirable to have a characterization of matrices for which this is possible. Theorem 11.2 gives such a characterization by a result of Vazirani and Yannakakis [74].

Another consequence of Theorem 11.2 is a solution of the even directed circuit problem [63,66,68,74]. The question is whether there exists a polynomial-time algorithm to decide if a digraph has a circuit of even length. Again, Theorem 11.3 provides such an algorithm by [74]. There are other equivalent formulations of the result in terms of 2-colouring of hypergraphs [29,60], and several others in terms of sign-nonsingular matrices [9,26,67].

Acknowledgements

The author acknowledges partial support by NSF under Grant No. DMS-9623031, and by NSA under Contract No. MDA904-98-1-0517.

References

[1] E. R. L. Aldred, D. A. Holton & B. Jackson, Uniform cyclic edge connectivity in cubic graphs, *Combinatorica*, **11** (1991), 81–96.

[2] K. Appel & W. Haken, Every planar map is four colorable, part I: discharging, *Illinois Journal of Mathematics*, **21** (1977), 429–490.

[3] K. Appel, W. Haken & J. Koch, Every planar map is four colorable, part II: reducibility, *Illinois Journal of Mathematics*, **21** (1977), 491–567.

[4] K. Appel & W. Haken, Every planar map is four colorable, *Contemporary Mathematics*, **98** (1989).

[5] D. Archdeacon, A Kuratowski theorem for the projective plane, PhD thesis, Ohio State University, 1980.

[6] S. Arnborg & A. Proskurowski, Characterization and recognition of partial 3-trees, *SIAM Journal of Algebraic and Discrete Methods*, **7** (1986), 305–314.

[7] D. Barnette, On generating planar graphs, *Discrete Mathematics*, **7** (1974), 199–208.

[8] T. Böhme, On spatial representations of graphs, in *Contemporary Methods in Graph Theory* (ed. R. Bodendieck), Bibliographisches Institut, Mannheim (1990), pp. 151–167.

[9] R. A. Brualdi & B. L. Shader, *Matrices of sign-solvable linear systems*, Cambridge Tracts in Mathematics, 116, Cambridge University Press (1995).

[10] J. W. Butler, A generation procedure for the simple 3-polytopes with cyclically 5-connected graphs, *Canadian Journal of Mathematics*, **26** (1974), 686–708.

[11] Y. Colin de Verdière, Sur un nouvel invariant des graphes et un critère de planarité, *Journal of Combinatorial Theory, Series B*, **50** (1990), 11–21.

[12] R. Diestel, *Graph Decompositions. A Study in Infinite Graph Theory*, Oxford University Press, Oxford (1990).

[13] G. A. Dirac, Homomorphism theorems for graphs, *Mathematische Annalen*, **153** (1964), 69–80.

[14] H. H. Glover, J. P. Huneke & C. S. Wang, 103 graphs that are irreducible for the projective plane, *Journal of Combinatorial Theory, Series B*, **27** (1979), 332–370.

[15] H. Hadwiger, Über eine Klassifikation der Streckenkomplexe, *Viertel Jahres Schrift der Naturforschenden Gesellschaft Zürich*, **88** (1943), 133–142.

[16] R. Halin, Über einen graphentheoretischen Basisbegriff und seine Anwendung auf Färbungsprobleme, Dissertation, University of Köln, 1962.

[17] R. Halin, Über einen Satz von K. Wagner zum Vierfarbenproblem, *Mathematische Annalen*, **153** (1964), 47–62.

[18] R. Halin, Zur Klassifikation der endlichen Graphen nach H. Hadwiger und K. Wagner, *Mathematische Annalen*, **172** (1967), 46–78.

[19] R. Halin & K. Wagner, Homomorphiebasen von Graphenmengen, *Mathematische Annalen*, **147** (1962), 126–142.

[20] A. Huck, Reducible configurations for the cycle double cover conjecture, manuscript, 1997.

[21] T. Johnson & R. Thomas, A splitter theorem for internally 4-connected graphs, manuscript 1999.

[22] L. Jorgensen, Contraction to K_8, *Journal of Graph Theory*, **18** (1994), 431–448.

[23] P. W. Kasteleyn, Dimer statistics and phase transitions, *Journal of Mathematical Physics*, **4** (1963), 287–293.

[24] P. W. Kasteleyn, Graph theory and crystal physics, in *Graph Theory and Theoretical Physics* (ed. F. Harary), Academic Press, New York (1967), pp. 43–110.

[25] A. K. Kelmans, On 3-connected graphs without essential 3-cuts or triangles, *Soviet Mathematics, Doklady Akademii Nauk, SSSR*, **33** (1986), 698–703.

[26] V. Klee, R. Ladner & R. Manber, Sign-solvability revisited, *Linear Algebra and its Applications*, **59** (1984), 131–158.

[27] K. Kuratowski, Sur le problème des courbes gauches en topologie, *Fundamenta Mathematicae*, **15** (1930), 271–283.

[28] C. H. C. Little, A characterization of convertible (0, 1)-matrices, *Journal of Combinatorial Theory, Series B*, **18** (1975), 187–208.

[29] L. Lovász, On chromatic number of finite set-systems, *Acta Mathematica Academiae Scientiarum Hungaricae*, **19** (1968), 59–67.

[30] L. Lovász & M. D. Plummer, *Matching Theory*, Annals of Discrete Mathematics, 29, North-Holland, Amsterdam (1986).

[31] L. Lovász & A. Schrijver, A Borsuk theorem for antipodal links and a spectral characterization of linklessly embeddable graphs, *Proceedings of the American Mathematical Society*, **126** (1998), 1275–1285.

[32] W. Mader, Über trennende Eckenmengen in homomorphiekritische Graphen, *Mathematische Annalen*, **175** (1968), 245–252.

[33] W. McCuaig, Edge-reductions in cyclically k-connected cubic graphs, PhD thesis, University of Waterloo, Waterloo, Ontario, 1987.

[34] W. McCuaig, Edge-reductions in cyclically k-connected cubic graphs, *Journal of Combinatorial Theory, Series B*, **56** (1992), 16–44.

[35] W. McCuaig, Pólya's permanent problem, manuscript (81 pages) 1997.

[36] W. McCuaig, Generating braces, manuscript (25 pages) 1998.

[37] G. Pólya, Aufgabe 424, *Archiv der Mathematik und Physik*, (1913), 271.

[38] N. Robertson, Minimal cyclic-4-connected graphs, *Transactions of the American Mathematical Society*, **284** (1984), 665–687.

[39] N. Robertson, D. P. Sanders, P. D. Seymour & R. Thomas, Efficiently four-coloring planar graphs, *Proceedings of the ACM Symposium on the Theory of Computing*, **28** (1996), 571–575.

[40] N. Robertson, D. P. Sanders, P. D. Seymour & R. Thomas, A new proof of the four-color theorem, *Electronic Research Announcements of the AMS*, **2** (1996), 17–25.

[41] N. Robertson, D. P. Sanders, P. D. Seymour & R. Thomas, The four-colour theorem, *Journal of Combinatorial Theory, Series B*, **70** (1997), 2–44.

[42] N. Robertson & P. D. Seymour, Graph minors XIII. The disjoint paths problem, *Journal of Combinatorial Theory, Series B*, **63** (1995), 65–110.

[43] N. Robertson & P. D. Seymour, Graph minors XVII. Excluding a nonplanar graph, submitted 1991.

[44] N. Robertson & P. D. Seymour, Graph minors XX. Wagner's conjecture, manuscript 1996.

[45] N. Robertson, P. D. Seymour & R. Thomas, Excluding infinite minors, *Discrete Mathematics*, **95** (1991), 303–319.

[46] N. Robertson, P. D. Seymour & R. Thomas, Linkless embeddings of graphs in 3-space, *Bulletin of the American Mathematical Society*, **28** (1993), 84–89.

[47] N. Robertson, P. D. Seymour & R. Thomas, Hadwiger's conjecture for K_6-free graphs, *Combinatorica*, **13** (1993), 279–361.

[48] N. Robertson, P. D. Seymour & R. Thomas, Kuratowski chains, *Journal of Combinatorial Theory, Series B*, **64** (1995), 127–154.

[49] N. Robertson, P. D. Seymour & R. Thomas, Petersen family minors, *Journal of Combinatorial Theory, Series B*, **64** (1995), 155–184.

[50] N. Robertson, P. D. Seymour & R. Thomas, Sach's linkless embedding conjecture, *Journal of Combinatorial Theory, Series B*, **64** (1995), 185–227.

[51] N. Robertson, P. D. Seymour & R. Thomas, Cyclically 5-connected cubic graphs, manuscript 1996.

[52] N. Robertson, P. D. Seymour & R. Thomas, Excluded minors in cubic graphs, manuscript 1995.

[53] N. Robertson, P. D. Seymour & R. Thomas, Tutte's edge-coloring conjecture, *Journal of Combinatorial Theory, Series B*, **70** (1997), 166–183.

[54] N. Robertson, P. D. Seymour & R. Thomas, Permanents, Pfaffian orientations, and even directed circuits, *Annals of Mathematics*, to appear.

[55] N. Robertson, P. D. Seymour & R. Thomas, Girth six graphs have Petersen minors, manuscript 1997.

[56] H. Sachs, On a spatial analogue of Kuratowski's theorem on planar graphs—an open problem, in *Proceedings of a Conference held in Lagów, Poland, February 10–13, 1981* (eds. M. Borowiecki, J. W. Kennedy & M. M. Syslo), *Lecture Notes in Mathematics*, 1018, Springer-Verlag, Berlin (1983), pp. 230–241.

[57] H. Sachs, On spatial representation of finite graphs, *Colloquia Mathematica Societatis János Bolyai*, 37. Finite and infinite sets, (eds. A. Hajnal, L. Lovász & V. T. Sós), North-Holland, Amsterdam (1984), pp. 649–662.

[58] D. P. Sanders, P. D. Seymour & R. Thomas, Edge three-coloring cubic doublecross graphs, in preparation.

[59] D. P. Sanders & R. Thomas, Edge three-coloring cubic apex graphs, manuscript 1998.

[60] P. D. Seymour, On the two-colouring of hypergraphs, *The Quarterly Journal of Mathematics*, Oxford Second Series, **25** (1974), 303–312.

[61] P. D. Seymour, Decomposition of regular matroids, *Journal of Combinatorial Theory, Series B*, **28** (1980), 305–359.

[62] P. D. Seymour, Nowhere-zero flows, in *Handbook of combinatorics* (eds. R. L. Graham, M. Grötschel & L. Lovász), North-Holland, Amsterdam (1995), pp. 289–299.

[63] P. D. Seymour & C. Thomassen, Characterization of even directed graphs, *Journal of Combinatorial Theory, Series B*, **42** (1987), 36–45.

[64] P. G. Tait, Note on a theorem in geometry of position, *Transactions of the Royal Society of Edinburgh*, **29** (1880), 657–660.

[65] R. Thomas, An update on the four-color theorem, *Notices of the American Mathematical Society*, **45** (1998), 848–859.

[66] C. Thomassen, Even cycles in directed graphs, *European Journal of Combinatorics*, **6** (1985), 85–89.

[67] C. Thomassen, Sign-nonsingular matrices and even cycles in directed graphs, *Linear Algebra and its Applications*, **75** (1986), 27–41.

[68] C. Thomassen, The even cycle problem for directed graphs, *Journal of the American Mathematical Society*, **5** (1992), 217–229.

[69] W. T. Tutte, A contribution to the theory of chromatic polynomials, *Canadian Mathematical Journal*, **6** (1954), 80–91.

[70] W. T. Tutte, Convex Representations of Graphs, *Proceedings of the London Mathematical Society*, **10** (1960), 304–320.

[71] W. T. Tutte, A theory of 3-connected graphs, *Indagationes Mathematicae*, **23** (1961), 441–455.

[72] W. T. Tutte, On the algebraic theory of graph colorings, *Journal of Combinatorial Theory*, **1** (1966), 15–50.

[73] L. G. Valiant, The complexity of computing the permanent, *Theoretical Computer Science*, **8** (1979), 189–201.

[74] V. V. Vazirani & M. Yannakakis, Pfaffian orientations, 0-1 permanents, and even cycles in directed graphs, *Discrete Applied Mathematics*, **25** (1989), 179–190.

[75] K. Wagner, Über eine Eigenschaft der ebenen Komplexe, *Mathematische Annalen*, **114** (1937), 570–590.

[76] K. Wagner, Über eine Erweiterung des Satzes von Kuratowski, *Deutsche Mathematik*, **2** (1937), 280–285.

[77] K. Wagner, Bemerkungen zu Hadwigers Vermutung, *Mathematische Annalen*, **141** (1960), 433–451.

[78] H. Whitney, 2-isomorphic graphs, *American Journal of Mathematics*, **55** (1933), 245–254.

School of Mathematics
Georgia Institute of Technology
Altanta, GA 30332-0160, USA
thomas@math.gatech.edu

Parity, Cycle Space, and K_4-Subdivisions in Graphs

C. Thomassen

Summary We survey some parity arguments and problems in graph theory, in particular some that can be attacked using the cycle space of a graph. We discuss some results on specific collections of cycles that generate the cycle space. We explain how the space generated by the cycles through two prescribed edges in a graph is used in a proof of the conjecture made by B. Toft in 1974 that every 4-chromatic graph contains a totally odd K_4-subdivision, that is, a subdivision of K_4 in which each edge of K_4 corresponds to an odd path. (Another proof of Toft's conjecture was found independently by W. Zang). We prove the new result that every 4-connected graph with at least three triangles contains a totally odd K_4-subdivision if and only if it does not contain a vertex whose deletion results in a bipartite graph. In particular, every 4-connected planar graph contains a totally odd K_4-subdivision. Finally, we offer some conjectures on path systems and subdivisions with parity constraints on the lengths.

1 Introduction

Parity arguments are often both elegant and powerful. An early parity result in graph theory is Redei's theorem [12] saying that the number of directed Hamiltonian paths in any tournament is odd. It implies, in particular, that every tournament has a directed Hamiltonian path. While this is an easy exercise, Redei's theorem inspired Forcade [4] to a parity result where the corresponding existence result is highly nontrivial. Forcade proved that every tournament whose number of vertices is a power of 2 has the following remarkable property: The number of Hamiltonian paths with any (fixed) orientation of the edges is odd. In Forcade's argument, the condition on the number of vertices is crucial but that condition is not needed in the corresponding existence result: Andrew Thomason [17] proved the deep result that every sufficiently large tournament has a Hamiltonian path with any prescribed edge orientations. Recently, Havet and Thomassé [7] announced that eight vertices are sufficient (while seven are not).

Another early parity result is Smith's theorem, see [22], that the number of Hamiltonian cycles through any fixed edge xy in any cubic graph is even. Tutte [22] gave a proof of Smith's theorem based on counting. Andrew Thomason [16] extended Smith's theorem to any graph G in which all degrees are odd. Thomason's beautiful argument is as follows: Consider any Hamiltonian path P whose first vertex is x and whose second vertex is y. We consider P as a vertex in a new graph H. Path P is joined to the vertex Q in H if Q can be obtained from P by adding an edge incident with the last vertex of P and then deleting an appropriate edge of P. Then H has an even number of vertices of odd degree, and those vertices of H are precisely those Hamiltonian paths in G that start with xy and can be extended to a Hamiltonian cycle of G. If G is cubic, then H is a particularly simple graph: All components of H are paths or cycles. Smith's theorem implies that, if G has a Hamiltonian cycle, then G has a second Hamiltonian cycle as well. Thomason's argument provides a very simple algorithm for finding it: Just walk along a path in H. However, the length of that path may be exponential in the size of G, as shown by Krawczyk [9], and it is a challenging open problem whether a second Hamiltonian cycle can be found in polynomial time.

Parity arguments can also be used for non-existence, as illustrated by the following argument, due to Horst Sachs (private communication), proving the easy part of Kuratowski's theorem saying that K_5 and $K_{3,3}$ have no planar embeddings (drawings) without edge crossings. It suffices to consider embeddings such that every edge is a simple polygonal arc (i.e. every edge is the union of a finite number of straight line segments, no two of which intersect except possibly at a common end). It is easy to see that, for any two embeddings (possibly with edge crossings) of the same graph G, one of the embeddings can be transformed into the other using a sequence of transformations where each transformation consists in moving a point p on G along a straight line and also moving the lines that are incident with p but keeping the rest of the embedding fixed. It is easy to ensure that two edges never have infinitely many points in common. Now Sachs observed that the number of edge crossings of two edges not incident with the same vertex of G is always odd when G is one of K_5 or $K_{3,3}$. Hence neither of these two graphs has planar embeddings with no edge crossings. One remarkable aspect of this argument is that it does not use Euler's formula or even the Jordan Curve Theorem.

2 The cycle space of a graph and generating sets of cycles

Our terminology is essentially that of Bondy and Murty [2]. If E is a set, then the set of subsets of E may be regarded as a vector space over the field of two elements where the sum of two sets, called the *modulo 2 sum*, is the symmetric difference of the two sets. If E is the edge set of a graph G, then the subspace generated by the cycles of G is called the *cycle space* of G and is denoted $Z(G)$. The subspace of $Z(G)$ generated by the even cycles is called the *even cycle space* of G.

Proposition 2.1 *If G is a connected graph with n vertices and m edges, then the cycle space of G has dimension $m - n + 1$. If G is 2-connected and nonbipartite, then the even cycle space has dimension $m - n$.*

Proof Consider first the case where G is 2-connected. If C is any cycle of G, then we may obtain G from C by successively adding a path having the ends but no other vertex in common with the current graph. If G is nonbipartite,

Parity, Cycle Space, and K_4-Subdivisions in Graphs

we may assume that C is odd. Whenever we add a path, the dimension of the cycle space increases by one. So does the dimension of the even cycle space if C is odd. This proves the Proposition when G is 2-connected. If G is not 2-connected, the first assertion is proved by induction on the number of blocks. ∎

Proposition 2.2 *If D is a strongly connected directed graph, then the directed cycles generate the cycle space of D*

Proof The directed graph D can be obtained from any directed cycle in D by successively adding a directed path having only its ends in common with the currect directed graph or by adding a directed cycle having only one vertex in common with the currect directed graph. In each step the dimension of the space generated by the directed cycles increases by one as does the dimension of the cycle space. ∎

The cycle space and the dual concept cocycle space (see [2]) are natural and attractive mathematical object which are useful in the study of electrical networks where the elements of these spaces describe currents and voltages, respectively. The cycle space is also useful for proving graph theorems that might be hard to prove otherwise, as first demonstrated by Bondy and Lovász [1]. An old result of Dirac [3] says that, for any set S of k vertices in a k-connected graph G, G has a cycle containing S. That statement is easily proved by induction on k: If C is a cycle containing all vertices of S except one, say x, then Menger's theorem, see [2, p. 46], implies that G has k paths from x to C having only x in common, pair by pair. Now the union of these paths and C contains the desired cycle. Toft [21] observed that Dirac's argument implies that a nonbipartite graph has an odd cycle through any k prescribed vertices provided the connectivity is at least $2k - 1$. Toft [21] also asked how much $2k - 1$ can be reduced. (It cannot be replaced by k as in Dirac's result because of the complete k-regular bipartite graph with one edge added.) Bondy and Lovász [1] settled Toft's question by proving the following result on the cycle space.

Theorem 2.3 *If S is a set of k vertices in a $(k + 1)$-connected nonbipartite graph G, then the set of cycles in G containing S generate the cycle space of G.*

This theorem settles Toft's question because the even cycle space of a nonbipartite graph G is a proper subspace of the cycle space of G, by Proposition 2.1. Therefore, every $(k + 1)$-connected nonbipartite graph has an odd cycle through any prescribed set of k vertices.

A result which says that any graph satisfying a certain condition has a cycle of a certain type raises the question if the cycles of that type generate the cycle space. Below we mention a few results of that nature.

The first is motivated by the fact that the facial cycles (that is, the cycles forming the face boundaries) of a 2-connected planar graph generate the cycle space. Tutte [23] proved the following.

Theorem 2.4 *If G is a 3-connected graph, then the cycle space of G is generated by the collection of those chordless cycles C in G such that $G - V(C)$ is connected.*

A graph of minimum degree $d > 1$ clearly contains a cycle of length at least $d + 1$. Hartman [6] proved the following.

Theorem 2.5 *If G is a 2-connected graph of minimum degree d, then the cycles of length at least $d + 1$ generate the cycle space of G, unless d is odd and G is the complete graph on $d + 1$ vertices.*

A directed graph D is *strongly k-connected* if $D - S$ is strong (that is, strongly connected) for any set S of fewer than k vertices. A *tournament* is a directed graph such that there is precisely one arc (directed edge) between any two vertices. The present author [19] proved the following two theorems.

Theorem 2.6 *If T is a strongly 4-connected tournament with n vertices, then the cycle space of T is generated by the directed cycles of length n or $n - 1$.*

Theorem 2.7 *If T is a strongly 10^8-connected tournament with n vertices, then the directed cycles of length $n - 1$ generate a space of co-dimension at most 1 in the cycle space.*

3 The cycle space and collections of cycles determining uniquely a graph up to isomorphism

As demonstrated by the theorem of Bondy and Lovász (Theorem 2.3), the cycle space is useful for existence results. The cycle space can also sometimes be used to show that a graph is uniquely determined from certain collections of cycles. Whitney [25] proved that a 3-connected graph is uniquely determined by its collection of cycles. More precisely, if G and H are graphs, G is 3-connected, and there is a bijection f of $E(G)$ onto $E(H)$ such that a subset E of $E(G)$ is the edge set of a cycle in G if and only if $f(E)$ is the edge set of a cycle in H, then G and H are isomorphic. (For a short proof of this theorem of Whitney, see [24, p. 83].) More generally, if a collection of cycles in G generate the cycle space of a 3-connected graph G, then G is uniquely determined by that collection, because the cycles are the minimal nonempty sets in the cycle space. There is a similar result for directed graphs D whose underlying graphs are 3-connected, as proved by the present author [19]. (In the theorem below, two directed graphs are *anti-isomorphic* if one is isomorphic to the directed graph obtained from the other by reversing all arcs.)

Theorem 3.1 *Let D and F be two strongly connected directed graphs whose underlying graphs are 3-connected. Let $C(D)$ and $C(F)$ be collections of directed cycles in D and F, respectively, and let f be a bijection of $E(D)$ onto $E(F)$ taking $C(D)$ onto $C(F)$. Then D and F are isomorphic or anti-isomorphic.*

This result, combined with cycle space results, was used to prove the following.

Theorem 3.2 *If T and F are strongly 10^{15}-connected tournaments, and there is a bijection of the arc set of T onto the arc set of F taking the directed Hamiltonian cycles of T onto the directed Hamiltonian cycles of F, then T and F are isomorphic or anti-isomorphic.*

We now turn to small directed cycles in tournaments. Let us call a directed cycle of length k a *k-cycle*. A tournament is not, in general, determined uniquely, up to isomorphism or anti-isomorphism, by its 3-cycles, even if the tournament is highly connected. To see this let us construct a tournament T by first taking three pairwise disjoint vertex sets A, B, C and adding all arcs from A to B, from B to C, and from C to A. The remaining arcs are added at random. If A, B, C are large, then T is highly connected. Let F be the tournament obtained from T by reversing all arcs in A. Then T and F have the same 3-cycles. But T can be chosen such that it is neither isomorphic nor anti-isomorphic to F. Instead Goldberg and Moon [5] proved the following.

Theorem 3.3 *Every strong tournament is uniquely determined, up to isomorphism or anti-isomorphism, by its collection of 3-cycles and 4-cycles.*

Goldberg and Moon [5] conjectured that a strong tournament is uniquely determined, up to isomorphism or anti-isomorphism, by its collection of 4-cycles. This was verified in [19] where the following two theorems were proved. If T is a tournament, then $T(4)$ denotes the subdigraph which is the union of all 4-cycles in T.

Theorem 3.4 *If T and F are strong tournaments with the same vertex set with at least six vertices such that $T(4) = F(4)$, then $T = F$.*

Theorem 3.5 *If T is a strong tournament with at least four vertices, then the 3-cycles and 4-cycles in $T(4)$ generate the cycle space of $T(4)$.*

Examples show that the underlying graph of $T(4)$ need not be 3-connected. However, as shown in [19], the underlying graph of $T(4)$ is almost 3-connected in the sense that any separating set of two vertices separates a vertex of degree 2 from the rest of the graph. Theorem 3.1 can be extended to such directed graphs. Therefore Theorems 3.4 and 3.5 imply Theorem 3.3. They were also used to prove the following result by the author [19], which settles the conjecture of Goldberg and Moon.

Theorem 3.6 *If T is a strong tournament with at least four vertices, then T is uniquely determined, up to isomorphism or anti-isomorphism, by its collection of 4-cycles.*

Using Theorems 3.4 and 3.5, it is not difficult to extend Theorem 3.3 to the infinite case. However, Theorem 3.6 does not extend to the infinite case, not even the countable case, as shown in [19].

4 The cycle space generated by the cycles through two fixed edges

This section describes the cycle space generated by the cycles through two fixed independent edges. (Two edges are *independent* if they have no end in common.) McCuaig and Rosenfeld [11] proved that a 3-connected graph has both an odd cycle and also an even cycle through any two prescribed edges in the graph if and only if the graph minus the two edges is nonbipartite. They pointed out that this result cannot be immediately proved by the method of Bondy and Lovász since the cycles through two prescribed edges in a 3-connected graph do not generate the cycle space. (Any cycle which contains precisely one of the edges cannot be written as a modulo 2 sum of cycles that contain both edges.) However, it was shown in [20] that the cycle space generated by the cycles through two prescribed edges is large enough to give a generalization of the result of McCuaig and Rosenfeld [11]. I proved the following in [20].

Theorem 4.1 *Let e_1, e_2 be independent edges in a 3-connected graph G with n vertices. Let H be a multigraph obtained from G by replacing some non-cutedges of $G - \{e_1, e_2\}$ by multiple edges. Let m denote the number of edges of H, and let Z denote the subspace of the cycle space of H generated by the cycles in H containing e_1, e_2. Then $\dim(Z) = m - n$.*

Theorem 4.1 implies immediately the following result of Lovász [10].

Corollary 4.2 *If G is a 3-connected graph and e_1, e_2, e_3, are independent edges in G, and $G - \{e_1, e_2, e_3\}$ is connected, then G has a cycle containing e_1, e_2, e_3.*

Proof Since e_3 is not a cut edge of $G - \{e_1, e_2\}$, we may apply Theorem 4.1 to the graph H obtained from G by replacing e_3 by a multiple edge consisting of, say, q edges. Since the dimension of Z tends to infinity as q tends to infinity, G must have a cycle through e_1, e_2, e_3. ∎

Corollary 4.3 *Let G be a 3-connected graph, and let e_1, e_2 be two independent edges. If C is a cycle in $G - \{e_1, e_2\}$, then C is the modulo 2 sum of cycles containing e_1, e_2.*

Proof Let Z (respectively Z') denote the subspace of the cycle space $Z(G)$ of G generated by the cycles containing both (respectively both or none) of e_1, e_2. Then
$$Z \subseteq Z' \subseteq Z(G), \text{ and } Z' \neq Z.$$
As $\dim(Z) = \dim(Z(G)) - 1$, it follows that $Z = Z'$. ∎

Corollary 4.4 *Let G be a 3-connected graph, and let e_1, e_2 be two independent edges. Let H be a subdivision of G. (If e_i becomes a path in H, then we also denote that path by e_i.) If $H - \{e_1, e_2\}$ has an odd cycle, then H has an odd cycle containing e_1, e_2, and H has an even cycle containing e_1, e_2.*

Proof Let C be an odd cycle in $H - \{e_1, e_2\}$. By Corollary 4.3, C is the modulo 2 sum of cycles $C_1, C_2, \ldots, C_q$ in H containing e_1, e_2. Since C is odd, one of $C_1, C_2, \ldots, C_q$ must be odd. Since C contains none of e_1, e_2, q is even. Hence at least one of $C_1, C_2, \ldots, C_q$ must be even. ∎

Corollary 4.5 *Let e_1, e_2, e_3 be three pairwise independent edges in a 3-connected nonbipartite graph G. Then G has an odd cycle containing at least two of e_1, e_2, e_3.*

Proof By Corollary 4.4, we may assume that $G - \{e_1, e_2, e_3\}$ is bipartite with bipartition A, B, say. Since G is nonbipartite, we may assume that e_1 has both ends in A. If e_2 joins A with B, then any cycle in $G - e_3$ through e_1 and e_2 is odd. So assume that e_2 (and similarly e_3) joins two vertices which are both in A or both in B. By Corollary 4.2, we may assume that $G - \{e_1, e_2, e_3\}$ is disconnected. Now it is easy to see that G is bipartite, a contradiction. ∎

Using Theorem 4.1 it is very easy to decide if two given independent edges in a 3-connected graph G are contained in an odd (respectively even) cycle. If G minus the two edges is nonbipartite, then the answer to both questions is affirmative. However, if G minus the two edges is bipartite, then we consider any cycle C through the two edges. All other cycles through the two edges have the same parity as C. A little more careful analysis yields the following result in [20].

Theorem 4.6 *There exists a polynomially bounded algorithm for the following problem. Given a graph G and two edges e_1, e_2 in G, do all cycles through e_1, e_2 have the same parity?*

5 The cycle space of a graph and K_4-subdivisions

In [18] it was shown that, for every natural number k, there exists a natural number $f(k)$ such that the following holds: If G is a graph of chromatic number at least $f(k)$, then G contains a subdivision of the complete graph K_k such that each edge of K_k corresponds to a path in the subdivision of any prescribed parity. Previously, Toft [21] asked if every 4-chromatic graph contains a totally odd K_4-subdivision (TOKS), that is a subdivision of K_4 in which each edge corresponds to an odd path, see also [8]. Toft's conjecture was proved by Zang [26] and, independently (but later) by the present author [20]. In this section we point out how the cycle space is utilized in the proof in [20].

Let v be a vertex in a graph G. If G has a TOKS in which v has degree i (where $i = 0$ or 2), then we say that v is of *type i* in G. We say that v is of *type* 1 if G has a TOKS in which each of the three paths starting at v has length 1. Finally, we say that v is of *type* 3 in G if G has two K_4-subdivisions such that v has degree 3 in both of them, the two K_4-subdivisions contain the same three edges incident with v, one of them is a TOKS, and in the other K_4-subdivision the three paths starting at v have even length, and the three other paths have odd length. The idea in [20] is to prove that every vertex in a 4-chromatic graph is of type $0, 1, 2$ or 3. However, this is not true, as shown by the graph obtained by applying Hajos' construction (see [8]) to two copies of K_4, that is, we delete an edge xy and uw, respectively in each K_4, and then we identify x and u and add the edge yw. Therefore we now introduce a *strong TOKS* recursively as follows: If v is a vertex of type 1 in a TOKS, then we say that TOKS is a strong TOKS in which v is a *distinguished vertex*. Suppose now that H and L each is a strong TOKS with distinguished vertex v and u, respectively. Then we delete an edge vx in H and an edge uy in L. We identify v and u, and we add the edge xy. The resulting graph is also called a strong TOKS with v as distinguished vertex. It is easy to see that a strong TOKS contains a TOKS. Also, a vertex of type 1 is the distinguished vertex in a strong TOKS. Now the following was proved in [20].

Theorem 5.1 *Let v be any vertex in any 4-chromatic graph G. Then either v is of type 0 or 2 or 3 in G, or else G has a strong TOKS in which v is a distinguished vertex.*

We sketch the proof of Theorem 5.1, which is by induction on the number of vertices of G. The subgraph of G induced by the neighbours of v is denoted $N(v)$. Now the proof of Theorem 5.1 is divided into two cases.

Case 1: $N(v)$ has no two independent edges.

Case 2: $N(v)$ has two independent edges e_1 and e_2.

In Case 1 we may assume that $N(v)$ has no triangle, since otherwise v is of type 1. So $N(v)$ has a vertex u such that each edge of $N(v)$ (if any) is incident with u. Now we form a new graph H by first deleting v and then identifying all vertices of $N(v) - u$ into a single vertex w, say. Clearly, H has chromatic number at least 4. We now apply the induction hypothesis to w and H. If w is of type 0 or 2 in H, then clearly v is of type 0 or 2 in G. If w is of type 3, then clearly v is of type 3 or 2 or 0 in G. There remains only the case where w is the distinguished vertex of a strong TOKS in H. That case is not easy. It is disposed of by establishing some connectivity properties of $G - v$ and then considering appropriate paths from the vertices in G corresponding to w to the rest of the strong TOKS.

In Case 2 we may assume that $G - v$ has no odd cycle containing e_1 and e_2 for otherwise v is of type 1. Therefore we may apply Theorem 4.1 and the ideas in the proof of Theorem 4.6. If $G-v$ is 3-connected, then $G-v$ is nearly bipartite, by Theorem 4.1. On the other hand, if $G - v$ is not 3-connected, then we form its so-called 3-*blocks* (see [20]). One of these 3-blocks is nearly bipartite. We then use this nearly bipartite subgraph to make a reduction (for example identifying almost all vertices in one of the bipartite classes) and complete the proof by induction.

6 Towards a characterization of the graphs containing no totally odd K_4-subdivisions

In this section we establish some sufficient conditions for a graph to contain a TOKS. We prove that every 4-connected graph G with at least three triangles contains a TOKS if and only if G has no vertex v such that $G-v$ is bipartite. As a corollary, every 4-connected planar graph contains a TOKS.

A vertex of degree at least 3 in a graph is called a *branch vertex*.

Proposition 6.1 *Let C_1: $x_1x_2x_3x_1$ and C_2: $y_1y_2y_3y_1$ be vertex-disjoint triangles in a 3-connected graph G. If G has three pairwise disjoint paths P_1, P_2, P_3 from C_1 to C_2, not all of the same parity, then G contains a TOKS.*

Proof Let the notation be such that P_i has ends x_i and y_i for $i = 1, 2, 3$. If two of P_1, P_2, P_3 are even, then $C_1 \cup C_2 \cup P_1 \cup P_2 \cup P_3$ contains a TOKS. So assume that P_2, P_3, say, are odd and that P_1 is even. Let z_1 be the neighbour of x_1 on P_1. As G is 3-connected, $G - x_1$ has two paths P_4, P_5 from z_1 to $P_2 \cup P_3 \cup C_2$ having only z_1 in common. Let P_1' be the path obtained by first walking from y_1 along P_1 until we hit one of P_4, P_5, say P_5, and then following P_5 to z_1. We may assume that the end of P_4 distinct from z_1 is a vertex z_2 on P_2. If P_1' is odd, then P_4 can be extended to an odd path from z_1 to either x_2 or y_2, and that odd path can be extended to a TOKS whose branch vertices are either z_1, x_1, x_2, x_3 or z_1, y_1, y_2, y_3. So we may assume that P_1' is even. We may also assume that the path in $P_2 \cup P_4$ from z_1 to x_2 is even since, otherwise, G has a TOKS with branch vertices z_1, x_1, x_2, x_3 (if $z_2 \neq y_2$) or y_2, x_1, x_2, x_3 (if $z_2 = y_2$). Since the parity of P_1' is distinct from the parity of the subpath of P_1 from z_1 to y_1, it follows that P_1 has a subpath P_6 such that P_6 has only its ends in common with $P_1' \cup P_4$ and such that the unique cycle C_3 in $P_1' \cup P_4 \cup P_6$ is odd. If P_6 has both ends on P_1' or both ends on P_4, then by the previous reasoning, G has a TOKS with branch vertices z_1, x_1, x_2, x_3 or z_1, y_1, y_2, y_3 or y_1, x_1, x_2, x_3. So we may assume that P_6 has one end on $P_1' - z_1$ and the other on $P_4 - z_1$. We may also assume that x_2 is not an end of P_4 since, otherwise, C_2 can be extended to a TOKS with branch vertices x_2, y_1, y_2, y_3. More generally, we may assume that the subpath of P_2 from z_2 to y_2 is even. We may assume that y_1 is an end of P_6 since, otherwise, G has a TOKS with branch vertices z_2, x_1, x_2, x_3. This implies in particular that $P_1' = P_5$ by the definition of P_1'. We may assume that $z_2 = y_2$ since otherwise, G has a TOKS with branch vertices y_1, x_1, x_2, x_3.

We have proved so far that, for any two paths P_4, P_5 in $G - x_1$ from z_1 to $P_2 \cup P_3 \cup C_2$ having only z_1 in common, one of them must have y_1 as an end, and the other must have one of y_2, y_3 as an end. We repeat this argument with y_1, u_1 instead of x_1, z_1 where u_1 is the neighbour of y_1 on P_1 (and P_6 which has the edge incident with y_1 in common with P_1). Then $G - y_1$ has two paths P_7, P_8 from u_1 to $P_2 \cup P_3 \cup C_1$ having only u_1 in common. Moreover, the two ends of P_7, P_8 distinct from u_1 must be in C_1 by the previous reasoning. But, it is easy to make one of P_7, P_8 end in y_2 using P_4 and P_6. This contradiction completes the proof. ∎

Theorem 6.2 *A 4-connected graph G containing at least three triangles contains a TOKS if and only if G has no vertex v such that $G - v$ is bipartite.*

Proof The necessity of the last condition is obvious. So we now assume that G is 4-connected, contains at least three triangles, and contains no vertex whose deletion results in a bipartite graph. Suppose (reductio ad absurdum) that G contains no TOKS. We derive several properties of G and finally reach a contradiction. Following [2] we write $G - x$ instead of $G - \{x\}$ and hence also $G - x - y$ instead of $G - \{x, y\}$ when x and y are vertices of the graph G.

(1) *G has no two triangles with a common edge.*

To prove (1), let us assume that C_1: $xyzx$ and C_2: $uyzu$ are triangles in G. Then $G - y - z$ is 2-connected and has no odd path between x and u. Hence $G - y - z$ is bipartite with bipartition A, B say, and x, u are in the same bipartite set of $G - y - z$, say A. As none of $G - y$ or $G - z$ are bipartite, G has edges yw and zr where w, r are in B. As $G - y - z$ is 2-connected, it has two paths, one from w to r and the other from w to one of x, u such that these two paths have only w in common. Now G has a TOKS with branch vertices w, y, z and one of x, u. This contradiction proves (1).

(2) *G has no three triangles with a common vertex.*

To prove (2), let us assume that G has three triangles all containing the vertex v. Let e_1, e_2, e_3 be the edges in these triangles not incident with v. Since G contains no TOKS, $G - v$ has no odd cycle containing two of e_1, e_2, e_3. By Corollary 4.5, $G - v$ is bipartite, a contradiction to the initial assumption of the present proof.

(3) *G has no two triangles with a common vertex.*

To prove (3), let us assume that $xyzx$ and $xuvx$ are triangles in G. Let T be a third triangle. By (1), y, z, u, v are distinct. By (2), T does not contain x.

Consider first the case where T contains none of y, z, u, v. Since G is 4-connected, $G - x$ has three pairwise disjoint paths from T to $\{y, z, u, v\}$. By Proposition 6.1, G contains a TOKS, a contradiction.

Consider next the case where T contains at least one of y, z, u, v and hence precisely one of y, z, u, v (by (1)), say y. Then $G - x - y$ has two disjoint paths P_1, P_2 from $T - y$ to $\{u, v\}$. By Proposition 6.1, the paths P_1, P_2 are both odd. If neither of P_1, P_2 contains z, then G contains a TOKS by Proposition 6.1. So assume that one of them, say P_1, contains z. As P_1 is odd, we may assume that the subpath of P_1 from z to u is odd. Now G contains a TOKS with branch vertices $\{x, u, v, z\}$. This contradiction proves (3).

By (3), G has three pairwise disjoint triangles C_1, C_2, C_3. We claim that the notation can be chosen such that

(4) G has three pairwise disjoint paths M_1, M_2, M_3 from C_1 to C_2 such that $M_1 \cup M_2 \cup M_3$ contains at most one vertex of C_3.

To prove (4), Let us consider any three pairwise disjoint paths P_1, P_2, P_3 from C_1 to C_2. Assume, without loss of generality, that P_1 contains a vertex x of C_3. If P_1 contains another vertex of C_3, then G contains three pairwise disjoint paths, not all of the same parity, between some two of C_1, C_2, C_3, a contradiction to Proposition 6.1. We may also assume that P_2 contains precisely one vertex y of C_3. Now let P_4 be a path in $G - x - y$ from $C_3 - x - y$ to $C_1 \cup C_2 \cup P_1 \cup P_2 \cup P_3$. If P_4 terminates at P_1 or P_2 we get a contradiction to Proposition 6.1. On the other hand, if P_4 terminates on P_3, then (4) is satisfied.

Consider three pairwise disjoint paths P_3, Q_3, R_3 from C_3 to $C_1 \cup C_2 \cup M_1 \cup M_2 \cup M_3$ (one of which may have length 0), and let H denote the graph $C_1 \cup C_2 \cup C_3 \cup M_1 \cup M_2 \cup M_3 \cup P_3 \cup Q_3 \cup R_3$. We may assume that P_3, Q_3, R_3 terminate at vertices p, q, r on M_1, M_2, M_3, respectively, since otherwise G contains a TOKS by Proposition 6.1. Let P_1, P_2 denote the subpaths of M_1 from p to C_1, C_2, respectively. The paths Q_1, Q_2, R_1, R_2 are defined analogously.

The path P_1 must be even since, otherwise, G has a TOKS whose branch vertices are p and the vertices of C_1. More generally,

(5) each of the paths $P_1, P_2, P_3, Q_1, Q_2, Q_3, R_1, R_2, R_3$ is even.

As G is 4-connected, $G - p - q - r$ has a path P' connecting distinct components of $H - p - q - r$. Assume without loss of generality the P' starts at a vertex p_1 in P_1. Then

(6) the other end p_2 of P' is at one of the paths P_2, P_3.

For clearly, p_2 is at one of the paths $P_2, P_3, Q_2, Q_3, R_2, R_3$. The vertex p_2 must be on one of P_2, P_3, say P_2, since, otherwise, G contains three pairwise disjoint paths, not all of the same parity, between some two of C_1, C_2, C_3, a contradiction to Proposition 6.1. Put $p_3 = p$. Then

(7) each of the vertices p_1, p_2, p_3 can play the role of p.

By (5) and (7), the three paths between p_1, p_2, p_3 are even. They form a 2-connected graph disjoint from $M_2 \cup M_3 \cup Q_3 \cup R_3$. We extend it to a maximal 2-connected subgraph $H(p)$ disjoint from $M_2 \cup M_3 \cup Q_3 \cup R_3$. Let p'_1, p'_2, p'_3 be the first vertices we meet in $H(p)$ when we traverse P_1, P_2, P_3 from C_1, C_2, C_3, respectively. Since $H(p)$ is 2-connected, we conclude that

(8) *each of the vertices p'_1, p'_2, p'_3 can play the role of p.*

Assume without loss of generality that Q_1 has positive length. Then we consider a path Q' in $G - p'_1 - q - r$ joining the two components of $M \cup H(p) - p'_1 - q - r$. The maximality of $H(p)$ implies that Q' does not start at a vertex of P_1. Assume therefore that Q' starts at a vertex of Q_1. We may assume that the other end of Q' is in Q_2 or Q_3. We then define $H(q)$ in a similar way as we defined $H(p)$, and we define q'_1, q'_2, q'_3 in a similar way as we defined p'_1, p'_2, p'_3. Now we consider a path R' in $G - p'_1 - q'_1 - r$ joining the two components of $M \cup H(p) \cup H(q) - p'_1 - q'_1 - r$, and we use that path to define $H(r)$ and r'_1, r'_2, r'_3. Consider finally a path in $G - p'_1 - p'_2 - p'_3$ joining the two components of $M \cup H(p) \cup H(q) \cup H(r) - p'_1 - p'_2 - p'_3$. Since each of the vertices p'_1, p'_2, p'_3 can play the role of p, and each of the vertices q'_1, q'_2, q'_3 can play the role of q, and each of the vertices r'_1, r'_2, r'_3 can play the role of r, we now get a contradiction to either (6) or the maximality of $H(p)$. This contradiction completes the proof. ∎

Corollary 6.3 *Let G be a 4-connected graph. If G can be embedded in the plane, then G contains a TOKS. If G can be embedded in the projective plane, then G contains a TOKS if and only if, for every vertex v in G, the graph $G - v$ is nonbipartite.*

Proof Euler's formula implies that G has at least four triangles. Moreover, if G is planar, then, for each vertex v, the graph $G - v$ has a triangle. We can therefore apply Theorem 6.2. ∎

A 4-connected graph embedded in the projective plane may contain a vertex whose deletion results in a bipartite graph. To see this take the Cartesian product of a path and a cycle of length 2 modulo 4. Draw this graph in the plane. There are two facial cycles of length greater than 4. In one of these we add a vertex v joined to all vertices of the facial cycle. In the other we add all diagonals. The resulting graph G is 4-connected and projective planar. Moreover, $G - v$ is bipartite.

7 Open problems

A deep result of Robertson and Seymour [13] provides, for each fixed natural number k, a polynomially bounded algorithm for the *k-path problem*: Let G be a graph and let $x_1, x_2, \ldots, x_k, y_1, y_2, \ldots, y_k$ be vertices in G. Does G contain k pairwise disjoint paths $P_1, P_2, \ldots, P_k$ such that P_i joins x_i, y_i for $i = 1, 2, \ldots, k$? Let us define the *odd k-path problem* as the k-path problem with the additional requirement that all paths $P_1, P_2, \ldots, P_k$ are odd. We suggest the following extension.

Conjecture 7.1 *For each fixed natural number k, there exists a polynomially bounded algorithm for the odd k-path problem.*

If true, Conjecture 7.1 implies the conjecture of Zang [26] that there exists a polynomially bounded algorithm for the *TOKS problem*, that is, the problem of deciding if a given graph G contains a TOKS. (In fact, if Conjecture 7.1 holds we can even decide, in polynomial time, which quadruples of vertices can play the role of the branch vertices.) The TOKS problem was mentioned by Sewell and Trotter [14] who gave a polynomially bounded algorithm for producing a maximum independent vertex set in a graph containing no TOKS. Conjecture 7.1 is open even for $k = 2$. The following weaker conjecture may also be of interest.

Conjecture 7.2 *For each fixed natural number k, there exists a polynomially bounded algorithm for the following problem: Given two disjoint sets A, B each of k vertices in a graph G, does G contain k pairwise disjoint paths from A to B not all of the same parity?*

For $k = 2$, Conjecture 7.2 reduces to Theorem 4.6. An affirmative answer for $k = 3$ is very likely to solve the TOKS problem restricted to those 3-connected graphs that contain two disjoint triangles, using Proposition 6.1. A 3-connected planar graph containing many pairwise disjoint triangles need not contain a TOKS, as demonstrated by the Cartesian product of a triangle and any path. Perhaps every 4-connected graph with two disjoint triangles contains a TOKS. Perhaps even the following holds.

Conjecture 7.3 *If G is a 4-connected graph containing two disjoint odd cycles, then G contains a TOKS.*

If true, Conjecture 7.3 solves the TOKS problem restricted to 4-connected graphs because Conjecture 7.3 is equivalent to Conjecture 7.4 below.

Conjecture 7.4 *If G is a 4-connected graph, then G contains a TOKS if and only if $G - v$ is nonbipartite for every vertex v.*

Clearly, Conjecture 7.4 implies Conjecture 7.3. To see that Conjecture 7.3 implies Conjecture 7.4, it suffices to consider those 4-connected graphs G such that G has no two disjoint odd cycles, and $G - v$ is nonbipartite for every vertex v. It has been shown by Lovász, see [15, p. 546], that any such graph G contains three edges forming a triangle such that the deletion of these edges leaves a bipartite graph. But then it is easy to extend the triangle to a TOKS.

References

[1] J. A. Bondy & L. Lovász, Cycles through specified vertices of a graph, *Combinatorica*, **1** (1981), 117–140.

[2] J. A. Bondy & U. S. R. Murty, *Graph Theory with Applications*, Macmillan, London (1976).

[3] G. A. Dirac, In abstrakten Graphen vorhandene 4-Graphen und ihre Unterteilungen, *Mathematische Nachrichten*, **22** (1960), 61–85.

[4] R. Forcade, Parity of paths and circuits in tournaments, *Discrete Mathematics*, **6** (1973), 115–118.

[5] M. Goldberg & J. W. Moon, Arc mappings and tournament isomorphisms, *Journal of the London Mathematical Society*, **3** (1971), 378–384.

[6] I. B. A. Hartman, Long cycles generate the cycle space of a graph, *European Journal of Combinatorics*, **4** (1983), 237–246.

[7] F. Havet & S. Thomassé, Paths in tournaments, a proof of Rosenfeld's conjecture, *Journal of Combinatorial Theory, Series B*, in press.

[8] T. Jensen & B. Toft, *Graph Coloring Problems*, John Wiley, New York (1995).

[9] A. Krawczyk, A note on finding a second hamiltonian cycle in cubic graphs, preprint, December 5, 1995.

[10] L. Lovász, Problem 5, *Periodica Mathematica Hungarica*, **4** (1974), 82.

[11] W. D. McCuaig & M. Rosenfeld, Parity of edges containing specified edges, in *Cycles in Graphs* (eds. B. R. Alspach & C. D. Godsil), *Annals of Discrete Mathematics*, 27, North-Holland, Amsterdam (1985), pp. 419–431.

[12] L. Redéi, Ein kombinatorischer Satz, *Acta Litterarum ac Scientiarum Szeged*, **7** (1934), 39–43.

[13] N. Robertson & P. D. Seymour, Graph minors XIII. The disjoint paths problem, *Journal of Combinatorial Theory, Series B*, **63** (1995), 65–110.

[14] E. C. Sewell & L. E. Trotter, Stability critical graphs and even subdivions of K_4, *Journal of Combinatorial Theory, Series B*, **59** (1993), 74–84.

[15] P. D. Seymour, Matroid minors, in *Handbook of Combinatorics* (eds. R. L. Graham, M. Grötschel & L. Lovász), North-Holland, Amsterdam (1985), pp. 419–431.

[16] A. Thomason, Hamiltonian cycles and uniquely edge colourable graphs, in *Advances in Graph Theory* (ed. B. Bollobás), *Annals of Discrete Mathematics*, 3, North-Holland, Amsterdam (1978), pp. 259–268.

[17] A. Thomason, Paths and cycles in tournaments, *Transactions of the American Mathematical Society*, **296** (1986), 167–180.

[18] C. Thomassen, Graph decomposition with applications to subdivisions and path systems modulo k, *Journal of Graph Theory*, **7** (1983), 261–271.

[19] C. Thomassen, Whitney's 2-switching theorem, cycle spaces, and arc mappings of directed graphs, *Journal of Combinatorial Theory, Series B*, **46** (1989), 257–291.

[20] C. Thomassen, Totally odd K_4-subdivisions in 4-chromatic graphs, Math Report No. 1998-04, Technical University of Denmark, 1998.

[21] B. Toft, Problems 10 and 11, in *Recent Advances in Graph Theory: Proceedings of the Symposium held in Prague, June 1974* (ed. M. Fiedler), Academia Praha, Prague (1975), pp. 543–544.

[22] W. T. Tutte, On Hamiltonian circuits, *Journal of the London Mathematical Society*, **21** (1946), 98–101.

[23] W. T. Tutte, How to draw a graph, *Proceedings of the London Mathematical Society*, **13** (1963), 743–768.

[24] D. J. A. Welsh, *Matroid Theory*, Academic Press, New York (1976).

[25] H. Whitney, Congruent graphs and the connectivity of graphs, *American Journal of Mathematics*, **54** (1932), 150–168.

[26] W. Zang, Proof of Toft's conjecture: Every graph containing no fully odd K_4 is 3-colorable, *Journal of Combinatorial Optimization*, **2** (1998), 117–188.

Department of Mathematics
Technical University of Denmark
DK-2800 Lyngby
Denmark
C.Thomassen@mat.dtu.dk

Models of Random Regular Graphs

N. C. Wormald

Summary This is a survey of results on properties of random regular graphs, together with an exposition of some of the main methods of obtaining these results. Related results on asymptotic enumeration are also presented, as well as various generalisations to random graphs with given degree sequence. A major feature in this area is the small subgraph conditioning method. When applicable, this establishes a relationship between random regular graphs with uniform distribution, and non-uniform models of random regular graphs in which the probability of a graph G is weighted according to the number of subgraphs that G has of a certain type. Information can be obtained in this way on the probability of existence of various types of spanning subgraphs, such as Hamilton cycles and decompositions into perfect matchings. Uniformly distributed labelled random regular graphs receive most of the attention, but also included are several non-uniform models which come about in a natural way. Some of these appear as spin-offs from the small subgraph conditioning method, and some arise from algorithms which use simple approaches to generating random regular graphs. A quite separate role played by algorithms is in the derivation of random graph properties by analysing the performance of an appropriate greedy algorithm on a random regular graph. Many open problems and conjectures are given.

1 Introduction

Random graphs first appeared in clever probabilistic proofs by Erdős of the existence of graphs with special properties such as arbitrarily large girth and chromatic number. These had not at that time been found constructively. Much later, the study of random regular graphs took off, beginning with the works of Bender and Canfield [5], Bollobás [10] and Wormald [117, 118]. This has since been fuelled in part by applications in other areas such as computer science. An interesting application occurs in biogeography, where random non-negative integer matrices with given row and column sums are of interest. See Wilson [113], for example.

This paper is a survey of that part of random graph theory in which the degrees of vertices are restricted. Such work concentrates on regular graphs as the most interesting examples, and the results on regular graphs often extend easily to more general degree sequences. A *d-regular* graph is one with all vertices of degree d. For $d = 3$ these are often called *cubic* graphs.

The asymptotic enumeration of objects of a given type frequently goes hand in hand with the problem of generating the same objects uniformly at random, and with finding some of the limiting probabilities in the resulting probability space (see Jerrum and Sinclair [58]). However, the two problems are not equivalent.

The distinction between the approaches of exact and asymptotic enumeration is important. Probably the first result on short cycles in random regular graphs of degree at least 3 was the determination of the expected number of triangles in random cubic graphs [114]. This was done by using recurrence relations, and the asymptotic result $\frac{4}{3}$ was obtained. However, this method of exact enumeration followed by asymptotic analysis has not been able to reach any further for the type of problems that we are concerned with here. For instance, for the numbers of 2- and 3-connected cubic graphs there are recurrence relations [116], but no asymptotic formulae have been obtained from these. A more direct probabilistic approach, with an initially asymptotic viewpoint, can do much more and leads to asymptotic enumeration results in any case. For instance, many of the results in the book on random graphs by Bollobás [16] can be translated this way. In spite of this, exact and asymptotic enumeration turns out to be used rather heavily for some problems here, albeit indirectly (for example, consider the proofs in Section 4.1).

Results on random regular graphs can of course be used to show that there exist graphs with particular combinations of properties; for instance d-regular graphs which are d-connected, have arbitrarily large girth, are Hamiltonian and have no non-trivial automorphisms. There are other ways of getting regular graphs with interesting properties in a constructive way, some of them quite sophisticated (surveyed by Chung [26]), and these supply graphs with some of the properties that can be found in random graphs. A notable exception is the diameter, where the best known constructions are only within a factor $c > 1$ of that of a random d-regular graph.

However, the study of random regular graphs is recently blossoming, and some pretty results are newly emerging, such as the almost sure property that the edges can be partitioned into disjoint Hamilton cycles (when the degree is even). This survey attempts to cover all the results, and show some relationships between old and new.

1.1 Overview and notation

Early in this article we encounter the uniform model of random d-regular graphs and various properties of it, in Section 2, and then uniform models similar to regular graphs (such as bipartite) in Section 3. Extensions to random non-regular graphs with given degree sequence are in the main mentioned at the end of the appropriate subsection. For some properties the methods extend easily in this way (short cycles, connectivity) but for others they seem not to at all. For example, it is known that a large random 3-regular graph is almost surely Hamiltonian, and so is a random 4-regular graph. However, present methods seem not to be strong enough to show that a large random graph with say half of its vertices of degree 3 and half of degree 4 is almost surely Hamiltonian.

In all this, we postpone detailed discussion of a general method called the small subgraph conditioning method until Section 4, because it requires considerable development and has interesting ramifications for models with non-uniform distribution. It is enough for now to know that this gives information on the distribution of random variables which count certain large subgraphs in the uniform model of random regular graphs. The problem of generating random regular graphs is considered in Section 5. We then visit some other models with non-uniform distributions, and some models which do not produce graphs with given degree sequence but are somehow similar. The final farewell is with comments on unsolved problems.

Almost all of the models discussed here assume that the graphs concerned have labelled vertices. Random unlabelled regular graphs can be accessed via the labelled model by studying the order of the automorphism group of the graphs (see Section 2.7).

For a non-negative integer j and real x, $[x]_j$ denotes $x(x-1)\cdots(x-j+1)$. We use the notation **P**, **E** and **Var** for probability, expectation and variance, sometimes subscripted as in $\mathbf{P}_\mathcal{G}$ to specify that the probability measure is as in the probability space $\mathcal{G}$. We say that an event H_n occurs a.a.s. (asymptotically almost surely) if $\mathbf{P}H_n \to 1$ as $n \to \infty$, with any obviously necessary parity restriction on n.

2 Uniform model for random regular graphs

We use $\mathcal{G}_{n,d}$ to denote the uniform probability space of d-regular graphs on the n vertices $\{1, 2, \ldots, n\}$ (where dn is even). So sampling from $\mathcal{G}_{n,d}$ is equivalent to taking such a graph uniformly at random (u.a.r.). We use $|\mathcal{G}|$ to denote the number of elements of a uniform space $\mathcal{G}$.

Another probabilistic space can be defined as follows. Suppose that dn is even, and for non-triviality take $d \geq 1$. Consider a set of dn *points* partitioned into n *cells* $v_1, v_2, \ldots, v_n$ of d points each. A perfect matching of the points into $\frac{1}{2}dn$ pairs is called a *pairing*. A pairing P corresponds to a multigraph (with loops permitted) $G(P)$ in which the cells are regarded as vertices and the pairs as edges: a pair (x, y) in P corresponds to an edge (v_i, v_j) of $G(P)$ where $x \in v_i$ and $y \in v_j$. Since each graph (which we assume is simple; i.e. has no loops or multiple edges) corresponds to precisely $(d!)^n$ pairings, a regular graph can be chosen u.a.r. by choosing a pairing u.a.r. and rejecting the result if it has loops or multiple edges. Non-simple graphs are *not* produced uniformly at random since each for each loop the number of corresponding pairings is divided by 2, and for each k-tuple edge it is divided by $k!$. This is the *pairing model* of random regular graphs, given in this form first by Bollobás, but see Section 2.1 for a description of pre-existing models. We denote the (uniform) probability space of pairings by $\mathcal{P}_{n,d}$. We can assume that the points are the elements of $\{1, \ldots, n\} \times \{1, \ldots, d\}$, so that $G(P)$ is induced by a projection.

The pairing model gives a basis for proving properties of graphs in $\mathcal{G}_{n,d}$, by doing computations in $\mathcal{P}_{n,d}$, and conditioning on the event that the corresponding multigraph has no loops or multiple edges. We call this event "Simple", and note that it is a function of n and d. Knowledge of the value of $\mathbf{P}_{\mathcal{P}_{n,d}}(\text{Simple})$ permits the following simple exploitation of the connection between $\mathcal{P}_{n,d}$ and $\mathcal{G}_{n,d}$.

Lemma 2.1 *Let H be an event (set of graphs) in $\mathcal{G}_{n,d}$ and H' the set of pairings in $\mathcal{P}_{n,d}$ that correspond to graphs in H. Then*

$$\mathbf{P}_{\mathcal{G}_{n,d}}(H) = \frac{\mathbf{P}_{\mathcal{P}_{n,d}}(H')}{\mathbf{P}(\text{Simple})}.$$

Proof This comes immediately from the uniformity of the two models, and the fact that each graph corresponds to the same number of pairings. ∎

A pairing can be selected u.a.r. in many different ways. In particular, the points in the pairs can be chosen sequentially. At any stage, the first point in the next random pair chosen can be selected using any rule whatsoever, as long as the second point in that pair is chosen u.a.r. from the remaining points. For example, one can insist that the next point chosen is the next one available in any pre-specified ordering of the points, or comes from a cell containing one of the points in the previous pair chosen (if any points such are still unpaired). We use this idea several times in this article and so give it a name: the *independence property* of the pairing model.

In addition, the pairing model provides a simple mechanism for enumerating d-regular graphs asymptotically. Since the number of pairings, or perfect matchings, of t points is

$$f(t) = \frac{t!}{(t/2)!2^{t/2}}, \qquad (1)$$

the number of d-regular graphs on n vertices is precisely

$$|\mathcal{G}_{n,d}| = \frac{(dn)!\mathbf{P}(\text{Simple})}{(dn/2)!2^{dn/2}(d!)^n}. \qquad (2)$$

Thus, an asymptotic formula for $|\mathcal{G}_{n,d}|$ can be found by estimating $\mathbf{P}(\text{Simple})$.

The pairing model for random graphs with given degree sequence is an immediate extension of the regular case: for a degree sequence $d_1, \ldots, d_n$, the cell v_i contains d_i points, and a perfect matching of all the points is selected u.a.r. Restricting to no loops or multiple edges produces u.a.r. graphs with degree sequence $\mathbf{d} = (d_1, \ldots, d_n)$. This uniform model of graphs we denote by $\mathcal{G}_{n,\mathbf{d}}$. In all asymptotic statements about this model we assume $\mathbf{d}$ is restricted to sequences with even sum.

We next discuss models for related combinatorial configurations, and then outline results on properties of $\mathcal{G}_{n,d}$.

2.1 History of uniform models

The pairing model was first given in its simple explicit form by Bollobás (beginning with [9]) and called the *configuration model*. (We prefer not to refer to pairings as configurations since the latter has a broad meaning which is useful in many contexts, for example in Section 4.2.) A little earlier, Bender and Canfield [5] used a model in their enumeration of graphs with given degrees which is implicitly a generalisation of the pairing model (see Section 4 of that paper). The difference is that they use involutions in place of perfect matchings of points, because they study 0-1 matrices with given row sums, which are the adjacency matrices of graphs with given degrees, and as a result the entries of the matrix on the main diagonal must be treated differently. (An entry 1 in the main diagonal does not correspond to a loop in the pairing model, since a loop uses up *two* points.) In addition, their analysis was made more general by allowing the possibility to specify a limited number of edges to be forbidden.

Even earlier, Békéssy et al. [4] studied random 0-1 matrices with given row and column sums by using essentially a more general model, in which the points in the cells are arranged into classes of arbitrary specified sizes and permuted at random. The argument in [4] gave asymptotically the number of such configurations in which no two points in the same cell are of the same class. In the general case, this gives a model of random bicoloured graphs with given vertex degrees. On the other hand, in the special case that every colour class has cardinality 2, this is very similar to the pairing model, the only difference being that the edges of the graph are labelled. The results in [4] give the number of pairings which induce no loops in the multigraph. Independently of [5], the author's PhD thesis [115] studied random regular graphs from the point of view of unlabelling the edges in the model of multigraphs arising from [4], which then becomes the pairing model. However, this model was studied only indirectly in [115] because enumeration results on multigraphs with loops were sought. Once achieved, these were built on to find the numbers of simple graphs (with given degree sequence), as well as properties of such random graphs. The methods used for all these variations of the model are equivalent to those required to study the pairing model.

2.2 Related enumeration results

Of course results on probabilities in a uniform model are inextricably associated with enumeration results. Most computations in this paper aim for asymptotic results: unless otherwise specified, all limits refer to $n \to \infty$ with n restricted to even integers if d is odd.

Of major interest is an asymptotic formula for the number $|\mathcal{G}_{n,d}|$ of labelled d-regular graphs, which by (2) amounts to estimating $\mathbf{P}(\text{Simple})$. Read [91] gave an exact formula for $|\mathcal{G}_{n,d}|$ which is unfortunately too complicated to be easily amenable to asymptotic evaluation for general d. For $d = 3$ however, his analysis leads to the asymptotic expression $\frac{(6n)!}{288^n (3n)! e^2}$, which is equivalent to (3) for $d = 3$. In addition, any number of terms in an asymptotic series expansion for the number of cubic graphs can be obtained from his result. However, the inelegance of his derivation dissuaded him from applying it to larger d.

Bender and Canfield's asymptotic formula for $|\mathcal{G}_{n,d}|$ can be stated as

$$\mathbf{P}(\text{Simple}) \sim \exp\left(\frac{1-d^2}{4}\right) \quad \text{for fixed } d, \tag{3}$$

where here and elsewhere in this paper $a(n) \sim b(n)$ means $a(n) = (1+o(1))b(n)$ as $n \to \infty$ (with dn even, of course). Thus evaluating (2) using Stirling's formula gives the following.

Theorem 2.2 (Bender and Canfield [5]) *For fixed d*

$$|\mathcal{G}_{n,d}| \sim \sqrt{2} e^{(1-d^2)/4} \left(\frac{d^d n^d}{e^d (d!)^2}\right)^{\frac{1}{2}n}. \tag{4}$$

This was found independently in [115], and then Bollobás [9, 10] gave the pairing model proof and showed that the formula applied for $d = d(n) \leq \sqrt{2\log n} - 1$. A version of this proof of (3) is in the next section.

From (3) it follows that $\mathbf{P}(\text{Simple})$ is bounded below for fixed d, a fact which combined with Lemma 2.1 immediately yields the following. First, for an event H in $\mathcal{P}_{n,d}$, define $G(H)$ to be the event in $\mathcal{G}_{n,d}$ containing precisely all simple graphs of the form $G(P)$ for some $P \in H$.

Corollary 2.3 *Let $d \geq 1$ be fixed, and let H be an event which is a.a.s. true in $\mathcal{P}_{n,d}$. Then $G(H)$ is a.a.s. true in $\mathcal{G}_{n,d}$.*

McKay [72] used switchings (as described in Section 2.4) to extend the range of d in (3) to $d = o(n^{1/3})$. McKay and Wormald [81] then used a new sort of switching to find the formula for $d = o(\sqrt{n})$:

$$\mathbf{P}(\text{Simple}) = \exp\left(\frac{1-d^2}{4} - \frac{d^3}{12n} + O\left(\frac{d^2}{n}\right)\right). \tag{5}$$

Corollary 2.4 (McKay and Wormald [81]) *For $d = o(\sqrt{n})$ the number of d-regular graphs on n vertices is*

$$\frac{(dn)!}{(\frac{1}{2}dn)! 2^{dn/2} (d!)^n} \exp\left(\frac{1-d^2}{4} - \frac{d^3}{12n} + O\left(\frac{d^2}{n}\right)\right).$$

McKay and Wormald also obtained a formula for $d \approx cn$ [80] which may have consequences for quite dense random regular graphs:

$$|\mathcal{G}_{n,d}| \sim \sqrt{2} \left(2\pi n \lambda^{d+1} (1-\lambda)^{n-d}\right)^{-n/2} \exp\left(\frac{-1 + 10\lambda - 10\lambda^2}{12\lambda(1-\lambda)}\right)$$

where $\lambda = d/(n-1)$, provided $d = d(n)$ is an integer-valued function such that, for n sufficiently large, dn is even and $\min\{d, n-d-1\} > cn/\log n$ for some $c > \frac{2}{3}$.

The formulae mentioned in this section have their counterparts for graphs with given degree sequences, the most widely applicable being in [81] and [80]. These formulae allowed McKay and Wormald [82] to create a useful model of the degree sequence of a random graph in $\mathcal{G}(n,p)$. Since it has proved so useful, the Bender–Canfield formula for bounded degrees d_i is included here:

$$|\mathcal{G}_{n,\mathbf{d}}| \sim \frac{(2m)! e^{-\lambda - \lambda^2}}{m! 2^m \prod_{i=1}^n d_i!} \tag{6}$$

where $2m = \sum d_i$ and $\lambda = \frac{1}{2m} \sum \binom{d_i}{2}$.

2.3 Short cycle distribution

The number of short cycles in random regular graphs of small degree has an asymptotically Poisson distribution, as is the usual rule with the sum of many nearly independent rare events. (This effect is called the "Poisson paradigm" by Alon and Spencer [3].) The usual method of proving this uses an asymptotic version of the fact that a Poisson variable is determined by its moments. This method is presented here, saving a much more powerful (in the case of regular graphs) switching method for Section 2.4.

We say that a set of variables $X_i = X_i^{(n)}$, for i in some finite set I, defined on a sequence of probability spaces indexed by n, are *asymptotically independent Poisson* with means λ_i if their joint distribution tends to that of independent Poisson variables whose means are fixed numbers λ_i. To be precise,

$$\lim_{n \to \infty} \mathbf{P}\left(\bigwedge_{i \in I} \{X_i = r_i\}\right) \to \prod_{i \in I} e^{-\lambda_i} \frac{\lambda_i^{r_i}}{r_i!} \tag{7}$$

for every fixed set of non-negative integers r_i, $i \in I$. (Recall that n is restricted to even numbers in all limits if d is odd.) Throughout this section, we use $Z(\lambda)$ to denote a Poisson random variable with expectation λ.

Theorem 2.5 (Bollobás [10], Wormald [115, 118]) *For d fixed, let $X_i = X_{i,n}$ ($i \geq 3$) be the number of cycles of length i in a graph in $\mathcal{G}_{n,d}$. For fixed $k \geq 3$, $X_3, \ldots, X_k$ are asymptotically independent Poisson random variables with means $\lambda_i = \frac{(d-1)^i}{2i}$.*

This was derived in [10] using the pairing model, and independently in [115,118] essentially from (6).

Theorem 2.5 plays a major role in the results on Hamiltonicity and related properties (see Section 4.2). As mentioned above, we also need to calculate limiting probabilities in the pairing model conditional on the event Simple. For this, a crucial step is the determination of **P**(Simple) asymptotically. Since loops and multiple edges are cycles of length 1 and 2, the short cycle distribution in the multigraphs corresponding to pairings provides a convenient common generalisation of **P**(Simple) and the distribution in $\mathcal{G}_{n,d}$. This leads to the following result, which was first derived in full by Bollobás.

Theorem 2.6 (Bollobás [10]) *For d fixed, let $X_i = X_{i,n}$ $(i \geq 1)$ be the number of cycles of length i in the random multigraph coming from a pairing in $\mathcal{P}_{n,d}$. For $k \geq 1$, $X_1, \ldots, X_k$ are asymptotically independent Poisson random variables with means $\lambda_i = \frac{(d-1)^i}{2i}$.*

A Corollary of this is (3), obtained by considering the event $X_1 = X_2 = 0$.

Before proceeding further, the reader deserves an explanation as to why small connected subgraphs other than cycles are not examined here. It is because unless they are trees or unicyclic, they do not exist (a.a.s., unless d is permitted to grow as a function of n, in which case the analysis gets more difficult; see Section 2.4). The following result has been explicitly or implicitly given since the first examinations of random regular graphs, and is easily proved by estimating the expected number of subgraphs of the given type in $\mathcal{P}_{n,d}$.

Lemma 2.7 *For fixed d and any fixed graph F with more edges than vertices, $G \in \mathcal{G}_{n,d}$ a.a.s. contains no subgraph isomorphic to F.* ∎

Thus, the neighbourhood of a random vertex in a huge random d-regular graph looks just like part of an infinite tree. It follows from the lemma that a.a.s. no two cycles of bounded length are joined by a path of bounded length.

Proof of Theorems 2.5 and 2.6 The plan is quite simple, following the method of moments. In this, the joint factorial moments of $X_{1,n}, \ldots, X_{k,n}$ are shown to tend in the limit to those of the independent Poisson variables $Z(\lambda_1), \ldots, Z(\lambda_k)$. That is, for every sequence of fixed non-negative integers $r_1, \ldots, r_k$,

$$\lim_{n \to \infty} \mathbf{E}\left(\prod_{i=1}^{k}[X_{i,n}]_{r_i}\right) \to \prod_{i=1}^{k} \lambda_i^{r_i}. \tag{8}$$

Theorem 2.6 follows from this by a well-known result (see Lemma 2.8). Theorem 2.5 comes from Theorem 2.6 by Lemma 2.1.

To compute the left hand side of (8), for each $1 \leq i \leq k$ distinguish the pairs of points corresponding to an ordered set of r_i cycles of length i. The required value is the number of pairings with such distinguished pairs, divided by the total number $f(dn)$ of pairings. Letting $s = r_1 + 2r_2 + \cdots + kr_k$, the number of ways to choose the s distinguished pairs in the correct configuration for the desired cycles is asymptotic to

$$\frac{(d(d-1)n)^s}{\prod_{i=1}^{k}(2i)^{r_i} r_i!}$$

Models of Random Regular Graphs

if exactly s different vertices are involved in the pairs, and is $O(n^{s-1})$ otherwise (since the vertices can be chosen in this many ways, and given the vertices, the distinguished pairs can form only a finite number of configurations). The cycles of each length can be ordered in $\prod_{i=1}^{k} r_i!$ ways, and the pairing can then be completed in
$$f(dn - 2s) \sim f(dn)/(dn)^s$$
ways, by (1). Multiplying these together gives the right side of (8), as required. ∎

That proof required the following multivariate generalisation of Brun's sieve, which is an implication of the Bonferroni inequalities (see [3] for example). This generalisation is well known in probability theory (see Chung [27]) and is given without explicit proof in [16]. We show here how it follows from Brun's sieve in a quite elementary way. This proof mimics the structure of the proof of Theorem 2.5 given in [118]. It is included especially because it uses alterations of probability measure, which will appear again in Section 4.1. Without loss of generality we could assume that the $X_{i,n}$ are indicator variables for random events.

Lemma 2.8 *Let $\lambda_1, \ldots, \lambda_k$ be some set of fixed non-negative reals, and let $X_{1,n}, \ldots, X_{k,n}$ be non-negative integer random variables defined on the same space $\mathcal{G}_n$ for each n. If (8) holds for each fixed set of non-negative integers $r_1, \ldots, r_k$, then the variables $X_{1,n}, \ldots, X_{k,n}$ are asymptotically independent Poisson with means λ_i.*

Proof This is by induction on k; for $k = 1$ it is Brun's sieve, so assume $k > 1$.

We take two cases: firstly $\lambda_k = 0$. Then (8) with $r_1 = \cdots = r_{k-1} = 0$ and $r_k = 1$ gives $\mathbf{E}X_k \to 0$, and so (7) holds whenever $r_k \neq 0$. On the other hand, it holds for $r_k = 0$ by the inductive hypotheses.

So now assume $\lambda_k > 0$. Let $\mathcal{G}_n'$ denote the probability space obtained from $\mathcal{G}_n$ by weighting the probability measure according to $[X_{k,n}]_{r_k}$. Let $\mathbf{P}'$ and $\mathbf{E}'$ denote probability and expectation in $\mathcal{G}_n'$, whereas $\mathbf{P}$ and $\mathbf{E}$ refer to $\mathcal{G}_n$. Then for any event R and variable X in $\mathcal{G}_n$,

$$\mathbf{P}'(R) = \frac{\mathbf{E}([X_{k,n}]_{r_k} \mid R)\mathbf{P}(R)}{\mathbf{E}[X_{k,n}]_{r_k}}, \qquad \mathbf{E}'(X) = \frac{\mathbf{E}(X[X_{k,n}]_{r_k})}{\mathbf{E}[X_{k,n}]_{r_k}}. \tag{9}$$

Thus from (8)

$$\mathbf{E}' \prod_{i=1}^{k-1} [X_{i,n}]_{r_i} \to \frac{\prod_{i=1}^{k} \lambda_i^{r_i}}{\lambda_k^{r_k}} = \prod_{i=1}^{k-1} \lambda_i^{r_i}.$$

So by the inductive hypothesis, $X_1, \ldots, X_{k-1}$ are asymptotically Poisson in $\mathcal{G}_n'$ with means λ_i. The same is true in $\mathcal{G}_n$. So if R is the event that $X_1 = r_1, \ldots, X_{k-1} = r_{k-1}$, we have $\mathbf{P}'(R) \to \Pi = \prod_{i=1}^{k-1} \mathbf{P}(Z(\lambda_i) = r_i)$, and also $\mathbf{P}(R) \to \Pi$. Hence from the left equation in (9),

$$\mathbf{E}([X_{k,n}]_{r_k} \mid R) - \mathbf{E}[X_{k,n}]_{r_k} \to 0.$$

But from (8) $\mathbf{E}[X_{k,n}]_{r_k} \to \lambda_k^{r_k}$, and so

$$\mathbf{E}([X_{k,n}]_{r_k} \mid R) \to \lambda_k^{r_k}.$$

That is, in the probability space obtained from $\mathcal{G}_n$ by conditioning on the event $R = R_n$, the factorial moments of $X_{k,n}$ tend to those of $Z(\lambda_k)$. Hence another application of Brun's sieve implies that $X_{k,n}$ is asymptotically Poisson with mean λ_k in this conditional space. Now

$$\begin{aligned}\mathbf{P}(X_1 = r_1, \ldots, X_k = r_k) &= \mathbf{P}(X_k = r_k \mid R)\mathbf{P}(R) \\ &\to \mathbf{P}(Z(\lambda_k) = r_k) \prod_{i=1}^{k-1} \mathbf{P}(Z(\lambda_i) = r_i)\end{aligned}$$

as required. ∎

Results on the short cycle distribution for $G \in \mathcal{G}_{n,\mathbf{d}}$ which generalise those given above were obtained in [10] and [118].

2.4 The switching method: subgraphs, eigenvalues and spanning trees

The switching method enables us to prove results about subgraphs of $G \in \mathcal{G}_{n,d}$ when d, or the size of the subgraph, grows much more quickly than is permitted in the proofs of Theorems 2.5 and 2.6 given above. It often applies for $d = o(n^c)$ where c is some "reasonable" number like $\frac{1}{2}$.

We define a *simple switching* in a pairing P to be the replacement of two pairs $\{p_1, p_2\}, \{p_3, p_4\}$ by $\{p_1, p_3\}, \{p_2, p_4\}$ or $\{p_1, p_4\}, \{p_3, p_2\}$. This induces a switching of two edges of the corresponding graph, and early switching results analysed this operation directly on graphs without reference to pairings. McKay [69] introduced switchings to the random regular graph scene to obtain general upper and lower bounds on the probability of a subgraph occurring. In [68] he used the method to obtain bounds on the probabilities of cycles of various (unbounded) lengths occurring in $G \in \mathcal{G}_{n,d}$. It follows that such G a.a.s. satisfies a certain condition which, as McKay showed in [67], implies that the distribution of eigenvalues of G tends towards a fixed function, which he determined. The argument exploits a connection between the eigenvalues and the cycle distribution via the number of closed walks of length k (noting that this is the trace of the kth power of the adjacency matrix, which is of course the sum of the kth powers of the eigenvalues). The same connection was used by Broder and Shamir [24] to obtain an asymptotic almost sure upper bound on the second-largest eigenvalue in absolute value, for d even. This bound was decreased by Friedman [39] to $2\sqrt{2d-1} + 2\log d + c'$ holding with probability $1 - O(n^{-c})$, where c' depends on c.

In [70] McKay used another eigenvalue connection, the matrix tree theorem, in a similar way to show that if a sequence of regular graphs G_i on $n_i \to \infty$

Models of Random Regular Graphs

vertices satisfies a certain asymptotic condition on numbers of cycles then the number of spanning trees in G_i is $(c_d + o(1))^{n_i}$ where

$$c_d = \frac{(d-1)^{d-1}}{(d^2 - 2d)^{\frac{1}{2}d-1}}.$$

The required condition is shown, using the switching results, to be true a.a.s. for $G \in \mathcal{G}_{n,d}$, implying that the nth root of the number of spanning trees in G is a.a.s. asymptotic to c_d (by which we mean it is a.a.s. in the interval $(c_d - \epsilon, c_d + \epsilon)$ for all $\epsilon > 0$). McKay also proved a variety of stronger versions and other variations of these statements.

The main power of switchings is the ability to obtain accurate estimates of very small probabilities. As an example, we illustrate with a use of the more modern switchings introduced in [79] and [81] which turn out to give results more easily and stronger than using simple switchings as in McKay [71, 72]. The following result is part of the proof of (5) (giving an asymptotic formula for $|\mathcal{G}_{n,d}|$) for $d = o(n^{1/3})$ in [79].

Lemma 2.9 (McKay and Wormald [79]) Let $S_{a,b}$ denote the set of pairings $P \in \mathcal{P}_{n,d}$ such that $G(P)$ has precisely a loops, b double edges, and no edges of multiplicity greater than 2, nor double loops. Then for $1 \le a < d + \omega(n)$ and $b < d^2 + \omega(n)$, where $\omega(n) \to \infty$ arbitrarily slowly, $|S_{a,b}|/|S_{a-1,b}| = \frac{d-1}{2a}(1 + O(\frac{d+\omega(n)}{n}))$.

Proof Denote a pair of points $\{p_i, p_j\}$ by $p_i p_j$. Given any pairing in $S_{a,b}$, choose a pair $p_1 p_2$ which projects onto a loop, choose two other pairs $p_3 p_4$ and $p_5 p_6$, and replace all three pairs by the pairs $p_1 p_3$, $p_2 p_5$ and $p_4 p_6$ to produce a pairing P'. This switching operation can be written as the composition of two simple switchings.

We count how many ways this can be done so that P' lands in $S_{a-1,b}$. There are $2a$ ways to choose p_1 and p_2 (as an ordered pair), and roughly dn ways to choose each of p_3 and p_5, which determine p_4 and p_6. After this there is only one way to perform the switching described. However, some choices of p_3 and p_5 lead to unwanted multiple edges created by the switching ($O(d^3 n)$ choices do this) or to triple edges ($O(bdn)$ do this) or unwanted loops ($O(adn)$ do this) or destroy loops or double edges unintentionally ($O(adn + bdn)$ do this). So the number of P' in $S_{a-1,b}$ corresponding to P is

$$2ad^2 n^2 \left(1 - O\left(\frac{d^2 + a + b}{dn}\right)\right). \tag{10}$$

On the other hand, in how many ways can $P' \in S_{a-1,b}$ be produced? To reverse the procedure, choose p_1 and p_2 in the same cell (in $d(d-1)n$ ways), and choose p_4 to be any other point (in $dn - 2$ ways). Then p_3, p_5 and p_6 are determined, in that order, as the points paired in P' with the three chosen points. Call this choice "bad" if it does not determine the reverse of the procedure as described. There are only a few causes of badness, but counting as before shows that the number of bad choices cannot exceed $O(d^3 an + d^2 bn + d^4 n)$. Hence the number of P in $S_{a,b}$ corresponding to P' is

$$d^2(d-1)n^2 \left(1 - O\left(\frac{d^2 + da + b}{dn}\right)\right),$$

and the result follows on division of (10) by this expression. ∎

This lemma can be used to obtain the asymptotic probability that $a = 0$, conditional upon a given value of b, by estimating $|S_{a,b}|/|S_{0,b}|$ (in the form of a telescoping product of the ratios in the lemma) and summing over all a which contribute significantly. (Large values of a can be ruled out by a simple expectation argument.) This gives the reciprocal of the desired probability for $d = o(n^{1/3})$. For d in this range, the restrictions in the hypotheses of the lemma are easily justified in almost all pairings by expectation arguments. To complete an estimation of $\mathbf{P}(\text{Simple})$, we can use other types of switchings for estimation of $|S_{0,b}|/|S_{0,0}|$, and combining the two, we can estimate $|S_{a,b}|/|S_{0,0}|$ for all the significant terms. Summing over a and b gives an estimate of $|\mathcal{P}_{n,d}|/|S_{0,0}|$, which gives the estimate of $1/\mathbf{P}(\text{Simple})$.

The superiority of the switching used in this proof over a simple switching (switch a loop-inducing pair with any other pair) is that the number of ways to reverse the procedure is almost independent of P. For the single switching, the number of valid ways to reverse the procedure, given P', depends heavily on the number of triangles in P'. McKay [72] argued about the average number of reversals, in order to reach $d = o(n^{1/3})$; similar averaging arguments in [81] for the "modern" switchings gave (5) (reaching $d = o(\sqrt{n})$). Possibly this averaging can be pushed further, but at the expense of a lot of work, and the next step would only be $n^{3/5}$.

Further uses of switchings are mentioned in Sections 2.5, 2.6, 2.7, 2.8, 3.2 and 5. Switchings also work fine on graphs in the model $\mathcal{G}_{n,\mathbf{d}}$. In addition, an argument using one simple switching implies that many variables defined on random regular graphs are sharply concentrated (Theorem 2.19). A general description of the switching argument and an application to $\mathcal{G}(n,p)$ can also be found in [123]. There the switchings give a relationship between probabilities of events rather than cardinalities of sets, but they still taste of counting.

2.5 Longer cycles

Theorem 2.5 implies an asymptotic formula for the probability that $G \in \mathcal{G}_{n,d}$ has girth k when k and d are fixed (see Theorem 2.12). By the switching method we can show that this asymptotic formula also applies when $k \to \infty$, simultaneously with $d \to \infty$. This is done for k close to $\frac{1}{2}\log_d n$ by McKay et al. [83]. The results on distribution of short cycles can also be extended in this way to cycles of unbounded length.

Garmo [50] used the small subgraph conditioning method to examine long cycles. He showed that provided $k = k(n) \to \infty$ faster than $c \log n$ but $k = o(n)$, the number of k-cycles in $G \in \mathcal{G}_{n,d}$ is concentrated near its expectation, and he found its limiting distribution on the appropriate scale. If $k/n \to q$, $0 < q < 1$, the number is not so concentrated, the variance is a non-zero constant times the square of the expectation, and the limiting distribution was found.

2.6 Connectivity and diameter

For fixed $d \geq 3$, $G \in \mathcal{G}_{n,d}$ is d-connected a.a.s., as was shown independently by Bollobás [11] and by Wormald [117] (originally in [115]). This result can readily be extended using switchings (Section 2.4) to d growing rather modestly with n. The following extension seems to be the strongest in this sense which has appeared in print, but the upper bound was just what was needed at the time and by no means represents the limit imposed by the method.

Theorem 2.10 (Luczak [63]) *For $3 \leq d(n) \leq n^{0.02}$, $G \in \mathcal{G}_{n,d}$ is d-connected a.a.s.*

This is proved below in the more restricted range $3 \leq d(n) \leq 3 \log n$, which is about as far as one can easily go without switchings or something similar. The proof below actually gives the bound $O(n^{2-d})$ on the probability that $G \in \mathcal{G}_{n,d}$ is not d-connected for d fixed. It is easy to see that this bound cannot be improved for fixed d (apart from the value of the constant implicit in the O). In [117] and [63], non-regular versions of this theorem were given. (See Theorem 2.16 below.)

Curiously, larger values of d seem to be more difficult to handle, but we can confidently conjecture that the same basic result holds even for d of the order of n.

Conjecture 2.11 *For $3 \leq d = d(n) < n - 3$, $G \in \mathcal{G}_{n,d}$ is d-connected a.a.s.*

This conjecture is trivially true also for $d = n - 1$ and $n - 2$. The exclusion of $n - 3$ is necessary, since the complement of G is then a 2-regular graph which has non-zero probability of having a 4-cycle. Deleting the other $n - 4$ vertices from G gives a disconnected graph. To prove the conjecture for d bigger than $n^{0.02}$ would not be too hard using switchings; something like $\sqrt{n}$ should be possible. However, to reach past $d = n^{1-\epsilon}$ for arbitrarily small ϵ would probably require a really new idea. On the other hand, for d very close to n, it would become easy to verify by analysing the complement.

For $d = 2$ the story is different: a random graph in $\mathcal{G}_{n,2}$ is a.a.s. disconnected. In fact, it is connected if and only if it is Hamiltonian, the probability of which is easy to compute asymptotically. The number of Hamiltonian graphs is $\frac{n!}{2n}$, and dividing by $|\mathcal{G}_{n,2}|$ (say from Theorem 2.2) gives

$$\mathbf{P}(G \in \mathcal{G}_{n,2} \text{ is Hamiltonian}) \sim \tfrac{1}{2} e^{3/4} \sqrt{\pi} n^{-1/2} \approx 1.876 n^{-1/2}. \tag{11}$$

The cyclic connectivity of regular graphs, especially 3-regular, has also been of considerable interest. A set S of vertices or edges in a connected graph G is *cycle-separating* if at least two components of $G - S$ contain cycles. G is *cyclically k-vertex-connected* (*cyclically k-edge-connected*) if it has no cycle-separating vertex (respectively, edge) set of cardinality less than k. For $d \geq 3$, a large d-regular graph with a j-cycle immediately has cycle-separating vertex and edge sets of cardinality $(d-2)j$. By Theorem 2.5, the probability that the shortest cycle has length j tends to a non-zero constant. This shortest cycle length, or girth, is asymptotically the determining factor for cyclic connectivity, as shown in [117].

Theorem 2.12 *Let $d \geq 3$ and $k \geq 3$ be fixed. The probabilities that $G \in \mathcal{G}_{n,d}$ is*

(i) cyclically k-vertex-connected

(ii) cyclically k-edge-connected

(iii) of girth at least $k/(d-2)$

are all asymptotic to

$$\prod_{3 \leq i < k/(d-2)} \exp\left(\frac{-(d-1)^i}{2i}\right)$$

as $n \to \infty$.

Theorem 2.5 determines this limiting probability from (iii).

Theorems 2.10 and 2.12 are both proved by showing that deleting a small subset of the vertices of $G \in \mathcal{G}_{n,d}$ a.a.s. does not produce two reasonably large components. This uses a simple expectation argument (and, essentially and surreptitiously, Markov's inequality, which can be found in many probability texts such as Grimmett and Stirzaker [53]). It can employ either the formula for the number of graphs with given degree sequence (as in [117]), or calculations directly in the pairing model and then making use of Lemma 2.1 (as by Bollobás [11] and [16, Thm VII.32], although the proof in the latter is oversimplified: the analysis is false for large values of the variable a). Use of the pairing model approach is demonstrated in the proofs given below.

A notion closely related to the usual proofs of the connectivity results (for example Lemma 2.14 below) is the *isoperimetric number* of a graph G, which is the minimum value of the ratio $|\partial U|/|U|$ over all $U \subseteq V(G)$, $|U| \leq |V(G)|/2$. Here ∂U is the set of edges with exactly one endpoint in U. Bollobás [17] gave an asymptotically almost sure lower bound on this number for $G \in \mathcal{G}_{n,d}$. It follows that the number is a.a.s. between $\frac{1}{2}d - \epsilon(d)$ and $\frac{1}{2}d + \epsilon(d)$ where $\epsilon(d) \to 0$ as $d \to \infty$.

The *diameter*, diam(G), of a graph G is the maximum, over all $u, v \in V(G)$, of the distance from u to v. For a d-regular graph G this clearly satisfies

Models of Random Regular Graphs

$|V(G)| \leq 1 + \sum_{i=0}^{\text{diam}(G)-1} d(d-1)^i$. Graphs realising this bound are called Moore graphs and are extremely rare. Roughly speaking, given n and d the inequality gives a lower bound on the diameter of roughly $\log_{d-1} n$. Using methods somewhat related to those used for connectivity, Bollobás and de la Vega [21] showed the following.

Theorem 2.13 *The diameter* $\text{diam}(G)$ *of* $G \in \mathcal{G}_{n,d}$ *a.a.s. satisfies* $1 + \lfloor \log_{d-1} n \rfloor + \left\lfloor \log_{d-1} \left(\frac{(d-2)}{6d} \log n \right) \right\rfloor \leq \text{diam}(G) \leq 1 + \left\lceil \log_{d-1}((2+\epsilon)dn \log n) \right\rceil$.

Another result involving distance was obtained by Bollobás [13], on the probability that for $G \in \mathcal{G}_{n,d}$, each vertex v is determined uniquely by the sequence $n_1(v), \ldots, n_k(v)$ where $n_i(v)$ is the number of vertices of distance i from v, and k is a function of n. This event implies that the automorphism group of G is trivial.

We turn now to the proofs of Theorems 2.10 and 2.12. These are included not only because the latter has not been proved directly using the pairing model before, but also as an example of using the independence property of the pairing model. This is in the following lemma. First a technical definition: let us define an (s, j)-*separating set* of a graph G to be a set $S \subseteq V(G)$ with $|S| = s$ such that $G - S$ has a component with exactly j vertices.

Lemma 2.14 *Let d and s be arbitrary. For $P \in \mathcal{P}_{n,d}$,*

$$\mathbf{P}(G(P) \text{ has an } (s,j)\text{-separating set}) < 3^{2+s/d} \left(\frac{j+s}{n} \right)^{j(\frac{1}{2}d-1)} n^{\frac{1}{2}s}(j+s)^{\frac{3}{2}s}$$

for all $j < \frac{2}{3}n - s$.

Proof Let X denote the expected number of (s, j)-separating sets. To show $\mathbf{P}(X \geq 1) \to 0$ it is enough, by Markov's inequality, to show $\mathbf{E}X \to 0$. If such a set S exists, let F be a component of $G - S$ with $|V(F)| = j$ and let $T = S \cap N(F)$. Put $t = |T|$. The j cells of P corresponding to $V(F)$ and the t cells for T can be chosen in at most $\binom{n}{j} n^t$ ways.

Since each vertex in T has degree at least 1 in the subgraph of G induced by $F \cup T$, the number r of edges with at least one end in F satisfies $r \geq (jd+t)/2$. Given r and the cells in $V(F) \cup T$, what is the probability that a random pairing has r pairs of the required type? By the independence property of the pairing model, such pairs can be selected by repeatedly selecting the pair containing the next unpaired point in a cell in $V(F)$. The probability of succeeding in the ith step of this random process, conditional upon succeeding in all earlier steps, is $\frac{(j+t)d-2i+1}{nd-2i+1}$. So the probability of selecting r such pairs is at most

$$\prod_{i=1}^{r} \frac{j+t-(2i-1)/d}{n-(2i-1)/d} \leq \left(\frac{[j+t]_{2r/d}}{[n]_{2r/d}} \right)^{\frac{1}{2}d} \tag{12}$$

where we define $[j]_{i+\epsilon}$ for $i \in \mathbb{Z}$ and $0 < \epsilon < 1$ to be $[j]_i(j-i)^\epsilon$. Thus

$$\mathbf{E}X \leq \sum_{t=0}^{s} \sum_{r} \left(\frac{[j+t]_{2r/d}}{[n]_{2r/d}}\right)^{\frac{1}{2}d} \binom{n}{j} n^t < 3 \sum_{t=0}^{s} \left(\frac{[j+s]_{j+t/d}}{[n]_{j+t/d}}\right)^{\frac{1}{2}d} \frac{[n]_j}{j!} n^t \quad (13)$$

since $j + t \leq j + s < \frac{2}{3}n$ and $2r/d \geq j + t/d$. The sum over $t \leq s$ can be conveniently bounded by 3 times the term with $t = s$. We have

$$\frac{[j+s]_{j+s/d}}{[n]_{j+s/d}} < \min\left(\frac{(j+s)^s j!}{[n]_j (n-j-s/d)^{s/d}}, \frac{(j+s)^{j+s/d}}{n^{j+s/d}}\right).$$

Apply the first term in the minimum to one factor in (13) along with $n-j-\frac{s}{d} > \frac{1}{3}n$, and use the second term on the other factors, to obtain the lemma. ∎

In the case of graphs, which by definition have no loops or multiple edges, the result of Lemma 2.14 can be improved for small j as follows.

Lemma 2.15 Let $d = d(n) < n^{1/3}$. For $P \in \mathcal{P}_{n,d}$,

$$\mathbf{P}(G(P) \in \mathcal{G}_{n,d} \text{ and has an } (s,j)\text{-separating set}) < 2 \left(\frac{j+s}{n}\right)^{j(d-\frac{1}{2}j+\frac{1}{2})} n^{j+s}.$$

Proof Firstly, we can assume $j < 2d$ and $s < d^2$ since otherwise there is nothing to prove. If G is a graph, the lack of loops and multiple edges permits replacement of the inequality $r \geq (jd+t)/2$ in the proof of Lemma 2.14 by $r \geq jd - \binom{j}{2}$. Repeating the first part of that argument, and bounding (12) by $(j+t)^{\frac{1}{2}d}/n^{\frac{1}{2}d}$, shows that the worst case is $t = s$ and $r = (jd+t)/2$. The sum over t and r can be bounded by say twice this largest term, which gives the lemma. ∎

Proof of Theorem 2.10 For $n \geq d+1$, a graph with n vertices which is not d-connected contains a $(d-1, j)$-separating set for some $1 \leq j \leq \frac{1}{2}n$. For d-regular graphs, $j = 1$ can be excluded immediately. For $2 \leq j \leq d$ the bound in Lemma 2.15 is $O(n^{2-d}(d+1)^{2d-1})$, with the worst case always occurring at $j = 2$, as well as $j = 3$ in the case $d = 3$, the other cases contributing negligibly. For $j > d$ we use the bound in Lemma 2.14, which is at most

$$\frac{27(j+d)^{j(\frac{1}{2}d-1)+\frac{3}{2}d}}{n^{j(\frac{1}{2}d-1)-\frac{1}{2}(d-1)}}. \quad (14)$$

This is $O(n^{2-d})$ for $j = d+1$ and $j = d+2$ (worst case $d = 3$), $O(n^{1-d})$ for $j = d+3$, and much smaller for $j = \frac{1}{2}n$. For the values in between, the second derivative with respect to j of the logarithm of (14) has the same sign as $j(d-2) + d(2d-7)$, which is positive for $j \geq 3$. Hence it is a convex function of j, and the sum of (14) over $d+3 \leq j \leq \frac{1}{2}n$ is $O(n^{2-d})$. So for $P \in \mathcal{P}_{n,d}$, the probability that $G(P)$ is a graph which is not d-connected is $O(n^{2-d}(d+1)^{2d-1})$.

The theorem now follows by Lemma 2.1 and the estimate for $\mathbf{P}(\text{Simple})$ in (5). ∎

Proof of Theorem 2.12 A cyclically k-edge-connected d-regular graph with sufficiently many vertices must have girth at least $k/(d-2)$. On the other hand, let G be d-regular, of girth at least $k/(d-2)$, and with a cycle-separating vertex set S such that $|S| = s < k$. Then $G - S$ has two components with cycles and hence one of them has j vertices for $k/(d-2) \leq j \leq \frac{1}{2}n$. Since s and d are bounded, the upper bound in Lemma 2.14 is

$$\frac{O(1)(j+s)^{j(\frac{1}{2}d-1)+\frac{3}{2}s}}{n^{\frac{1}{2}j(d-2)-\frac{1}{2}s}}.$$

Since $j(d-2) \geq s+1$, this is $O(n^{-\frac{1}{2}})$ for j near its lower bound, and analysis as for (14) shows the sum over all relevant j is $O(n^{-\frac{1}{2}})$. Hence for $P \in \mathcal{P}_{n,d}$, $G(P)$ a.a.s. either has girth less than $k/(d-2)$ or is cyclically k-vertex-connected and hence (for n large enough) also cyclically k-edge-connected. It follows by Lemma 2.1, (3) and Theorem 2.5 that the probability that $G \in \mathcal{G}_{n,d}$ is cyclically k-vertex-connected, cyclically k-edge-connected, or of girth at least $k/(d-2)$, all tend to the same non-zero constant as $n \to \infty$. ∎

Note The proof of Theorem 2.12 is easily adapted to permit d and k to grow slowly with n, using the result mentioned in Section 2.5 to estimate the probability of having large girth. However, even for d fixed, it is only easy to reach $k = o(\log \log n)$. To go past this point would require more careful estimation of the probability of having a cycle-separating set in the conditional space of large girth, which would appear to be even more difficult than extending the range of d in Theorem 2.10. Again, the switching method would presumably be required here.

We close this subsection with mention of a similar result for $\mathcal{G}_{n,\mathbf{d}}$.

Theorem 2.16 ([117]) *Fix $3 \leq d \leq D$, and for each n let $\mathbf{d} = \mathbf{d}(n) = (d_1, \ldots, d_n) \in [d, \ldots, D]^n$ with $\sum d_i$ even. The probability that $G \in \mathcal{G}_{n,\mathbf{d}}$ is not d-connected is $O(n^{2-d})$.*

Luczak's theorem in [63] shows that this is also valid with the upper bound $d \leq n^{0.02}$, but with the bound $O(n^{2-d})$ weakened to $o(1)$. Theorem 2.16 is the degree-restricted counterpart of the result of Erdős and Rényi [36] that a random graph with n vertices and $\frac{1}{2}n \log n + O(n)$ edges a.a.s. has connectivity equal to its minimum degree. Without degree restrictions, a much higher edge density is required before higher connectivity occurs, but for both models of random graphs the basic effect is the same: vertices of degree less than d are a.a.s. the only obstruction to d-connectedness.

In [63], Luczak also considers random graphs with given degree sequence where the minimum entry is 2, obtaining results on the number and nature of the components. Molloy and Reed [85] gave a nice extension of these results, to determine asymptotically necessary and sufficient conditions on the degree sequence $\mathbf{d}$ (with minimum degree 2) for a random graph in $\mathcal{G}_{n,\mathbf{d}}$ to have a.a.s. a giant component.

2.7 Automorphisms and unlabelled graphs

What is the number $U_{n,d}$ of unlabelled d-regular graphs on n vertices? Let $\mathcal{S}$ be any set of unlabelled graphs on n vertices, and $\mathcal{S}'$ the set of all labelled versions of graphs in $\mathcal{S}$. A simple application of the Frobenius–Burnside lemma gives

$$|\mathcal{S}| = \frac{1}{n!} \sum_{\sigma \in S_n} |\text{fix}(\sigma)| \qquad (15)$$

where S_n denotes the group of permutations of the label set $\{1, 2, \ldots, n\}$ and $\text{fix}(\sigma)$ denotes the set of graphs in S' fixed by σ. (See Harary and Palmer [54] for a variety of applications of this principle to the enumeration of unlabelled graphs.) With id denoting the identity permutation, $|\text{fix}(\text{id})| = |\mathcal{G}_{n,d}|$ and so

$$\sum_{\sigma \in S_n \setminus \{\text{id}\}} |\text{fix}(\sigma)| = o(|\mathcal{G}_{n,d}|) \qquad (16)$$

is equivalent to

$$U_{n,d} \sim |\mathcal{G}_{n,d}|/n!. \qquad (17)$$

Rewrite $\text{fix}(\sigma) = \sum_{G \in S'} x(G, \sigma)$, where $x(G, \sigma)$ is 1 if $G \in \text{fix}(\sigma)$ and 0 otherwise. Reversing the order of summation in (15) then gives

$$|\mathcal{S}| = \frac{1}{n!} \sum_{G \in S'} \phi(G),$$

where $\phi(G)$ denotes the number of $\sigma \in S_n$ which fix G.

Hence (16), via (17), is equivalent to

$$\mathbf{E}(\phi(G)) \to 1 \quad \text{for} \quad G \in \mathcal{G}_{n,d}. \qquad (18)$$

Bollobás [14] showed that this condition holds for fixed $d \geq 3$. In a more general investigation of McKay and Wormald, this was extended to the following result.

Theorem 2.17 ([78]) *Let $\epsilon > 0$ and $3 \leq d = d(n) = O(n^{\frac{1}{2}-\epsilon})$. Then $\mathbf{E}(\phi(G)) \to 1$ for $G \in \mathcal{G}_{n,d}$.*

This and Corollary 2.4 imply the following.

Corollary 2.18 *For $\epsilon > 0$ and $3 \leq d = d(n) = O(n^{\frac{1}{2}-\epsilon})$,*

$$U_{n,d} = \frac{(dn)!}{(dn/2)! 2^{dn/2} (d!)^n n!} \exp\left(\frac{1-d^2}{4} - \frac{d^3}{12n} + o(1)\right).$$

Versions of Theorem 2.17 are invariably proved by going back to (16) and noting that if $\sigma \in S_n \setminus \{\text{id}\}$, then for any graph in $\text{fix}(\sigma)$, the arrangement of edges incident with vertices moved by σ is severely restricted. The various possible subgraphs induced by these edges have a low probability of occurring in $G \in \mathcal{G}_{n,d}$, and the sum of this probability over all such subgraphs and σ is $o(1)$. Bounding $|\text{fix}(\sigma)|$ is achieved by classifying $G \in \text{fix}(\sigma)$ according to the numbers of edges joining vertices in cycles of σ of given small lengths. This leads to optimisation over several variables. A proof is given for bounded d in [120] which is significantly simpler than the others because of the way in which it handles these variables.

Computations to bound the probability of subgraphs occurring can be done directly in the pairing model as in [14] and [120] for fixed d, but for d growing significantly with n as in [78] the switching method (Section 2.4) is required. No doubt Theorem 2.17 is still valid with the upper bound on d increased to $d \leq n - 4$, but a proof of this seems out of reach at present.

The results of McKay and Wormald [78] actually treat the model $\mathcal{G}_{n,\mathbf{d}}$ with varying degrees, and results are obtained like Theorem 2.17, but the upper bound on degrees depends on how many vertices of low degree are permitted.

2.8 Packing, covering and colouring

The independence property of the pairing model has been used heavily in packing and covering problems (including colouring, which is equivalent to covering with independent sets). Some algorithms (say, for finding independent sets) can be analysed by generating the pairs of the pairing in the order in which they are examined by the algorithm, as long as the algorithm is "greedy" in the sense that it does not have to know too much about the part of the pairing not yet generated. An upper or lower bound can then be obtained a.a.s. on the graph parameter of interest, by determining the behaviour of the algorithm a.a.s. Several results along these lines have been obtained by showing that the parameter of interest a.a.s. closely tracks a function given as the solution of differential equations throughout the generation process. Convergence is proved using supermartingale inequalities. This is referred to below as the *differential equation method*. It is described in [122], but see [124] for a much more comprehensive and up to date exposition.

Quite apart from this, there is a general concentration result using martingales which applies to many random regular graph parameters which are large. A general statement of the result seems to be lacking so far (though a particular case was referred to by Frieze and Suen [45]), and so we give the following. We write $P \sim P'$ to denote that two pairings P and P' differ by a simple switching.

Theorem 2.19 *If X_n is a random variable defined on $\mathcal{P}_{n,d}$ such that $|X_n(P) - X_n(P')| \leq c$ whenever $P \sim P'$, then*

$$\mathbf{P}(|X_n - \mathbf{E}X_n| \geq t) \leq 2\exp\left(\frac{-t^2}{dnc^2}\right)$$

for all $t > 0$.

Proof Hold n fixed and write $X = X_n$. Generate an element P of $\mathcal{P}_{n,d}$ by choosing the pairs consecutively. After each pair is completed, canonically choose the next point to be paired by some convention (such as the lowest-numbered unpaired point) and choose its mate randomly. Let P_0 be any pairing with the order of pairs specified according to the order of generation, let $P_0(m)$ denote the set of the first m pairs in P_0, and define

$$Y_m(P_0) = \mathbf{E}(X(P) \mid P_0(m) \subseteq P)$$

for $P \in \mathcal{P}_{n,d}$. Then $Y_0(P), Y_1(P), \ldots, Y_{dn/2}(P)$ is a martingale (in fact a martingale constructed this way is called a Doob martingale; see McDiarmid [65] for example), and we have $Y_0(P) = \mathbf{E}X$ and $Y_{dn/2}(P) = X(P)$. The theorem now follows from Hoeffding's inequality (also called Azuma's; see [65]), once we show that $|Y_{m+1}(P) - Y_m(P)| < c$ for all m; that is, the martingale has differences bounded by c.

For given P_0, let i denote the canonical next point after the pairs in $P_0(m)$ are chosen, and let S_j denote those pairings in $\mathcal{P}_{n,d}$ containing all of $P_0(m)$ as well as the pair $\{i,j\}$. Then for any j and k which occur in no pairs in $P_0(m)$, each $P \in S_j$ corresponds to a unique $P' \in S_k$ by the simple switching $\{\{i,j\},\{k,l\}\} \leftrightarrow \{\{i,k\},\{j,l\}\}$, where l is determined as the mate of k in P. This gives a bijection between S_j and S_k. By assumption $|X(P') - X(P)| \leq c$, and the required bound on the differences now follows because $\mathcal{P}_{n,d}$ is a uniform space. ∎

One way of deriving results on $\mathcal{G}_{n,d}$ using Theorem 2.19 is to consider Y_n defined on $\mathcal{G}_{n,d}$ such that $Y(G(P)) = X(P)$ for all $P \in $ Simple where X_n satisfies the hypotheses of the theorem. Then

$$\mathbf{P}(|Y_n - \mathbf{E}Y_n| \geq t) \leq \frac{2}{p} \exp\left(\frac{-t^2}{dnc^2}\right)$$

in $\mathcal{G}_{n,d}$, where p is the probability (in $\mathcal{P}_{n,d}$) of Simple. This gives concentration of X_n and normally implies that $\mathbf{E}X_n$ is well approximated by $\mathbf{E}Y_n$ (but $\mathbf{P}(\text{Simple})$ and the range of values of Y_n can affect quantification of this).

Independent sets, dominating sets and star forests An *independent set* or stable set in a graph G is a set of vertices no two of which are adjacent, and the *independence number* $\alpha(G)$ of G is the cardinality of a maximum independent set in G. A *dominating set* is a set S of vertices such that every vertex not in S has at least one neighbour in S, and the *dominating number* $\alpha'(G)$ of G is the cardinality of a minimum dominating set in G.

Since a simple switching in a pairing P can change $\alpha(G(P))$ by at most 1, Theorem 2.19 and Corollary 2.3 imply that $\alpha(G)/|V(G)|$ (called the *independence ratio*) is concentrated when d is fixed. For instance, it is a.a.s. within $\log n/\sqrt{n}$ of its expected value for $G \in \mathcal{G}_{n,d}$. A similar statement holds for the dominating number. In spite of this, no argument has been forthcoming to show that the expected value of $\alpha(G)/n$ tends to some constant for $G \in \mathcal{G}_{n,d}$ as $n \to \infty$. This (and a similar result for $\alpha'(G)/n$) would imply the following.

Conjecture 2.20 *For fixed d, there are constants $\beta(d)$ and $\beta'(d)$ such that for all $\epsilon > 0$, $|\alpha(G) - n\beta(d)| < \epsilon n$ and $|\alpha'(G) - n\beta'(d)| < \epsilon n$ a.a.s. for $G \in \mathcal{G}_{n,d}$.*

What bounds are known on $\alpha(G)/n$? Bollobás [12] used the expected number of independent sets of a given size to give an upper bound on $\alpha(G)/n$ a.a.s. for $G \in \mathcal{G}_{n,d}$, the upper bound being given in terms of the solution of an equation and being strictly less than $\frac{2\log d}{d}$ for all d. It can be checked that this bound is asymptotic to

$$\gamma_1(d) = \frac{2n}{d}(\log d - \log\log d + 1 - \log 2) \tag{19}$$

as $d \to \infty$. By sharpening the method, McKay [73] improved these slightly to functions $\gamma_2(d)$, a.a.s. for G of sufficiently large (but fixed) girth. In view of the concentration mentioned above and Theorem 2.5, $\gamma_2(d) + \epsilon$ applies, for any $\epsilon > 0$, with no girth restriction. (Beware of McKay's sign error in the terms $1 - \log 2$ in his expression for $\gamma_1(d)$ defined above.) The definition of γ_2 is complicated but numerical computations give $\alpha(G)/n < \gamma_2(3) < 0.4554$ a.a.s. for $G \in \mathcal{G}_{n,3}$, and similarly $\gamma_2(4) < 0.4164$ and $\gamma_2(5) < 0.3845$. McKay expressed disappointment in the small improvement he obtained over $\gamma_1(d)$ as $d \to \infty$, but as we mention below this was inevitable because $\gamma_1(d)$ is quite sharp to order $o(n/d)$ in this asymptotic sense.

Lower bounds were obtained by several authors; the best (at least for small d) comes from analysing what is called in [124] the degree-greedy algorithm: choose vertices to put in the growing independent set randomly; the first, v_1, is chosen uniformly at random, and in general after adding a vertex v_i to the set, delete v_i and its neighbours from the graph and randomly choose the next vertex v_{i+1} (uniformly) from those vertices now of minimal degree. The analysis uses the differential equation method. This was done first by Frieze and Suen [45] for $d = 3$, and independently for arbitrary $d \geq 3$ in [122]. It gives asymptotically almost sure lower bounds $\beta_1(d)$ on $\alpha(G)/n$. For example, $\beta_1(3) = 6\log\frac{3}{2} - 2 = 0.4328$, $\beta_1(4) = 0.3901$ and $\beta_1(5) = 0.3566$. These constants are computed numerically for $d \geq 4$ by solving systems of first order nonlinear differential equations, and the asymptotic behaviour for large d is unknown. Further values are in [122] and [124]. The best known simple formula for a lower bound when $d \geq 4$ is $\frac{1}{2} - \frac{1}{2}\left(\frac{1}{d-1}\right)^{2/(d-2)}$ from [122]. However, this is far from $\beta_1(d)$ for d small, and approaches only $\frac{1}{2}\gamma_1(d)$ for large d. It is obtained from the simpler algorithm which randomly greedily chooses vertices for the independent set regardless of their neighbours, analysed again by the differential equation method. Weaker lower bounds were obtained earlier by Bollobás [12] through lower bounds on the independent set size of all d-regular triangle-free graphs, and arguing that deleting all vertices in short cycles a.a.s. decreases the independent set size only marginally. Interestingly, the non-deterministic bounds used here are obtained by analysing the greedy algorithm on triangle-free graphs of given average degree. That line of reasoning was also improved by Shearer [105] but with only an iterative formula for arbitrary d which seems always to be less than $\beta_1(d)$.

Frieze and Łuczak [44] proved that functions $\gamma_1(d)+o(n/d)$ for $d \to \infty$, $d < n^{1/3} - \epsilon$, provide upper and lower bounds a.a.s. on $\alpha(G)$, $G \in \mathcal{G}_{n,d}$. However, this result implies nothing for small d. They used an argument partitioning the pairing into large pieces which are then analysed almost separately, and part of the proof relies on the results of McKay and Wormald [79] which were obtained using switchings.

The dominating number seems to have been well studied only in $\mathcal{G}_{n,3}$. Molloy and Reed [86] showed that the obvious lower bound $\frac{1}{4}n$ on $\alpha'(G)$ can be improved slightly to $\alpha'(G) > 0.2636n$ a.a.s. by simply computing the expected number of dominating sets of that size. On the other hand they gave the almost sure upper bound $0.3126n$ by analysing a greedy algorithm which walks along a Hamilton cycle. The analysis uses the differential equation method and relies on Theorem 4.5 which implies that when proving events a.a.s., one can assume $G \in \mathcal{G}_{n,3}$ is a random Hamilton cycle plus a random matching. This is easier to analyse than the algorithm corresponding to the degree-greedy algorithm for independent sets; Duckworth and Wormald [34] applied the differential equation method to such an algorithm and improved the upper bound to $0.2746n$. To summarise, $0.2636 < \alpha'(G)/n < 0.2746$ a.a.s. for $G \in \mathcal{G}_{n,3}$.

As a slight variation of the last result, by analysing almost the same algorithm, it is shown in [34] that $G \in \mathcal{G}_{n,3}$ a.a.s. has a forest of stars containing at least $0.7178n$ edges. The obvious upper bound is $0.75n$, which can be reduced to $0.7390n$ using the usual argument of computing the expected number of star packings with a given number of edges.

Chromatic number Frieze and Łuczak [44] also gave functions asymptotic to $\frac{d}{2\log d}$ for $d \to \infty$ which a.a.s. are upper and lower bounds on the chromatic number $\chi(G)$ for $G \in \mathcal{G}_{n,d}$. Like their bounds on independent set size, these give no restriction for small d. A related result by Bollobás and Clark [18] considers a generalised version of chromatic number.

The chromatic number of G is at least $n/\alpha(G)$, and so almost sure lower bounds for small d are given by the almost sure upper bounds on $\alpha(G)$ mentioned in Section 2.8. Better lower bounds come from considering the expected number of k-colourings. For instance, counting bicoloured graphs shows that for fixed $d \geq 3$, $G \in \mathcal{G}_{n,d}$ a.a.s. cannot be 2-coloured. For $r \geq 3$, Molloy and Reed [84] performed calculations of the expected number of r-colourings of $G \in \mathcal{G}_{n,d}$ which show that its nth root is at most $r(\frac{r-1}{r})^{d/2} + o(1)$ (for fixed r and d). Hence, for instance, a.a.s. $\chi(G) \geq 4$ for $G \in \mathcal{G}_{n,d}$ when $d \geq 6$.

Brooks' Theorem (see Diestel [33] for example) gives the upper bound

$\chi(G) \leq d$ a.a.s. Non-trivial asymptotically almost sure upper bounds for small d seem to be harder to obtain. Borodin and Kostochka [23], Catlin [25] and Lawrence [62] all showed that for K_4-free graphs G, $\chi(G) \leq \frac{3}{4}(\Delta(G) + 2)$. Hence by Lemma 2.7, $\chi(G) \leq \frac{3}{4}(d+2)$ a.a.s. for $G \in \mathcal{G}_{n,d}$. For triangle-free graphs G there is the bound $\chi(G) \leq \frac{2}{3}\Delta(G) + 2$, which has been attributed to Kostochka, and by Theorem 2.5 implies $\chi(G) \leq \frac{2}{3}d + 2$ with probability at least $e^{-(d-1)^3/6} - o(1)$ for $G \in \mathcal{G}_{n,d}$.

These results still leave open the question of whether a random 4-regular or 5-regular graph is a.a.s. 3-colourable. An answer to this would be a significant breakthrough.

Knowledge of the edge-chromatic number $\chi'(G)$ is in a much more satisfactory state. It is shown below (Corollary 4.16) that $G \in \mathcal{G}_{n,d}$ a.a.s. has a partition of the edge set into perfect matchings if n is even. Hence $\chi'(G) = d$ a.a.s. for n even, and clearly for n odd it must attain the upper bound $d+1$ of Vizing's theorem.

Linear arboricity The *linear arboricity* of a graph is the minimum number of edge-disjoint linear forests (i.e. forests in which each component is a path) which cover the edges of the graph. The linear arboricity conjecture [1] is that every d-regular graph has linear arboricity precisely equal to the obvious lower bound of $\lceil \frac{1}{2}(d+1) \rceil$. Alon [2] proved using the Lovász Local Lemma that all regular graphs with sufficiently large girth satisfy this. Based on his approach, McDiarmid and Reed proved that a random d-regular graph a.a.s. satisfies the conjecture.

Theorem 2.21 (McDiarmid and Reed [66]) *For fixed d, the linear arboricity of a random d-regular graph is a.a.s. $\lceil \frac{1}{2}(d+1) \rceil$.*

This theorem is proved easily from the more recent results on the structure of random regular graphs, as shown in Section 4.3. It is also shown there that the paths in all but one of the factors can be forced to be Hamilton paths.

2.9 Perfect matchings and Hamilton cycles

The aim of this section is to whet the appetite with some of the results on the existence of certain types of spanning subgraphs of random regular graphs. The method of proof of the strongest results in this direction is presented in Section 4, along with many extensions of these results which assert the existence of factorisations of regular graphs into subgraphs of prescribed types.

Every d-regular $(d-1)$-edge-connected graph of even order has a perfect matching (see, for example, Berge [8]). Hence Theorem 2.10 has an interesting consequence.

Corollary 2.22 (Bollobás [16]) *For fixed $d \geq 3$, $G \in \mathcal{G}_{n,d}$ a.a.s. has a perfect matching if dn is even.*

This fails for $d = 2$ since in that case $G \in \mathcal{G}_{n,d}$ a.a.s. has an odd cycle, and hence no perfect matching. One way to see this is to observe from Theorem 2.5 that, for fixed $k \geq 2$, the probability that $G \in \mathcal{G}_{n,2}$ has no odd cycle of length less than $2k$ is $\exp(-\prod_{i=2}^{k} \frac{1}{4i-2}) + o(1)$, which is $\exp(-\frac{1}{4} \log k) + o(1) \to 0$ as $k \to \infty$. (Here k goes "slowly" to ∞ while n goes "quickly".) Alternatively, this probability can be computed using generating function methods and then evaluated asymptotically (see Wilf [112, Section 5.3] for similar examples). In $\mathcal{P}_{n,2}$ there is the following exact expression, which is asymptotic to $Cn^{-1/4}$ for a constant C. It then follows from Lemma 2.1 that the probability is also $o(1)$ for $G \in \mathcal{G}_{n,d}$.

Lemma 2.23 *For $P \in \mathcal{P}_{n,2}$, the probability that $G(P)$ has no odd cycles is precisely*
$$\prod_{i=1}^{\lfloor \frac{1}{2} n \rfloor} \frac{4i-2}{4i-1}.$$

Proof By the independence property of the pairing model, the random pairing $P \in \mathcal{P}_{n,2}$ can be generated in such a way that the next point paired is always a point in a cell which already contains a paired point (if such a cell exists). Then $G(P)$ is generated by tracing along paths, not starting a new path until the old one becomes a cycle. Let F_i denote the event that the i'th pair chosen completes a cycle in $G(P)$ with the earlier pairs. In order to create no odd cycle, the requirement is that F_{2i-1} must be false for each $1 \leq i \leq \frac{1}{2}n$. (Conditioning on F_{2i-1} being false for all $i \leq j$, if F_{2j} holds then the cycle created will have even length and if F_{2j+1} holds the cycle will have odd length.) Given all earlier choices of pairs, the conditional probability of F_i is $(2n - 4i + 3)^{-1}$ regardless of the earlier choices. The product of $(1 - (2n - 4i + 3)^{-1})$ over $1 \leq i \leq \frac{1}{2}n$ gives the stated probability. ∎

The expected number and variance of the number of perfect matchings were computed asymptotically by Bollobás and McKay.

Theorem 2.24 (Bollobás and McKay[20]) *For $3 \leq d = d(n) \leq (\log n)^{1/3}$, let M_n denote the number of perfect matchings in $G \in \mathcal{G}_{n,d}$. Then with n restricted to even integers,*
$$\mathbf{E}M_n \sim \sqrt{2} e^{1/4} \left(\frac{(d-1)^{d-1}}{d^{d-2}} \right)^{\frac{1}{2}n}, \qquad \frac{\mathbf{E}M_n^2}{(\mathbf{E}M_n)^2} \to e^{-\frac{1}{4}(2d-1)/(d-1)^2} \sqrt{\frac{d-1}{d-2}}.$$

This result on the first moment is readily derived in the pairing model by counting how many ways one can first choose a perfect matching of the n vertices, then lay down a set of pairs in P which create the chosen matching in $G(P)$, and then complete these pairs to a full pairing. After this, divide by the total number of pairings. To obtain the result in $\mathcal{G}_{n,d}$ one can condition these counts on having no loops or multiple edges, using for instance the method of moments as in Section 2.3. (Lemma 4.4 below performs this task in general.) The second moment $\mathbf{E}M_n^2$ is calculated in a similar fashion by laying down the pairs corresponding to an ordered pair of perfect matchings, which is much more complicated than the first moment calculation due to an extra variable denoting the number of edges in common between the two perfect matchings.

Since the variance is $\mathbf{Var} M_n = \mathbf{E}M_n^2 - (\mathbf{E}M_n)^2$, it tends to a non-zero constant multiple of $(\mathbf{E}M_n)^2$. This gives some information on the distribution of M_n via *Chebyshev's inequality*

$$\mathbf{P}(|X - \mathbf{E}X| \geq t) \leq \frac{\sigma^2}{t^2},$$

valid for any real-valued random variable X with variance σ^2 and for any $t > 0$. Putting $t < \mathbf{E}X$ gives an upper bound on $\mathbf{P}(X = 0)$, which is loosely called the *second moment method*. This is not sharp enough in this case to imply Corollary 2.22. However, the elaborate refinement of the second moment method presented in Section 4 is sufficient to obtain not only Corollary 2.22, but also the full asymptotic shape of the distribution of the number of perfect matchings.

Ever since the earliest results on random regular graphs, the natural conjecture on Hamiltonicity was (by Bollobás [11], for example) that $G \in \mathcal{G}_{n,d}$ is a.a.s. Hamiltonian for all fixed $d \geq 3$. For $d = 2$ this is false, by (11) for example. The second moment method implies that asymptotically at least a certain percentage of large random 3-regular graphs are Hamiltonian, which was first shown by Robinson and Wormald in [98]. In fact, it was shown there that by restricting to graphs with no triangles, a higher proportion $(2 - 3e^{-13/12})$ is obtained, which can be shown by bijective type methods to be valid back in the unrestricted space $\mathcal{G}_{n,3}$. This idea evolved into the small subgraph conditioning method for estimating the distribution of the number of large subgraphs. The results in the following theorem were derived for $d = 3$ in [98], stated in general in [100] without being used, and then derived and used by Frieze et al. [42].

Theorem 2.25 *For fixed $d \geq 3$, let H_n denote the number of Hamilton cycles in $G \in \mathcal{G}_{n,d}$. Then with dn restricted to even integers,*

$$\mathbf{E}H_n \sim e\sqrt{\frac{\pi}{2n}} \left(\frac{(d-2)^{d-2}(d-1)^2}{d^{d-2}} \right)^{\frac{1}{2}n}, \qquad \frac{\mathbf{E}H_n^2}{(\mathbf{E}H_n)^2} \to \frac{d}{(d-2)e^{2/(d-1)}}.$$

The calculations are similar to those for Theorem 2.24 but derivation of the second moment in particular is considerably more difficult. Again, two copies of the subgraph in question (this time a Hamilton cycle) are laid down and the rest of the cubic graph is completed. (All calculations can be made in the pairing model or directly in the graph model.) The number of ways of doing the laying down is computed as a function of the structure of the graph which is the intersection of the two Hamilton cycles. Two variables are required: the number of paths in this intersection and the number of isolated vertices. The resulting expression is summed over these two variables and evaluated asymptotically. (In the case $d = 3$ the number of isolated vertices must be 0, and so that case is much simpler than for general d.)

Further sharpening of the small subgraph conditioning method, as described in Section 4, produced the following.

Theorem 2.26 (Robinson and Wormald [99, 100]) *For fixed $d \geq 3$, $G \in \mathcal{G}_{n,d}$ is a.a.s. Hamiltonian.*

This was proved first for large d only, using an entirely different method, by Fenner and Frieze [38] ($d \geq 796$) and independently by Bollobás [15] ($d \geq 10^7$). The method used was developed by Fenner and Frieze [37] for a model of non-regular graphs. It combines the technique which Pósa [90] and Komlós and Szemerédi [61] successfully applied to determine the threshold for Hamiltonicity for $G \in \mathcal{G}(n,p)$ and a colouring argument by which the random graph is split into semi-random pieces of different colours. (This resembles the partitioning argument mentioned in Section 2.8 regarding independence number.) Neither of these works was attempting to achieve a sharpest possible result from this method. Frieze [41] later made improvements to cover all $d \geq 85$, giving a polynomial time constructive proof, before the appearance of [99] and [100] settled the matter for all $d \geq 3$.

Theorem 2.26 has now been improved in various ways to give much stronger results. It is now known that for $d \geq 3$, $G \in \mathcal{G}_{n,d}$ is a.a.s. decomposable into edge-disjoint Hamilton cycles (plus a matching if d is odd). In addition, Hamilton cycles are in a certain sense dense in random regular graphs: for $d \geq 3$ there is a.a.s. a Hamilton cycle containing a randomly selected set of $o(\sqrt{n})$ edges of $G \in \mathcal{G}_{n,d}$. These results and others are discussed more fully in Section 4.3.

By putting vertices of degree 2 in each selected edge, the last-mentioned result translates to a result about random graphs with all vertices of degree 3 except for a small number of degree 2. This is possibly the only non-trivial result known on the existence of perfect matchings or Hamilton cycles in $G \in \mathcal{G}_{n,\mathbf{d}}$, and yet the following is certain to be true.

Conjecture 2.27 *Restricting $\mathbf{d} = \mathbf{d}(n)$ to those sequences with elements d_i in the range $3 \leq d_i \leq D$ for some fixed d and D, $G \in \mathcal{G}_{n,\mathbf{d}}$ is a.a.s. Hamiltonian.*

It would be very interesting to prove this even for $d = 3$ and $D = 4$; probably all the difficulties for this are encountered in the special case where half the vertices are of degree 3 and half of degree 4 (mentioned in Section 1.1).

3 Other uniform models

3.1 Unlabelled regular graphs

How does one select an unlabelled graph uniformly at random? As in Section 2.7, let $\mathcal{S}$ be a set of unlabelled graphs on n vertices and let $\mathcal{S}'$ be the set of all labelled versions of them. By applying (15) to each singleton subset of $\mathcal{S}$, it follows that an unlabelled graph can be selected uniformly at random from $\mathcal{S}$ by selecting a labelled $G \in \mathcal{S}'$ with weight proportional to $|\phi(G)|$, and then ignoring the labels on the vertices. From Theorem 2.17 it follows that for $3 \leq d = o(n^{\frac{1}{2}-\epsilon})$, a structural property is a.a.s. true of $G \in \mathcal{G}_{n,d}$ if and only if it is a.a.s. true of a random unlabelled d-regular graph n-vertex graph.

3.2 Random bipartite graphs

For a model of random bipartite regular graphs we assume that n is even and that the vertices of one colour are labelled $1, 2, \ldots, \frac{1}{2}n$, as are the vertices of the other colour. The natural pairing model for d-regular bipartite graphs is obvious: the cells containing points are the same as for ordinary graphs, but the random perfect matching is equivalent to a bijection between the points in cells of one colour and those in cells of the other colour. As a result, calculations are usually even easier than in the graph case. O'Neil [89] found that the number of bicoloured d-regular graphs on n vertices (n even) is asymptotic to

$$\frac{(\frac{1}{2}dn)! e^{-\frac{1}{2}(d-1)^2}}{(d!)^n} \tag{20}$$

for $3 \leq d < (\log n)^{\frac{1}{4}-\epsilon}$ with $\epsilon > 0$, and the same result (at least for d fixed) was found in [4] using a model equivalent to the pairing model. O'Neil also obtained a few results on random bipartite graphs.

More recently McKay [71] extended (20) to higher d using switchings, and then McKay and Wang [76] extended to $d = o(\sqrt{n})$ using the new type of switchings exemplified by the proof of Lemma 2.9.

For simplicity when we refer to *the* model of random bipartite regular graphs we mean the model described above, which is actually uniform on *bicoloured* graphs. It is *not* uniform on bipartite d-regular labelled graphs because the number of 2-colourings of a bipartite graph G is 2^k where k is the number of components of G. To deal with graphs which are not bicoloured but are rather bipartite, one notes (see below) that, at least for $d \geq 3$, the bicoloured graphs are a.a.s. connected. This implies that the bipartite graphs (taken as $\mathcal{G}_{n,d}$ restricted to bipartite graphs) are similarly a.a.s. connected. For $d = 2$ the situation is a bit more complicated but results can be obtained by exact enumeration and we do not pursue this here. So, in this article, random regular bipartite graphs have the uniform bicoloured distribution.

The distribution of short cycles can be derived just as for graphs. The expected number of cycles of length i is asymptotically $\frac{(d-1)^i}{i}$ for all even

$i \geq 4$, and the numbers of cycles of given bounded lengths are asymptotically independent Poisson. This was all derived in [115].

Ellingham [35] showed the analogues of Theorems 2.10 (for d bounded) and 2.12 on connectivity of random regular bipartite graphs. The main result on the expected number of automorphisms of a random regular graph, Theorem 2.17, also applies in the bipartite case as shown by the very general results of McKay and Wormald [78, Note before Corollary 3.5].

The expectation and variance of the number of perfect matchings in random d-regular bipartite graphs were given asymptotically in [89] and [20] (for quite small d). For Hamilton cycles these quantities were given in [98], where the case $d = 3$ of the following result was proved simply from Chebyshev's inequality.

Theorem 3.1 (Robinson and Wormald [100]) *For fixed $d \geq 3$, a random d-regular bipartite graph is a.a.s. Hamiltonian.*

Godsil and McKay [51] obtained asymptotics for the number of $k \times n$ Latin rectangles for $k = o(n^{6/7})$. As part of the argument they find an asymptotic formula for the number of perfect matchings (respecting the bipartition) in a random edge-coloured bicoloured k-regular graph whose edges have been coloured such that each colour class is a perfect matching. This is equivalent to the bicoloured analogue of the random graph model $k\mathcal{G}_{n,1}$ defined in Section 4.3. Their derivation uses an integral formula to convert the problem into one about short cycles, and from here switchings are used (necessarily, for such a large k).

3.3 Directed graphs

Let $\mathcal{DG}_{n,d}$ denote the uniform model of random d-regular digraphs (in which each vertex has in- and out-degree d).

Cooper [28] showed for $D \in \mathcal{DG}_{n,2}$ that D is a.a.s. strongly 2-connected and [29] that a.a.s. every pair of vertices lies on a common directed cycle. For graphs, 2-connectedness implies such a property, but not so for digraphs.

For $d = 2$ the expected number of directed Hamilton cycles goes to 0, so digraphs in $\mathcal{DG}_{n,d}$ are a.a.s. not Hamiltonian. For larger d the situation is different.

Theorem 3.2 (Cooper et al. [31]) *For $d \geq 3$ a random digraph in $\mathcal{DG}_{n,d}$ a.a.s. has a directed Hamilton cycle.*

The proof of this employs the ideas from Section 4.3, and is outlined near the end of that section. A different but related proof was given by Janson using the small subgraph conditioning method directly (see Theorem 4.12).

The d-regular bipartite graph corresponding to a d-regular digraph has a perfect matching by Hall's theorem (see [22] for example), and so *all* d-regular digraphs have an edge decomposition into 1-regular factors.

3.4 Multigraphs

Nothing much has been done with uniformly distributed multigraphs. The asymptotic number with a given degree sequence (with an upper bound on the degrees) is found in [118], with loops permitted as well as multiple edges. Bender and Canfield [5] have a similar result, but they effectively count a loop as contributing degree 1 to a vertex. These results should be possible to extend to higher degrees.

Of course many results on the uniform model can be deduced from the pairing model, which is not uniform but is close. This can be quantified as follows (here multigraphs permit loops as well).

Theorem 3.3 (Janson [56]) *For d fixed, an event is true a.a.s. for $G(P)$, $P \in \mathcal{P}_{n,d}$, if and only if it is true a.a.s. for d-regular multigraphs chosen u.a.r.*

This does not follow from Theorem 2.6, because multigraphs with many multiple edges have a reduced probability of arising as $G(P)$. Hence Theorem 2.6 leaves open the possibility that a random d-regular multigraph a.a.s. has at least say $\log n$ multiple edges. Some further analysis of the appropriate moments was required to prove this theorem.

3.5 Hypergraphs

There are some results on uniformly distributed d-regular, r-uniform hypergraphs with d and r fixed, as $n \to \infty$ (subject to dn being a multiple of r). For many of the results on random regular graphs, similar techniques will suffice. For instance, Bollobás [10] stated an asymptotic formula for the number of r-uniform d-regular hypergraphs, and Cooper et al. [32] applied the small subgraph conditioning method to show that for fixed positive integers r and s,

$$\lim \mathbf{P}(G_{n,r,s} \text{ has a perfect matching}) = \begin{cases} 0 & s > \sigma_r \\ 1 & s < \sigma_r \end{cases}$$

where

$$\sigma_r = \frac{\log r}{(r-1) \log\left(\frac{r}{r-1}\right)} + 1.$$

3.6 Tournaments

In 1974 Spencer [106] used an argument akin to switchings to prove properties of subgraphs of random regular tournaments. This gave the logarithm of the number of n-vertex regular tournaments asymptotically. This result was superseded by McKay [74], who found an asymptotic formula for the number itself using saddle-point techniques. This method, also used for asymptotic enumeration of tournaments with given score sequence by McKay and Wang [77], is useful for studying random regular tournaments. In this way Gao et al. [46] obtained some results on subgraphs of random tournaments with a given score sequence, and have shown that the expected number of automorphisms tends to 1.

4 The small subgraph conditioning method, contiguity, and superposition models

When it was finally shown in [99] and [100] that almost all d-regular graphs are Hamiltonian for $d \geq 3$ (Theorem 2.26), the central work involved two applications of essentially the same general method. The method has since proved useful for gaining information on a variety of spanning subgraphs, as well as other large subgraphs, of random regular graphs. The strongest consequences are given in Section 4.3.

The way in which the method works can be described as follows. (A precise development of how to use it comes in the following subsections.) The random variable $Y = Y_n$ of interest (for instance, the number of Hamilton cycles in $G \in \mathcal{G}_{n,d}$) has a variance which is of a size comparable with the square of its expectation (as in, for example, Theorems 2.24 and 2.25). Thus, Chebyshev's inequality is not strong enough to show that a.a.s. $Y > 0$. However, the explanation for the large variance lies in the fact that the distribution of Y is affected by the presence of certain small but not too common subgraphs in the random graph—usually the short cycles of given lengths. (By not too common, we mean the expected number is bounded.) It turns out (in the cases where the method works) that conditioning on the small subgraph counts (up to some preselected size of small subgraphs) affects $\mathbf{E}Y$, altering it by some constant factor. However, luckily and yet mysteriously, such conditioning reduces the variance of Y, to the point that conditioning on the numbers of enough small subgraphs reduces the variance to any desired small fraction of $(\mathbf{E}Y)^2$.

Perhaps this is not a total mystery. For instance, one can imagine that the number of ways in which a Hamilton cycle or perfect matching can "go through" a short cycle depends on the length of the short cycle. Probably the main mystery is not that conditioning on the small cycle counts affects $\mathbf{E}Y$, but that the variance can be made *so* small by conditioning on the small subgraph counts. The computations which are performed in carrying out the small subgraph conditioning method demonstrate this as a fact but leave the user with no underlying explanation of the effect. These computations can be described as follows. The set of all regular graphs can be divided into groups according to the small subgraph counts. The variance $\mathbf{Var}Y$ of Y can be written as the expected value of the variance within a group, plus the variance of the group mean. The calculations show that the latter consumes almost all of the variance of Y, so the variance within any group is small. One deduction which can be made from this is that $Y > 0$ a.a.s. This explains the basic argument used in [99], where Y counts Hamilton cycles, to show that $G \in \mathcal{G}_{n,3}$ is a.a.s. Hamiltonian.

Moreover, the asymptotic distribution of Y is determined from the distribution of the group means and the sizes of the groups. The fact that the group means of the most common groups are close to each other implies that the value of Y does not vary too much in the random graphs (except for a rare event). For instance, taking Y as the number of perfect matchings in $G \in \mathcal{G}_{n,d}$, this shows that the number of perfect matchings in such a random graph usually varies by only a constant factor. This was used in [100] to deduce that adding a random perfect matching to an edge-disjoint random $(d-1)$-regular graph produces a random d-regular graph with something close to the uniform distribution. This relationship is called contiguity of the two models—perfect matching plus $(d-1)$-regular on the one hand, and d-regular on the other hand—as defined precisely below. Hence, if random $(d-1)$-regular graphs are a.a.s. Hamiltonian, so are random d-regular graphs, and so Theorem 2.26 follows by induction on d from the case $d = 3$. This is the second major use of the small subgraph conditioning method, and it leads naturally to the establishment of relationships between various non-uniform models of random regular graphs which are called here superposition models.

4.1 Contiguity of models

One of the most important conclusions from the small subgraph conditioning method is well described in terms of altered probabilistic models. Suppose that Y is a non-negative integer random variable defined on a space $\mathcal{G}$ with $\mathbf{E}Y \neq 0$. We define[1] a new model $\mathcal{G}^{(Y)}$ with the same underlying set as $\mathcal{G}$ by weighting the probability of each element G by $Y(G)$. That is, the probability of G in $\mathcal{G}^{(Y)}$ equals the probability in $\mathcal{G}$ multiplied by $Y(G)/\mathbf{E}Y$. Thus the probability of an event H in $\mathcal{G}^{(Y)}$ is $\mathbf{E}_{\mathcal{G}}(Y \wedge 1_H)/\mathbf{E}_{\mathcal{G}}Y$, where 1_H is the indicator function of H and $\wedge$ denotes the conjunction of events. We have already encountered this idea in the proof of Lemma 2.8 with $Y = [X_{k,n}]_{r_k}$.

It is a very interesting property of two different random graph models if all events true a.a.s. in one model are true a.a.s. in the other. To make this precise, suppose that $(\mathcal{G}_n)_{n \geq 1}$ and $(\widehat{\mathcal{G}}_n)_{n \geq 1}$ are two sequences of probability spaces such that $\mathcal{G}_n$ and $\widehat{\mathcal{G}}_n$ differ only in the probabilities. We say that these sequences are *contiguous*[2] if a sequence of events A_n is a.a.s. true in $\mathcal{G}_n$ if and only if it is a.a.s. true in $\widehat{\mathcal{G}}_n$, in which case we write

$$\mathcal{G}_n \approx \widehat{\mathcal{G}}_n.$$

Contiguity is clearly an equivalence relation on these sequences of spaces.

---

[1] This definition is for discrete probabilistic spaces, a context which can be assumed for this article. The method of this section holds in continuous spaces as well, by defining $dQ/dP = Y/\mathbf{E}_{\mathcal{G}}Y$ where P and Q are the probability measures in $\mathcal{G}$ and $\mathcal{G}^{(Y)}$ respectively, and the σ-algebra is the same for both spaces.

[2] Formally, two sequences P_n and Q_n of probability measures are contiguous if P_n and Q_n are defined on the same measurable space $(\mathcal{G}_n, \mathcal{F}_n)$ for each n, and for every sequence of measurable sets A_n, $\lim_{n \to \infty} P_n(A_n) = 0$ iff $\lim_{n \to \infty} Q_n(A_n) = 0$. Here I have defined the (sequences of) spaces to be contiguous if the probabilities are.

The original approach in [99] and [100] is sufficiently strong for all the applications of the small subgraph conditioning method which have occurred to date and was presented in general by Molloy et al. [87, Theorem 1]. However Janson [56] was able to streamline the procedure by adjusting the method of proof so as to eliminate several of the conditions required in [87]. This simplification has made little difference in applications of the method: it turns out that the extra conditions have been trivially satisfied in every application so far. However, the simplification is definitely advantageous.

Theorem 4.1 ([56], see also [87]) *Let $\lambda_i > 0$ and $\delta_i \geq -1$, $i = 1, 2, \ldots$, be real numbers and suppose that for each n there are random variables $X_i = X_i(n)$, $i = 1, 2, \ldots$, and $Y = Y(n)$ defined on the same probability space $\mathcal{G} = \mathcal{G}(n)$ such that X_i is non-negative integer valued, Y is non-negative and $\mathbf{E}Y > 0$ (for n sufficiently large). Suppose furthermore that*

(a) *For each $k \geq 1$ X_i, $i = 1, 2, \ldots, k$ are asymptotically independent Poisson random variables with $\mathbf{E}X_i \to \lambda_i$;*

(b)
$$\frac{\mathbf{E}(Y[X_1]_{j_1} \cdots [X_k]_{j_k})}{\mathbf{E}Y} \to \prod_{i=1}^{k} (\lambda_i(1+\delta_i))^{j_i}$$

for every finite sequence $j_1, \ldots, j_k$ of non-negative integers;

(c) $\sum_i \lambda_i \delta_i^2 < \infty$;

(d) $\dfrac{\mathbf{E}Y_n^2}{(\mathbf{E}Y_n)^2} \leq \exp\left(\sum_i \lambda_i \delta_i^2\right) + o(1)$ *as $n \to \infty$.*

Then
$$\mathbf{P}(Y_n > 0) = \exp\left(-\sum_{\delta_i = -1} \lambda_i\right) + o(1),$$

and, provided $\sum_{\delta_i = -1} \lambda_i < \infty$,

$$\overline{\mathcal{G}}^{(Y)} \approx \overline{\mathcal{G}}$$

where $\overline{\mathcal{G}}$ is the probability space obtained from $\mathcal{G}$ by conditioning on the event $\bigwedge_{\delta_i = -1}(X_i = 0)$.

The proof of this theorem is based on the ideas in the discussion at the start of Section 4. The role of condition (b) is to deduce by the method of moments the joint distribution of the X_i in $\mathcal{G}^{(Y)}$, which then gives the expected value of Y_n conditioned on $X_1, \ldots, X_k$. For the proof the factor $(1 + \delta_i)$ need not be interpreted. It is an adjustment to the expected value of X_i in the space $\mathcal{G}^{(Y)}$, so that $\delta_i = -1$ only if conditioning on the event $X_i > 0$ in $\mathcal{G}$ suppresses the value of Y asymptotically. Condition (d) has been at other times stated with

"=" instead of "≤", but "≤" implies "=" anyway (see Janson [56, Theorem 1, Note 4]), basically because variance is always non-negative.

The hypotheses of the theorem are normally verified in much the same way on each occasion. Most of the times this method has been used so far, the X_i are short cycle counts and so (a) is simply Theorem 2.5 or Theorem 2.6. In all other cases, the method of moments has sufficed for verifying (a). Part (d) is verified like Theorems 2.24 or 2.25, and (b) is usually like a combination of the two, harder than (a) but easier than (d).

It follows immediately from contiguity that the value of Y in $\overline{\mathcal{G}}$ is usually "close" to its expected value.

Corollary 4.2 *If the conditions of the theorem are satisfied then*
$$\lim_{\epsilon \to 0^+} \left\{ \lim_{n \to \infty} \mathbf{P}_n \left(\epsilon < \frac{Y_n}{\mathbf{E} Y_n} < \frac{1}{\epsilon} \right) \right\} = 1$$
where $\mathbf{P}_n$ denotes probability in $\overline{\mathcal{G}}$.

From the arguments in [99] and [100] it was clear that when the hypotheses of the theorem are satisfied, the distribution of Y is determined asymptotically. Janson made this explicit.

Theorem 4.3 (Janson [56]) *With the same hypotheses as Theorem 4.1,*
$$\frac{Y}{\mathbf{E} Y} \xrightarrow{d} W = \prod_{i=1}^{\infty} (1+\delta_i)^{Z_i} e^{-\lambda_i \delta_i} \qquad \text{as } n \to \infty,$$
where the variables Z_i are independent Poisson variables with $\mathbf{E} Z_i = \lambda_i$ for $i \geq 1$. Moreover, this convergence and the convergence of the X_i to the Z_i expressed in Theorem 4.1 (a) all hold jointly.

Thus in each case where contiguity is proved using Theorem 4.1, we can if so desired deduce the asymptotic distribution of the appropriate variable immediately by Theorem 4.3. Moreover, the joint convergence mentioned here means that the Z_i are linked to the X_i. Thus the distribution of Y conditioned on any values of a finite number of X_i is also determined asymptotically, for example the distribution of Y conditioned on $X_i = k$ is asymptotically that of W conditioned on $Z_i = k$. This makes the contiguity assertion of Theorem 4.1 easy to verify in view of the proof of [87, Corollary 1].

One more result is often useful when the X_i are short cycle counts in $\mathcal{P}_{n,d}$. It is readily proved by applying Lemma 2.8 to the space $\mathcal{G}^{(Y)}$.

Lemma 4.4 (Janson [56]) *If the conditions of Theorem 4.1 hold and $Y'(n)$ has the distribution of $Y(n)$ conditioned on $X_1(n) = X_2(n) = 0$ then*
$$\frac{\mathbf{E} Y'(n)}{\mathbf{E} Y(n)} \to \exp(-\lambda_1 \delta_1 - \lambda_2 \delta_2), \quad \frac{\mathbf{E}(Y'(n)^2)}{(\mathbf{E} Y'(n))^2} \to \exp(-\lambda_1 \delta_1 - \lambda_2 \delta_2) \frac{\mathbf{E}(Y(n)^2)}{(\mathbf{E} Y(n))^2}.$$

We mention one application of contiguity immediately. Janson showed in [56] that $G(P)$, for $P \in \mathcal{P}_{n,d}$, is contiguous to a uniformly chosen random regular multigraph (with loops permitted). Theorem 3.3 follows directly from this.

4.2 Applications to regular spanning subgraphs and long cycles

In order to discuss in some detail how to verify the hypotheses of Theorem 4.1, we consider the result of the arguments in [99], expressed in the language of contiguity as follows.

Theorem 4.5 *Let H_n denote the number of Hamilton cycles in $G \in \mathcal{G}_{n,3}$. Then $\mathcal{G}_{n,3}^{(H_n)} \approx \mathcal{G}_{n,3}$.*

Proof The first and second moments of H_n are given by Theorem 2.25 with $d = 3$. However, although the final result will be in $\mathcal{G}_{n,3}$, it is a little simpler to work with Theorem 4.1 where $\mathcal{G}(n) = \mathcal{P}_{n,3}$. In this case $Y = Y(n)$ is defined as the number of Hamilton cycles in the multigraph corresponding to $P \in \mathcal{P}_{n,3}$. We have

$$\frac{\mathbf{E}Y^2}{(\mathbf{E}Y)^2} \to 3. \tag{21}$$

(In $\mathcal{P}_{n,d}$ for $d \geq 3$ the limit is $\frac{d}{d-2}$, as shown in [42].) This second moment calculation is the most difficult part of this whole proof, and is virtually the same as the calculation required for $\mathcal{G}_{n,d}$.

Let $X_i = X_i(n)$ denote the number of cycles of length i in $G(P)$ for $P \in \mathcal{P}_{n,3}$. The condition of Theorem 4.1 (a) follows from Theorem 2.6 with

$$\lambda_i = \frac{2^{i-1}}{i} \tag{22}$$

($i \geq 1$). Working towards Theorem 4.1 (b), we next show that for any fixed $i \geq 1$

$$\frac{\mathbf{E}(YX_i)}{\mathbf{E}Y} \to \lambda_i(1 + \delta_i) \tag{23}$$

where

$$\delta_i = \frac{(-1)^i - 1}{2^i}.$$

Let D be some fixed set of pairs of points corresponding to a Hamilton cycle in pairings in $\mathcal{P}_{n,3}$. By symmetry all copies of D are equivalent and so in $\mathcal{P}_{n,3}$

$$\frac{\mathbf{E}(YX_i)}{\mathbf{E}Y} = \mathbf{E}(X_i \mid D \subseteq P).$$

If C is the set of pairs corresponding to an i-cycle (in which case we also call C itself an i-cycle), we classify C according to the configuration of paths in $G(D \cap C)$. We can assume that there must be at least one such path as all pairs in C cannot be in D provided $n > i$. Give these paths a consistent orientation along C (which multiplies counts by 2) and distinguish one path as first (which multiplies counts by the number of paths and induces a linear ordering of paths around C). Thus

$$\frac{\mathbf{E}(YX_i)}{\mathbf{E}Y} = \sum_Q \frac{1}{2|Q|} \mathbf{E}(X_i(Q) \mid D \subseteq P) \tag{24}$$

where Q denotes the sequence of lengths of paths, $|Q|$ is the number of paths in Q and $X_i(Q)$ is the number of i-cycles in P consistent with such a configuration Q. Fix on such a Q with k paths. There are asymptotically $(2n)^k$ ways to choose the starting points of the paths on D together with their directions along D. Almost all such points are well spaced for n large, and once they are chosen the pairs in C, if it is to correspond to an i-cycle yielding Q, are determined. The probability that these pairs all occur in $P \in \mathcal{P}_{n,3}$ conditional upon $D \subseteq P$ is asymptotically n^{-k}. Hence $\mathbf{E}(X_i(Q) \mid D \subseteq P) \to 2^{|Q|}$ and so (24) becomes

$$\frac{\mathbf{E}(YX_i)}{\mathbf{E}Y} \to \sum_{k \geq 1} \frac{2^k}{2k} \left| \{Q : |Q| = k\} \right|. \qquad (25)$$

The ordinary generating function for the number of configurations Q with x marking the total number of vertices involved and y marking the number of paths is $\frac{g(x,y)}{1-g(x,y)}$ where $g(x,y)$ is the generating function for one path; that is, $\frac{yx^2}{1-x}$. Thus, with square brackets denoting extraction of coefficients,

$$\begin{aligned}
\frac{\mathbf{E}(YX_i)}{\mathbf{E}Y} &\to \sum_{k \geq 1} \frac{2^k}{2k} [x^i y^k] \frac{yx^2}{1-x-yx^2} \\
&= [x^i] \sum_{k \geq 1} \frac{1}{2k} [y^{k-1}] \frac{2x^2}{1-x-2yx^2} \\
&= [x^i] \frac{1}{2} \int_0^1 \frac{2x^2}{1-x-2yx^2} \, dy \\
&= -[x^i] \tfrac{1}{2} \log(1-x-2x^2) + \tfrac{1}{2} \log(1-x),
\end{aligned}$$

and (23) follows. To obtain the condition of Theorem 4.1 (b), one observes that this argument works in general for higher moments and gives the required result.

Noting $\sum_{i \geq 1} \lambda_i \delta_i^2 = \log 3$ and recalling (21), we see that the hypotheses of Theorem 4.1 are satisfied. As $\delta_i = -1$ only for $i = 1$, $\overline{\mathcal{P}}_{n,3}$ is the restriction of $\mathcal{P}_{n,3}$ to pairings P with $X_1 = 0$, i.e. for which $G(P)$ has no loops. The conclusion is that $\overline{\mathcal{P}}_{n,3}^{(Y)} \approx \overline{\mathcal{P}}_{n,3}$. The theorem now follows using the definition of contiguity by restricting to $X_2 > 0$, since in $\mathcal{P}_{n,3}$ the probability of this event tends to a non-zero constant. ∎

Notes (i) An alternative way of proceeding from (25) to (23), which in some applications leads to significantly simpler computations (see Janson [56]), is the following. Each intersection $D \cap C$ corresponds to a sequence in $\{0,1\}^i$ determined by walking along C in a given direction from an arbitrary starting edge (which multiplies counts by $2i$) and writing 0 for an edge not in $D \cap C$ and 1 for an edge in $D \cap C$. Each 0 must be followed by 1 and contributes a factor 2 (by an argument analogous to (25)). Regarding 0 and 1 as two states in a Markov chain (or something similar), and imposing the condition that the final state is equal to the initial state (since the first edge must be returned to) we see that the sum over sequences is $\text{Tr}(A^i)$ where

$$A = \begin{pmatrix} 0 & 2 \\ 1 & 1 \end{pmatrix}.$$

As the sequence of all 1's is impossible, it follows that $\mathbf{E}(YX_i)/\mathbf{E}Y \to \frac{1}{2i}(\text{Tr}(A^i) - 1)$. Then by noting that A has eigenvalues 2 and -1, we obtain $\text{Tr}(A^i) = 2^i + (-1)^i$ and then (23) as required.

(ii) Working the proof in $\mathcal{G}_{n,3}$ instead of $\mathcal{P}_{n,3}$ merely requires eliminating loops and multiple edges in a couple of places, which can always be done using the method of moments. Alternatively, as in [99, 100], one can appeal directly to the enumeration results of Bender and Canfield [5] to count the ways of adding edges to an existing graph without creating multiple edges, since their main theorem allows arbitrary forbidden edges as long as a bounded number are incident with a vertex.

From the proof of Theorem 4.5 and Theorem 4.3 we obtain the following.

Corollary 4.6 (Janson[56]) *Let H_n denote the number of Hamilton cycles in $G \in \mathcal{G}_{n,3}$. Then*

$$\frac{H_n}{\mathbf{E}H_n} \xrightarrow{d} W = \prod_{i=3}^{\infty} (1+\delta_i)^{Z_i} e^{-\lambda_i \delta_i} \qquad \text{as } n \to \infty,$$

where $\lambda_i = \frac{2^{i-1}}{i}$, $\delta_i = \frac{(-1)^i - 1}{2^i}$, and Z_i are independent Poisson variables with $\mathbf{E}Z_i = \lambda_i$ for $i \geq 3$.

Proof These are the λ_i and δ_i obtained in the proof of Theorem 4.5, so Theorem 4.3 gives the claimed result for $G(P)$, $P \in \mathcal{P}_{n,3}$ but with $i \geq 1$. By the joint convergence of the variables, restricting to $\mathcal{G}_{n,3}$ (i.e. $X_1 = X_2 = 0$) is asymptotically equivalent to restricting to $Z_1 = Z_2 = 0$. ∎

Similar arguments, using [42] for the variance calculation, give the following.

Theorem 4.7 (Frieze et al. [42], Janson [56]) *Let H_n denote the number of Hamilton cycles in $G \in \mathcal{G}_{n,d}$. Then $\mathcal{G}_{n,d}^{(H_n)} \approx \mathcal{G}_{n,d}$ for $d \geq 3$.*

To obtain a result generalising Corollary 4.6, one only needs to note the values $\lambda_i = \frac{(d-1)^i}{2i}$, $\delta_i = \frac{(-1)^i - 1}{(d-1)^i}$ and that X_i is again the number of i-cycles (so the product in the definition of W is again over $i \geq 3$ for distribution in $\mathcal{G}_{n,d}$). This was done by Janson [56].

Proof of Theorem 2.26 This now follows immediately by the definition of contiguity, since the event $\{H_n > 0\}$ has probability 1 in $\mathcal{G}_{n,d}^{(H_n)}$, so it is true a.a.s. in $\mathcal{G}_{n,d}$. ∎

The proof of Theorem 2.26 in [100] was different. It involved proving the analogous property of perfect matchings given in Theorem 4.8, and then argued using contiguity inductively as developed in Section 4.3. In the case of n odd and d even this result could not be used directly, but instead another result was proved which related to the distribution of the number of matchings which miss just one vertex. (However, Janson [56] was the first to state the contiguity result for perfect matchings explicitly.) Other results on the distribution of numbers of spanning regular subgraphs in $\mathcal{G}_{n,d}$ have been obtained using the same method, with the conclusions summarised in the next theorem.

A *k-factor* of a graph is a k-regular spanning subgraph. (A perfect matching is the edge set of a 1-factor.) A *1-factorisation* of a d-regular graph is a set of d edge-disjoint 1-factors, equivalent to a partition of the edge set into perfect matchings. For enumeration purposes, we assume these matchings are ordered. (For $d = 3$ the number of 1-factorisations is then equal to the number of ordered pairs of disjoint 1-factors, called double 1-factors in [87].) For n even, let M_n (defined in $\mathcal{G}_{n,d}$) denote the number of perfect matchings and let T_n (defined in $\mathcal{G}_{n,3}$) denote the number of 1-factorisations of G. For all n let D_n denote the number of 2-factors in $G \in \mathcal{G}_{n,d}$ (D comes from dwa, which is Polish for 2), and let B_n denote the number of ordered pairs of edge-disjoint Hamilton cycles in $G \in \mathcal{G}_{n,4}$ (a good name for these is Hamilton bicycles; hence the letter B). The following result on M_n essentially appears in [100] and was given explicitly by Janson [56], that on T_n was obtained independently by Janson [56] and by Molloy et al. [87], that on D_n is by Robalewska [95], and the one on B_n is by Kim and Wormald [60].

Theorem 4.8 *Restricting to n even, $\mathcal{G}_{n,d}^{(M_n)} \approx \mathcal{G}_{n,d}$ for all $d \geq 3$, and $\mathcal{G}_{n,3}^{(T_n)} \approx \mathcal{G}_{n,3}$. With no restrictions on n, $\mathcal{G}_{n,d}^{(D_n)} \approx \mathcal{G}_{n,d}$ for $n \geq 4$ and $\mathcal{G}_{n,4}^{(B_n)} \approx \mathcal{G}_{n,4}$.*

Proving these results is like proving Theorem 4.7, but the calculations are much easier in the case of M_n, somewhat easier for T_n and D_n. The result for 2-factors (D_n) when $d = 4$ was conjectured in [56]. (More precisely, the stronger multigraph version was conjectured for $G(P)$ where $P \in \mathcal{P}_{n,d}$ restricted to loopless multigraphs, which no doubt follows in the same way. As noted in Section 2, this space of multigraphs is not uniform.) For edge-disjoint Hamilton cycles (B_n) the same method was again used, but a really new idea was required to compute the variance, including a proof of Theorem 4.11 below. In fact, it was a conjecture in [100] that a random 4-regular graph a.a.s. decomposes into two edge-disjoint Hamilton cycles, and the related contiguity result was later conjectured by Janson [56]. (Actually, the multigraph version was conjectured, which also follows from the results in [60].)

In each of these applications, the X_i are the short cycle counts as before, and so $\lambda_i = \frac{(d-1)^i}{2i}$ as in (22). Other relevant values (all in the graph space $\mathcal{G}_{n,d}$) are given in the following table. Hamilton cycles (H_n) are included for comparison.

Y_n	$\mathbf{E}Y_n \sim$	$\mathbf{E}Y_n^2/(\mathbf{E}Y_n)^2 \to$	δ_i
H_n	$e\sqrt{\frac{\pi}{2n}}\left(\frac{(d-2)^{d-2}(d-1)^2}{d^{d-2}}\right)^{\frac{1}{2}n}$	$\frac{d}{(d-2)e^{2/(d-1)}}$	$\frac{(-1)^i-1}{(d-1)^i}$
M_n	$\sqrt{2}e^{1/4}\left(\frac{(d-1)^{d-1}}{d^{d-2}}\right)^{\frac{1}{2}n}$	$e^{-\frac{1}{4}(2d-1)/(d-1)^2}\sqrt{\frac{d-1}{d-2}}$	$\frac{(-1)^i}{(d-1)^i}$
T_n	$2\sqrt{e}(\frac{4}{3})^{n/2}$	$4e^{-5/4}$	$\frac{(-1)^i}{2^{i-1}}$
D_n	$e^{1/4}\sqrt{2}\left(\frac{(d-1)(d-2)^{(d-2)/2}}{d^{(d-2)/2}}\right)^n$	$e^{-\frac{1}{4}(2d-1)/(d-1)^2}\sqrt{\frac{d-1}{d-2}}$	$\frac{(-1)^i}{(d-1)^i}$
B_n	$e^{7/4}\frac{\pi}{\sqrt{8n}}(\frac{3}{2})^n$	$e^{-55/36}\sqrt{24}$	$\frac{-2+(-1)^i}{3^i}$

Theorem 2.24 is used for the second moment of M_n. The values of the first two moments of M_n and T_n for $G(P)$, $P \in \mathcal{P}_{n,d}$ (and hence the corresponding contiguity results for this multigraph space) are given in [56]; in all applications examined so far these are obtained by omitting the powers of e, since these and only these factors arise when sieving to remove loops and multiple edges (see Lemma 4.4). Similarly, the contiguity relating to B_n for $G(P)$, $P \in \mathcal{P}_{n,d}$ restricted to loopless multigraphs, follows from the results in [60]. For general d, Janson also derives the value $\delta_i = \frac{(-1)^i}{(d-1)^{i-1}}$ relevant to 1-factorisations of d-regular graphs, but it seems to be too difficult to obtain the variance for general d.

In all cases, the limiting distribution of Y_n is given immediately from Theorem 4.3 (and this is done in [56] for M_n and T_n, and in [95] for D_n).

The obvious corollary of the result for T_n is that $G \in \mathcal{G}_{n,3}$ a.a.s. has a 1-factorisation for even n, but this already follows from Theorem 2.26 with $d = 3$. The result for M_n gives another proof of Corollary 2.22. The result for D_n shows that $G \in \mathcal{G}_{n,4}$ a.a.s. has a 2-factor, but this event has probability 1 anyway by Petersen's theorem (see [33] for example). So the only new result we obtain of this type is with B_n.

Corollary 4.9 *A random 4-regular graph a.a.s. decomposes into two edge-disjoint Hamilton cycles.*

The surprising equality of the values in the last two columns of the table for M_n and D_n has the following consequence.

Corollary 4.10 (Robalewska [95]) *For even n,*

$$\frac{M_n}{\mathbf{E}M_n} \quad \text{and} \quad \frac{D_n}{\mathbf{E}D_n}$$

have the same limit distribution in $\mathcal{G}_{n,d}$ ($d \geq 3$).

Moreover, from the joint convergence given by Theorem 4.1 and the fact that the X_i are the same variables in each of these two cases, it follows that M_n and D_n are far from independent: they are asymptotically linked, and we can write

$$\frac{M_n}{\mathbf{E}M_n} \sim \frac{D_n}{\mathbf{E}D_n} \quad \text{a.a.s.}$$

in $\mathcal{G}_{n,d}$.

Computing the variance of B_n in $\mathcal{G}_{n,4}$ required the following result, which interestingly relates to the models of random regular graphs in Section 4.3.

Theorem 4.11 (Kim and Wormald [60]) *Let n be even. Given four independent random matchings M_i, $i = 0, \ldots, 3$, of n vertices, the probability that $M_i \cup M_{i+1}$ induces a Hamilton cycle for each i (with subscripts mod 4) is asymptotic to p_H^4 as $n \to \infty$, where $p_H \sim \sqrt{\frac{\pi}{2n}}$ is the probability that two independent random matchings of n vertices induce a Hamilton cycle.*

There are recent extensions of Theorem 4.5 in other directions. Robinson and Wormald [101] established the analogue of this theorem in which $o(\sqrt{n})$ edges of $G \in \mathcal{G}_{n,3}$ are chosen u.a.r., and the variable H_n counts only Hamilton cycles which pass through these edges. A corollary is that if that number of edges of $G \in \mathcal{G}_{n,3}$ are chosen at random, then a.a.s. there is a Hamilton cycle passing through them all. The limiting distribution of the probability was also found, if $c\sqrt{n}$ edges are chosen. This is an interesting example in that the variables X_i in the use of Theorem 4.1 are not just the short cycles counts, but also the counts of short paths joining distinguished edges. A similar result is by Janson and Wormald [57] that if the edges of $G \in \mathcal{G}_{n,d}$ are randomly coloured in n colours, $\frac{1}{2}d$ of each colour, then a.a.s. there is a Hamilton cycle using precisely one of each colour (provided $d \geq 8$).

Moving away from regular spanning subgraphs, in [98] the expectation and variance of the number of decompositions of $G \in \mathcal{G}_{n,3}$ into a tree and a cycle were computed asymptotically. (The length of cycle is necessarily $\frac{1}{2}n + 1$.) Janson [55] computed the joint distribution with short cycles and verified that the small subgraph conditioning method performs as expected; that is, almost all cubic graphs have such a decomposition. The difficult part is as usual the variance, covered in [98], but the short cycle part of the calculation is perhaps more complicated than all the other examples done so far.

Finally, as an example of use of this method with non-spanning subgraphs, there are the results of Garmo mentioned in Section 2.5.

Several results have appeared on the random regular bipartite model. The calculations for Hamilton cycles in bipartite graphs for the case $d = 3$ are given in [98], in which case the simple application of Chebyshev's inequality to the expectation and variance is enough to deduce that the probability of Hamiltonicity tends to 1. The same is true of the general case $d \geq 4$ (but this does not seem to have been published). The bipartite analogue of Theorem 2.26 was proved in [100] *not* by proving the bipartite analogue of Theorem 4.7, but it did involve verifying the bipartite analogue of the statement in Theorem 4.8 about M_n. The variance of H_n in the bipartite case requires a function to be maximised virtually the same as that encountered for $\mathcal{G}_{n,d}$ in [42]. This calculation is almost the same as in the general graph case, and no doubt calculations for D_n also follow a similar pattern in bipartite graphs. The bipartite analogue for T_n (1-factorisations) of the statement in Theorem 4.8 is verified in [87].

Let H_n denote the number of (directed) Hamilton cycles in a digraph in $\mathcal{DG}_{n,d}$ (defined in Section 3.3). Janson [56] used the small subgraph conditioning method to obtain the following.

Theorem 4.12 (Janson [56]) $\mathcal{DG}_{n,d}^{(H_n)} \approx \mathcal{DG}_{n,d}$ for all $d \geq 3$.

Theorem 3.2 is a corollary of this.

Although the results discussed in this section are strong in some ways, the method is extremely difficult to extend to examination of $\mathcal{G}_{n,d}$ in general, and no good results in this direction seem to be known. It is not known for instance if Conjecture 2.27 has any chance of being proved by contiguity.

4.3 The superposition arithmetic of contiguity classes

Originating from the argument in [100], the notion of the union of two random regular graphs on the same vertex set is very useful for proving asymptotic properties of $\mathcal{G}_{n,d}$. If $\mathcal{G}$ and $\widehat{\mathcal{G}}$ are two probability spaces of random graphs or multigraphs on the same vertex set, we define their *sum* $\mathcal{G} + \widehat{\mathcal{G}}$ to denote the space whose elements are defined by the random multigraph $G \cup \widehat{G}$ (called the *superposition* of G and $\widehat{G}$) where $G \in \mathcal{G}$ and $\widehat{G} \in \widehat{\mathcal{G}}$ are generated independently. Similarly, define the *graph-restricted sum* of $\mathcal{G}$ and $\widehat{\mathcal{G}}$, denoted by $\mathcal{G} \oplus \widehat{\mathcal{G}}$, to be the space which is the restriction of $\mathcal{G} + \widehat{\mathcal{G}}$ to simple graphs (i.e. with no multiple edges—we have no cause to use this operation when loops are present). This is defined only if $\mathcal{G} + \widehat{\mathcal{G}}$ contains at least one simple graph. In order to ensure that all our probability spaces can be sensibly related, we assume when using these operations that the underlying sets are extended to cover all graphs (or multigraphs, as the case may be) on that vertex set. Thus any n-vertex graph which cannot be formed by the operation in question is included in the space, but with probability 0. We loosely call models defined by sums *superposition models*.

The two sum operations $+$ and $\oplus$ are clearly commutative and associative. We define
$$k\mathcal{G} = \mathcal{G} \oplus \cdots \oplus \mathcal{G},$$
with k terms on the right. (The analogous product of k by $\mathcal{G}$ using $+$ instead of $\oplus$ is not so interesting for our purposes here.)

It is also straightforward to derive the following result.

Lemma 4.13 (Janson[56]) *Suppose that $\mathcal{G}_n \approx \mathcal{G}'_n$ and $\widehat{\mathcal{G}}_n \approx \widehat{\mathcal{G}}'_n$ where all four spaces $\mathcal{G}_n$, $\mathcal{G}'_n$, $\widehat{\mathcal{G}}_n$ and $\widehat{\mathcal{G}}'_n$ are of graphs on n vertices. Then*
$$\mathcal{G}_n + \widehat{\mathcal{G}}_n \approx \mathcal{G}'_n + \widehat{\mathcal{G}}'_n.$$

Janson [56] discussed contiguity of multigraph models defined by both the sum $+$ and the graph-restricted sum $\oplus$, but in that paper the emphasis is on the sum. Results for the graph-restricted sum can be obtained using the multigraph results, and it is desirable to put this on a systematic base. A crucial property required of the graph models in order for this idea to work is that the probability that a simple graph is created in the sum space must be bounded away from 0 (which incidentally ensures that the graph-restricted sum is defined). To ensure this, we use the property of every natural model, that it is *label-independent*, by which we mean that the probability of any multigraph is the same as that of any particular relabelled version of that multigraph.

Lemma 4.14 *Suppose that $\mathcal{G}_n \approx \mathcal{G}'_n$ and $\widehat{\mathcal{G}}_n \approx \widehat{\mathcal{G}}'_n$ where all four spaces $\mathcal{G}_n$, $\mathcal{G}'_n$, $\widehat{\mathcal{G}}_n$ and $\widehat{\mathcal{G}}'_n$ are label-independent and all graphs in the spaces have bounded degree (and n vertices). Then*
$$\mathcal{G}_n \oplus \widehat{\mathcal{G}}_n \approx \mathcal{G}'_n \oplus \widehat{\mathcal{G}}'_n.$$

Proof Let A_n be any sequence of events which is a.a.s. true in $\mathcal{G}_n \oplus \widehat{\mathcal{G}}_n$. We can regard A_n as an event in $\mathcal{G}_n + \widehat{\mathcal{G}}_n$ also (after all, it is just a set of graphs). Let B_n be the event that $G \in \mathcal{G}_n + \widehat{\mathcal{G}}_n$ has a multiple edge. Then $A_n \vee B_n$ is a.a.s. true in $\mathcal{G}_n + \widehat{\mathcal{G}}_n$ (where $\vee$ denotes the union of events) and hence also in $\mathcal{G}'_n + \widehat{\mathcal{G}}'_n$ by Lemma 4.13. It then follows that A_n is a.a.s. true in $\mathcal{G}'_n \oplus \widehat{\mathcal{G}}'_n$ once we show that the probability of the complement of B_n in $\mathcal{G}'_n + \widehat{\mathcal{G}}'_n$ is bounded below by a positive constant.

For this it suffices to treat each pair of graphs $G_1 \in \mathcal{G}'_n$ and $G_2 \in \mathcal{G}'_n$ separately, and consider a random relabelling of each. We can use the method of moments. It is easily verified that the expected number λ of edges in common between two such random relabellings is exactly half the product of the average degrees of vertices in G_1 and G_2. The other moments are calculated and satisfy the relation (8) in Section 2.3 (with $k = 1$). Hence (by Brun's sieve or Lemma 2.8) the probability that the superposition of the relabelled copies of G_1 and G_2 creates no multiple edges is asymptotically $e^{-\lambda}$, which is bounded below in view of the upper bound on vertex degrees.

The reverse argument, from $\mathcal{G}'_n \oplus \widehat{\mathcal{G}}'_n$ to $\mathcal{G}_n \oplus \widehat{\mathcal{G}}_n$, is identical. ∎

For bicoloured graphs the analogue of Lemma 4.14 holds if the definition of label-independence is specified to include only those relabellings which rearrange the labels within each of the two label sets.

If $Y = Y(n)$ is the number of spanning k-regular subgraphs of $G \in \mathcal{G}_{n,d}$ which lie in some specified set S_n, then $G \in \mathcal{G}_{n,d}^{(Y)}$ has the distribution of $F(n) \cup F'(n)$ where the ordered pairs $(F(n), F'(n))$ are sampled uniformly such that $F(n) \in S_n$, $F'(n)$ is a $(d-k)$-regular graph on the same vertex set, and $F(n)$ and $F'(n)$ are edge-disjoint. Thus $\mathcal{G}_{n,d}^{(Y)} = \mathcal{U}_n \oplus \mathcal{G}_{n,d-k}$ where $\mathcal{U}_n$ is the uniform space on S_n. Similarly, if Y is the number of decompositions of the edge set of $G \in \mathcal{G}_{n,d}$ into j spanning k-regular subgraphs of $G \in \mathcal{G}_{n,d}$ each of which lie in a uniform space $\mathcal{U}_n$ (so $jk = d$) then $\mathcal{G}_{n,d}^{(Y)} = j\mathcal{U}_n$. So, for example, $\mathcal{G}_{n,3}^{(T_n)} = 3\mathcal{G}_{n,1}$ for n even. We can now restate Theorems 4.7 and 4.8 in the following form. Let $\mathcal{H}_n$ denote a uniformly random Hamilton cycle (n even), on the same n vertices as $G \in \mathcal{G}_{n,d}$ for all d. The only item in the following not explicitly covered already is (iv) for $d = 3$, which is equivalent to (ii) for $d = 3$.

Theorem 4.15

(i) $\mathcal{G}_{n,d-2} \oplus \mathcal{H}_n \approx \mathcal{G}_{n,d}$ for $d \geq 3$.

(ii) $\mathcal{G}_{n,d-1} \oplus \mathcal{G}_{n,1} \approx \mathcal{G}_{n,d}$ for $d \geq 3$ and n even.

(iii) $3\mathcal{G}_{n,1} \approx \mathcal{G}_{n,3}$ (n even).

(iv) $\mathcal{G}_{n,d-2} \oplus \mathcal{G}_{n,2} \approx \mathcal{G}_{n,d}$ for $d \geq 3$.

(v) $2\mathcal{H}_n \approx \mathcal{G}_{n,4}$.

We next examine implications of this theorem, taking an interesting special case first. From (ii) (used repeatedly), and (iii), and Lemma 4.14,

$$\mathcal{G}_{n,d} \approx d\mathcal{G}_{n,1} \qquad (26)$$

for n even and $d \geq 3$. Since every element of $d\mathcal{G}_{n,1}$ has a 1-factorisation, this implies the following.

Corollary 4.16 (Robinson and Wormald [100]) *For $d \geq 3$, $G \in \mathcal{G}_{n,d}$ a.a.s. has a 1-factorisation when n is even.*

Another way to say this is that the edge-chromatic number of these graphs is a.a.s. equal to d when n is even, which is best possible since it must be $d + 1$ by Vizing's theorem when n is odd.

The main problem with $d = 2$ is that $\mathcal{G}_{n,1} \oplus \mathcal{G}_{n,1} \not\approx \mathcal{G}_{n,2}$, since, as seen in Section 2.9, $G \in \mathcal{G}_{n,2}$ a.a.s. has an odd cycle, but $G \in 2\mathcal{G}_{n,1}$ clearly does not. Similarly, $\mathcal{G}_{n,1} \oplus \mathcal{G}_{n,1} \not\approx \mathcal{H}_n \not\approx \mathcal{G}_{n,2}$, but these three form the only exceptions in the rather pleasant arithmetic of contiguity classes of regular graph models. Earlier, weaker versions of this result [56, 87] appeared at the times when various parts of Theorem 4.15 were proved.

Corollary 4.17 Let $d \geq 3$, and suppose $d = 2j + \sum_{i=1}^{d-1} ik_i$ with all terms non-negative. Then
$$\mathcal{G}_{n,d} \approx j\mathcal{H}_n \oplus k_1\mathcal{G}_{n,1} \oplus \cdots \oplus k_{d-1}\mathcal{G}_{n,d-1},$$
with n restricted to even integers if $k_i \neq 0$ for any odd i.

Proof Lemma 4.14 is used abundantly throughout. From (i), (iv) and (v),
$$\mathcal{G}_{n,d} \approx j\mathcal{H}_n \oplus k\mathcal{G}_{n,2} \oplus \mathcal{G}_{n,d-2j-2k} \tag{27}$$
for any $k \leq \frac{1}{2}d - 2j$. If $k_i = 0$ for all odd i, take $k = \frac{1}{2}d - 2j$ and combine the copies of $\mathcal{G}_{n,2}$ into the desired spaces $\mathcal{G}_{n,i}$ using the same result in reverse (with $j = 0$) for each space. If not, we can assume n is even, use (27) with $k = k_2$ and (26) to get
$$\mathcal{G}_{n,d} \approx j\mathcal{H}_n \oplus k_2\mathcal{G}_{n,2} \oplus (d - 2j - 2k_2)\mathcal{G}_{n,1}$$
(unless $d - 2j - 2k_2 = 2$). Then recombine the copies of $\mathcal{G}_{n,1}$ into all the other terms required using (26) in reverse for each term $\mathcal{G}_{n,i}$, $i \geq 3$. The required k_1 copies of $\mathcal{G}_{n,1}$ will be surplus. The only case left is $d - 2j - 2k_2 = 2$ and $k_1 = 2$, whence either $j > 0$ or $k_2 > 0$. From above we have $\mathcal{G}_{n,d} \approx j\mathcal{H}_n \oplus (k_2+1)\mathcal{G}_{n,2}$, and any two of these spaces can be recombined to give $\mathcal{G}_{n,4}$. This can then be split as desired since $\mathcal{G}_{n,4} \approx \mathcal{G}_{n,3} \oplus \mathcal{G}_{n,1} \approx \mathcal{G}_{n,2} \oplus 2\mathcal{G}_{n,1}$ by (ii) (twice) and $\mathcal{G}_{n,3} \approx \mathcal{H}_n \oplus \mathcal{G}_{n,1}$ by (i). ∎

A *complete Hamiltonian decomposition* of a d-regular graph is a partition of its edge set into the edges of $\frac{d}{2}$ Hamilton cycles (for d even), or $\frac{d-1}{2}$ Hamilton cycles and a perfect matching (for d odd).

Corollary 4.18 (Kim and Wormald [60]) *For fixed $d \geq 3$, $G \in \mathcal{G}_{n,d}$ a.a.s. has a complete Hamiltonian decomposition.*

One restatement of Corollary 4.9 is that a random 4-regular graph a.a.s. has four 1-factors such that a certain pairwise union gives two Hamilton cycles. The following strengthening of this has not even been proved yet for $d = 3$. A *perfect 1-factorisation* of a d-regular graph is an edge-decomposition into d pairwise disjoint perfect matchings such that the union of any two of them gives a Hamilton cycle.

Conjecture 4.19 *For $d \geq 3$, $G \in \mathcal{G}_{n,d}$ a.a.s. has a perfect 1-factorisation when n is even.*

I also conjecture the contiguity version of this; i.e. that $\mathcal{G}_{n,d}^{Y_n} \approx \mathcal{G}_{n,d}^{Y_n}$ where Y_n is the number of perfect 1-factorisations of G.

There is strong evidence for this conjecture from the results in [60], where it is shown that imposing Hamiltonicity conditions on pairs of perfect matchings does not asymptotically alter the probability of such conditions holding between other pairs (at least, not for a small number of pairs). If this is true in general then the expected number of perfect 1-factorisations will be large. I believe that Theorem 4.11 extends as follows.

Conjecture 4.20 *Let n be even. The probability that $d \geq 4$ random matchings of n vertices create a perfect 1-factorisation of a d-regular graph is asymptotic to $(\frac{\pi}{2n})^{d(d-1)/4}$ as $n \to \infty$.*

The case $d = 3$ is not conjectured here, because it is proved in [60], which means that the expected number of perfect 1-factorisations in $G \in \mathcal{P}_{n,3}$ or $\mathcal{G}_{n,3}$ is known. The variance, as usual, is much harder to compute. The expected number of 1-factorisations of $G \in \mathcal{G}_{n,d}$ grows exponentially with n, so by the contiguity expressed in (26) it is a.a.s. exponentially large. It seems unlikely that none of these 1-factorisations is perfect, even if Conjecture 4.20 is in error by a large factor. This supports Conjecture 4.19.

Corollary 4.18 is quite a strong statement about the structure of $G \in \mathcal{G}_{n,d}$ and enables quick proof of some results already known. For even d it immediately shows that $G \in \mathcal{G}_{n,d}$ is a.a.s. d-edge-connected, giving the edge connectivity version of Theorem 2.10 for d fixed. Here are some other examples.

Proof of Theorem 2.21 Take the decomposition in Corollary 4.18, and remove one edge from each Hamilton cycle. Since there are only $\lfloor \frac{1}{2}d \rfloor$ such cycles, for n sufficiently large we can easily arrange that the removed edges form a matching (d even) or form a linear forest with the perfect matching in the decomposition (d odd). The Hamilton paths form the other linear forests, to obtain linear arboricity $\lfloor \frac{1}{2}d \rfloor + 1 = \lceil \frac{1}{2}(d+1) \rceil$. ∎

This proof also shows that the conclusion of Theorem 2.21 can be considerably strengthened: all but one of the forests in this proof are Hamilton paths, and the other has maximum path length 1 (even d) or 3 (odd d).

The model $\mathcal{H}_n \oplus \mathcal{G}_{n,1}$ has already been studied for its diameter, but without the convenience of contiguity. A weaker version of the following result was obtained by Bollobás and Chung [19], with lower bound $\log_2 n - 10$ and upper bound $\log_2 n + \log_2 \log n + 10$.

Corollary 4.21 *A Hamilton cycle plus a random matching a.a.s. gives a graph with diameter between $\log_2 n + \log_2 \log n - 4$ and $\log_2 n + \log_2 \log n + 4$.*

Proof This comes immediately from Theorems 4.5 and 2.13 with $d = 3$. ∎

Bollobás and Chung [19] also investigated the diameter of a fixed regular graph plus a random matching in order to show that it is possible to find large regular graphs with small diameter using randomisation for only a small part of the graph.

Janson [56] conjectured that the model of random 2-regular graphs obtained by taking a random permutation digraph and ignoring the directions of edges fits in with the arithmetic of models just like the models $\mathcal{H}_n$ and $\mathcal{G}_{n,d}$ (again requiring $d \geq 3$). However, no results in this direction have been forthcoming, probably because of the lack of a suitable framework for combining this model with $\mathcal{G}_{n,d}$. We conjecture that the model obtained from $\mathcal{G}_{n,2}$ by restricting to graphs with even cycles fits in the same way. (This is *not* the same as $2\mathcal{G}_{n,1}$—the probabilities are different.)

The theory of the superposition arithmetic of contiguity classes has progressed further for random regular graph models than for related models. For bipartite graphs, Theorem 3.1 is proved in [100] in the same way as Theorem 2.26 by treating 1-factors in random $(d-1)$-regular bipartite graphs, and examination of the argument will verify the analogue of Theorem 4.15 (ii) in the bipartite case. The proof of Theorem 3.2 in [31] also uses contiguity in just the same way: the bipartite regular graph related to a regular digraph is contiguous to the graph-restricted sum of a perfect matching and a bipartite $(d-1)$-regular graph. The result is proved first for $d=3$ working with the two random parts of this decomposition, and the results for larger d again follow by contiguity.

Some of the other necessary base results in this area are missing (see the comments on this near the end of Section 4.2). But this is due mainly to the fact that they are almost the same as the corresponding graph case. For directed graph models, even more results are missing, but this is due to real difficulties. Permutation digraphs, or 1-regular digraphs (with loops permitted) are equivalent to matchings in the bipartite graphs corresponding to d-regular digraphs, so contiguity is known for these. On the other hand, in view of Corollary 4.18 and Theorem 4.12 we have the following.

Conjecture 4.22 *A random digraph in $\mathcal{DG}_{n,d}$ a.a.s. has d edge-disjoint directed Hamilton cycles.*

We also of course conjecture the contiguity version of this. From Theorem 4.12 this would imply contiguity of sums of directed Hamilton cycles.

5 The generation problem

The difficulty of sampling from $\mathcal{G}_{n,d}$, i.e. of generating regular graphs u.a.r., and the related problem for $\mathcal{G}_{n,d}$, are part of the motivation for the "algorithmic" models given in Section 6. Those models do not have uniform, or even well understood, distributions. In this section we focus on uniform and near-uniform distribution.

5.1 Uniform generation for d large

Generating an element of $\mathcal{G}_{n,d}$ is easy in the following sense: just generate an element P of $\mathcal{P}_{n,d}$ and use $G(P)$ if it has no loops or multiple edges. If it is not suitable, repeat. This is implicit in the presentation of the pairing model by Bollobás [10], and in the model used by Bender and Canfield in [5]. It was discussed explicitly in [119], along with a non-probabilistic algorithm for generating 3-regular graphs (which is much more complicated).

The difficulty of this simple procedure is the number of repetitions required before success. The expected number is exactly $1/\mathbf{P}(\text{Simple})$, which from (3) and (5) is prohibitively large for even quite small d, say $d = 8$. What can we do for larger d?

McKay and Wormald [79] used the modern version of switchings to give an algorithm for generating a random d-regular graph for $d = O(n^{1/3})$ in which the expected time per graph is polynomial ($O(n^2 d^4)$ in a version that would be implementable, or $O(nd^3)$ in a version that qualifies as the programmer's nightmare). The idea is to generate a random pairing in the pairing model, where those pairings corresponding to multigraphs with given numbers of loops and multiple edges are all equally likely, and then use a random switching to get to a random pairing with fewer loops and multiple edges. An accept/reject procedure is used to make sure the resulting random pairing is uniformly distributed with those parameters. Iterating this procedure eventually reaches pairings P such that $G(P)$ is a simple d-regular graph uniformly distributed. The "old" switchings do not help with this problem even for $d = n^\epsilon$, and the idea definitely cannot be extended in any easy way to get an algorithm for degree $n^{1/3+\epsilon}$ with polynomial expected time per graph generated.

Earlier Tinhofer [109] considered choosing edges consecutively, each choice a uniform choice from some restricted set, so that a regular graph hopefully results. This is combined with an accept/reject procedure to produce the uniform distribution, which is based on *a posteriori* computation of the probability $p(G)$ of the generated graph. The probability of acceptance is the ratio of $p(G)$ to an upper bound on $p(G')$ for all G'. Unfortunately basically nothing is known about such upper bounds, with the result that such algorithms do not seem to be of practical use for uniform generation, except perhaps for very small graphs.

The description of how to generate a random unlabelled graph given in Section 3.1 was applied directly in [121] to generating random unlabelled d-regular graphs. The expected time complexity per graph is linear in n for d-regular graphs on n vertices if d is fixed. Achieving this requires judicious use of accept/reject procedures, as well as the simplification given in [120] of the multidimensional optimisation problem involved in Theorem 2.17. However, in an uncommon twist, the algorithm is not at all practicable for $d = 3$ and 4 unless n is quite *large*, say $n \geq 200$. This is because symmetries are more common for smaller n than for larger n.

For random graphs in $\mathcal{G}_{n,\mathbf{d}}$, the result in [79] applies to degree sequences for which the maximum degree is $O(m^{1/4})$, where m is the sum of the degrees. The result is a polynomial time algorithm for uniform generation. Those in [109] apply to any degree sequence (but tell us little).

5.2 Near-uniform generation

When uniform generation is difficult, it is worth looking at near-uniform generation. If the approximation to uniform can be made to a given accuracy, then probabilities can be obtained experimentally to the same accuracy.

Jerrum and Sinclair [58] gave a *fully polynomial almost uniform generator* for d-regular graphs on n vertices; that is, a generation algorithm which, for $\epsilon > 0$, runs in time polynomial in n and $\log(1/\epsilon)$, such that all graphs are generated with probabilities varying by a factor of at most $1 + \epsilon$. This result is remarkable in that it applies for *all d*. The analysis required to prove it uses sophisticated eigenvalue techniques for estimating the rate of convergence of a Markov chain to the stationary distribution. Unfortunately the polynomial does not have very low order so this algorithm seems to be of little practical significance.

Luczak and Wormald [64] studied a generalised form of the random processes described below in Section 6 which suggested the algorithm studied by Steger and Wormald [107]. This seems to be a very practical algorithm (time $O(nd^2)$ expected per graph for small d, and good practical results for all $d \leq \frac{1}{2}n$, which suffices by complementation). Although it does not give the uniform distribution, for $d < n^c$ where c is a small positive constant, it produces all graphs with asymptotically the same probability. That is, for such d there are upper and lower bounds on the probabilities of a graph on n vertices being generated which are asymptotically equal as $n \to \infty$. Unfortunately, there are indications that this algorithm does not give such a close approximation to uniform distribution when d gets past $n^{1/3}$.

This algorithm has useful properties even though it is equivalent to the quite naive idea of generating (non-uniformly) at random an element of the pairing model one pair at a time, at each step selecting only from those pairs which will not create loops or multiple edges with those already selected. This process occasionally gets stuck, with unpaired points remaining and not legally able to be paired, but experimental evidence strongly supports the following conjecture, with the value of the constant approximately $\frac{1}{3}$.

Conjecture 5.1 *Let $f(n,d)$ denote the probability that this procedure terminates with a regular graph. Then $f(n,d)$ is bounded below by a constant for all $d \leq \frac{1}{2}n$.*

Jerrum et al. [59] showed that the algorithm in [58] applies to a very wide class of degree sequences. The proof relies on showing that if the degrees do not vary by too much, the number of graphs does not changes radically with small perturbations of the degrees (maintaining the same number of vertices). This incidentally gives an explanation of the difficulty of enumerating graphs or generating random graphs with given degrees when the degrees vary wildly.

6 Algorithmically defined models

As explained in Section 5, it can be difficult to efficiently generate members of $\mathcal{G}_{n,d}$ uniformly at random. There are no good practical schemes for large d, and the state of affairs for $\mathcal{G}_{n,\mathbf{d}}$ is even worse. It is substantially these difficulties which motivate the "algorithmic" models given here, but aside from this they are often also of intrinsic interest due to their simplicity.

Tinhofer's scheme (see Section 5.1) is based on selection of edges by repeating a uniform selection from a restricted set. Wilson [113] uses a similar approach (ignoring questions of distribution). A simple form of such an algorithm is the following. Given d and n, start with n isolated vertices and repeatedly add edges joining vertices of degree strictly less than d. Each time, the edge added is chosen uniformly at random from all unfilled positions. This is the *degree restricted graph process* with parameter d, or d-process for short. The d-process stops when no more edges can be added, i.e. when the graph induced by the vertices of degree less than d is a clique.

This process was considered by Erdős, who asked for the asymptotic distribution of the number of vertices of degree less than d in the final graph (d fixed, $n \to \infty$). This question was settled in [102] using the differential equation method together with some other arguments. It was shown that a.a.s. the final graph is regular if dn is even, and almost regular, with one vertex of degree $d-1$ and the rest of degree d, otherwise.

The final graph of the d-process, conditioned on it being d-regular, gives a model of d-regular graphs which we denote here by $\mathcal{G}_{n,d}^{\text{deg}}$. Using the fact that the numbers of vertices of given degree follow close to the solutions of differential equations, Ruciński and Wormald [103] also determined the limiting distribution of the numbers of short cycles in $G \in \mathcal{G}_{n,2}^{\text{deg}}$. These are asymptotically independent Poisson, but the expected number of cycles of length i involves an i-dimensional integral. For $i=3$ this simplifies to $\frac{1}{2}\int_0^\infty \frac{(\log(1+x))^2\,dx}{xe^x} = 0.1887\ldots$, which is a little different from the expected number $\frac{1}{6}$ of 3-cycles in the uniform model $\mathcal{G}_{n,2}$. More recently, the same authors [104] showed that $G \in \mathcal{G}_{n,d}^{\text{deg}}$ is a.a.s. connected for $d \geq 3$, and the following is conjectured there.

Conjecture 6.1 *For $G \in \mathcal{G}_{n,3}^{\text{deg}}$, $\mathbf{P}(G \text{ is disconnected}) \sim cn^{-2}$ where $c \approx 0.25$.*

The corresponding statement in $\mathcal{G}_{n,3}$ holds with $c = \frac{2}{27}$.

If we do not condition on the final graph of the d-process being regular, the probability of disconnectedness seems, from simulation results, to be much higher than this. Nevertheless, the following seems quite plausible.

Conjecture 6.2 *Let $d \geq 3$ be fixed. For $G \in \mathcal{G}_{n,d}^{\text{deg}}$, G is a.a.s. d-connected.*

Telcs and Wormald [108], using a delicate application of the differential equation method, found the probability that $G \in \mathcal{G}_{n,2}^{\text{deg}}$ is a single Hamilton cycle is asymptotic to $\sqrt{\frac{\pi e^\tau}{2n}} \approx 1.819 n^{-1/2}$ where $\tau = \int_0^\infty \frac{(\log(1+x))^2\,dx}{xe^x}$, and also determined the expected number of cycles in $G \in \mathcal{G}_{n,2}^{\deg}$ quite accurately. These values are also close to those in the uniform model $\mathcal{G}_{n,2}$ (see (11) for example). Although the evidence is not very strong, the following is suspiciously hard to disprove.

Conjecture 6.3 *For fixed $d \geq 1$ and dn even, $\mathcal{G}_{n,d} \approx \mathcal{G}_{n,d}^{\deg}$.*

Motivated by [109] and [113], in which the aim was to conveniently generate random graphs with given degrees, Robalewska [96] studied a random process by which stars are added to fill the required vertex degrees. We consider the regular case here, and take d fixed. Beginning with n isolated vertices, at each step choose u.a.r. a vertex v of minimum degree, and choose u.a.r. $d - d(v)$ other vertices of degree strictly less than d. Edges are added from v to these vertices, then the step is repeated. As with the d-process, stop when the required edges do not exist or the graph is d-regular. This is called the *star d-process*. Multiple edges are impossible by the degree-filling nature of the process.

In the case $d = 2$ the short cycle distribution of the final graph was obtained in [96], as well as the asymptotic probability of Hamiltonicity and the distribution of the number of cycles. It was also shown that the final graph is a.a.s. 2-regular. The latter result was extended to d-regular for fixed d, provided dn is even, in [97]. The methods are similar to those used for d-processes. Again, we can consider restricting to those processes in which the final graph to d-regular, to obtain a probability space which we call here $\mathcal{G}_{n,d}^{\text{stardeg}}$.

Conjecture 6.4 *For fixed $d \geq 1$ and dn even, $\mathcal{G}_{n,d} \approx \mathcal{G}_{n,d}^{\text{stardeg}}$.*

7 A wider perspective

There are several interesting random graph models which involve either special regular graphs or in some sense near-regular graphs. After mentioning some cubic examples, we look at results on planar graphs and near-regular graphs in a bit more detail.

McKay et al. [75] found the asymptotic number of claw-free cubic graphs. Together with the result in [101], their result implies that almost all of these graphs are Hamiltonian.

Garmo [48, 49] defined random railways to be random cubic multigraphs with one of the three half-edges at each vertex distinguished, and studies special connectivity properties of these (properties defined with respect to the distinguished half-edges).

7.1 Random regular planar graphs

For enumeration purposes, planar graphs are normally studied embedded in the sphere and with a rooting which consists of distinguishing an edge and a vertex incident with that edge. For 3-connected graphs the enumeration is equivalent to labelled planar graphs (just multiply the number of rooted graphs by $\frac{n!}{4n}$ to get the corresponding number of labelled planar graphs).

Many enumeration results and several on random planar graphs were obtained in this way, for instance Tutte [110] showed that a random 3-connected planar graph a.a.s. has no automorphisms. Then a more general theory of random planar graphs began with the paper of Richmond et al. [92] showing that a random 3-connected cubic planar graph has exponentially small probability of being Hamiltonian. Richmond and Wormald [93] extended this result to more classes of planar maps. These papers also give results on the frequency of subgraphs in such graphs.

Up to this point, results were quite dependent on generating function analysis, but Bender et al. [6] freed the topic from generating functions to a large extent (although their argument to show this uses a modification of the generating function singularity arguments in [92, 93]). Building on this, Richmond and Wormald [94] showed that many classes of planar maps are almost all asymmetric, thus including a simpler proof of the result of Tutte mentioned above (amongst other things).

Bender et al. [7], returning to generating function methods, showed that in almost all 3-connected cubic planar graphs with n edges, the largest cyclically 4-edge-connected cubic component has about $n/2$ edges. See also Gao and Wormald [47] for more general results.

7.2 Non-regular models

Some random graph models cannot easily be specialised to models of regular graphs, and yet have features in common, in particular vertices of low degree but with a guaranteed lower bound on the degrees. Here is only a brief mention of some of the results of this type.

Several authors have considered the random graph coming from a random digraph in which k arcs are chosen out of each vertex at random, and then the orientations are suppressed and multiple edges are coalesced. Fenner and Frieze [37] showed for $k \geq 23$ that this is a.a.s. Hamiltonian. The big open problem in this area is whether this is true for all $k \geq 3$. For $k = 2$ the opposite is true, as sketched in an exercise in Bollobás [16, Section VIII.5], because of the asymptotically almost sure occurrence of small subgraphs called spiders which kill Hamilton cycles. The result for $k \geq 23$ has been far superseded by the recent result of Cooper and Frieze [30] that a random 3-in, 3-out digraph (in which from each vertex three arcs are randomly chosen in and three out) is a.a.s. Hamiltonian.

An early result of Walkup [111], subsequently used by others, is that a random d-out directed bicoloured graph with n vertices in each part (in which from each vertex d arcs are randomly chosen out) a.a.s. contains a perfect matching provided $d \geq 2$ (and a.a.s. does not if $d = 1$). Frieze [40] later showed that the superposition of two random 1-out-regular digraphs a.a.s. has a perfect matching (n even). Also in the spirit of the models in Section 4, Frieze et al. [43] showed that the superposition of five random trees is a.a.s. Hamiltonian. The big open problem in this direction is whether three is enough.

Recently a model of random regular graphs with edge faults has been studied. Here the edges of $G \in \mathcal{G}_{n,d}$ are randomly deleted independently with probability p each. For example, Goerdt [52] looked at the giant component, and Nikoletseas and Spirakis [88] obtained concentration results on the second eigenvalue of the adjacency matrix of the giant component.

7.3 Further unsolved problems

Besides the various conjectures sprinkled through this paper, there is one issue which benefits from emphasising. Large values of d are in general a problem. For many of the results mentioned in this article, the obvious conjecture is that the result holds for much higher values of d (such as Conjecture 2.11). Can the generating function method in [80] help here? It certainly does permit computation of the probability of sets of edges being present, to some extent. For another example, can one in this way show that for $d \sim \frac{1}{2}n$, $G \in \mathcal{G}_{n,d}$ a.a.s. has no non-trivial automorphisms? At least it would be good to extend Theorem 2.17 past $d = o(n^{1/2})$. Also can a practical near-uniform generation algorithm be found for very large d (say $d \approx \sqrt{n}$ or $d = cn$)? In addition, further results on random tournaments (Section 3.6) would be interesting, but this suffers for one thing from a problem similar to the case of $\mathcal{G}_{n,d}$ with large d: high edge density restricts switchings.

For some of these questions, asymptotic enumeration of $|\mathcal{G}_{n,d}|$ in the range $\sqrt{n} < d < cn/\log n$ may help. This is still wide open, and has strong implications for the model of the degree sequence of a random graph in $\mathcal{G}(n, p)$ which is given in [82].

Acknowledgements

This research was supported by the Australian Research Council. I am grateful to Brendan McKay for reading a version of the manuscript carefully, and to the anonymous referee for many suggestions and corrections to the presentation.

References

[1] J. Akiyama, G. Exoo & F. Harary, Covering and packing in graphs III, cyclic and acyclic invariants, *Mathematica Slovaca*, **30** (1980), 405–417.

[2] N. Alon, The linear arboricity of graphs, *Israel Journal of Mathematics*, **62** (1988), 311–325.

[3] N. Alon & J. H. Spencer, *The Probabilistic Method*, Wiley, New York (1992).

[4] A. Békéssy, P. Békéssy & J. Komlós, Asymptotic enumeration of regular matrices, *Studia Scientiarum Mathematicarum Hungarica*, **7** (1972), 343–353.

[5] E. A. Bender & E. R. Canfield, The asymptotic number of non-negative integer matrices with given row and column sums, *Journal of Combinatorial Theory, Series A*, **24** (1978), 296–307.

[6] E. A. Bender, Z. C. Gao & L. B. Richmond, Submaps of Maps I. General 0-1 Laws, *Journal of Combinatorial Theory, Series B*, **55** (1992), 104–117.

[7] E. A. Bender, L. B. Richmond & N. C. Wormald, Largest 4-connected components of 3-connected planar triangulations, *Random Structures & Algorithms*, **7** (1995), 273–285.

[8] C. Berge, *Graphes et Hypergraphes*, Dunod, Paris (1970).

[9] B. Bollobás, A probabilistic proof of an asymptotic formula for the number of labelled regular graphs, Preprint Series, Matematisk Institut, Aarhus Universitet, 1979.

[10] B. Bollobás, A probabilistic proof of an asymptotic formula for the number of labelled regular graphs, *European Journal of Combinatorics*, **1** (1980), 311–316.

[11] B. Bollobás, Random graphs, in *Combinatorics* (ed. H. N. V. Temperley), *London Mathematical Society Lecture Note Series*, 52, Cambridge University Press, Cambridge (1981), pp. 80–102.

[12] B. Bollobás, The independence ratio of regular graphs, *Proceedings of the American Mathematical Society*, **83** (1981), 433–436.

[13] B. Bollobás, Distinguishing vertices of random graphs, in *Graph Theory* (ed. B. Bollobás), *North-Holland Mathematics Studies*, 62, North-Holland, Amsterdam (1982), pp. 33–49.

[14] B. Bollobás, The asymptotic number of unlabelled regular graphs, *Journal of the London Mathematical Society*, **26** (1982), 201–206.

[15] B. Bollobás, Almost all regular graphs are Hamiltonian, *European Journal of Combinatorics*, **4** (1983), 97–106.

[16] B. Bollobás, *Random Graphs*, Academic Press, London (1985).

[17] B. Bollobás, The isoperimetric number of a random graph, *European Journal of Combinatorics*, **9** (1988), 241–244.

[18] B. Bollobás & L. Clark, Generalized chromatic numbers of random regular graphs, in *Graph Theory, Combinatorics and Algorithms, Vol. 1, 2* (eds. Y. Alavi & A. Schwenk), Wiley, New York (1995), pp. 209–219.

[19] B. Bollobás & F. R. K. Chung, The diameter of a cycle plus a random matching, *SIAM Journal on Discrete Mathematics*, **1** (1988), 328–333.

[20] B. Bollobás & B. D. McKay, The number of matchings in random regular graphs and bipartite graphs, *Journal of Combinatorial Theory, Series B*, **41** (1986), 80–91.

[21] B. Bollobás & W. Fernandez de la Vega, The diameter of random regular graphs, *Combinatorica*, **2** (1982), 125–134.

[22] J. A. Bondy & U. S. R. Murty, *Graph Theory with Applications*, Elsevier, New York (1976).

[23] O. Borodin & A. Kostochka, On an upper bound of a graph's chromatic number depending on the graph's degree and density, *Journal of Combinatorial Theory, Series B*, **23** (1977), 247–250.

[24] A. Broder & E. Shamir, On the second eigenvalue of random regular graphs, in *28th Annual Symposium on Foundations of Computer Science*. IEEE Computer Society Press, Washington DC (1987), pp. 286–294.

[25] P. Catlin, A bound on the chromatic number of a graph, *Discrete Mathematics*, **22** (1978), 81–83.

[26] F. R. K. Chung, Constructing random-like graphs, in *Probabilistic Combinatorics and Its Applications* (ed. B. Bollobás), *Proceedings of Symposia in Applied Mathematics*, 44, American Mathematical Society, Providence, Rhode Island (1991), pp. 21–55.

[27] K. L. Chung, *A Course in Probability Theory*, second edition, Academic Press, New York (1974).

[28] C. Cooper, A note on the connectivity of 2-regular digraphs, *Random Structures & Algorithms*, **4** (1993), 469–472.

[29] C. Cooper, On the 2-cyclic property in 2-regular digraphs, *Random Structures & Algorithms*, **6** (1995), 439–448.

[30] C. Cooper & A. M. Frieze, Hamilton cycles in a class of random directed graphs, *Journal of Combinatorial Theory, Series B*, **62** (1994), 151–163.

[31] C. Cooper, A. Frieze & M. Molloy, Hamilton cycles in random regular digraphs, *Combinatorics, Probability and Computing*, **3** (1994), 39–50.

[32] C. Cooper, A. Frieze, M. Molloy & B. Reed, Perfect matchings in random r-regular, s-uniform hypergraphs, *Combinatorics, Probability and Computing*, **5** (1996), 1–14.

[33] R. Diestel, *Graph Theory*, Springer, New York (1997).

[34] W. Duckworth & N. C. Wormald, Maximum star forests and minimum dominating sets in random cubic graphs, in preparation.

[35] M. N. Ellingham, The asymptotic connectivity of labelled coloured regular bipartite graphs, in *Combinatorial Mathematics X.* (ed. L. R. A. Casse), *Springer Lecture Notes in Mathematics*, 1036, Springer, Berlin (1983), pp. 177–188.

[36] P. Erdős & A. Rényi, On the strength of connectedness of a random graph, *Acta Mathematica Academiae Scientiarum Hungaricae*, **12** (1961), 261–267.

[37] T. I. Fenner & A. M. Frieze, On the existence of hamiltonian cycles in a class of random graphs, *Discrete Mathematics*, **45** (1983), 301–305.

[38] T. I. Fenner & A. M. Frieze, Hamiltonian cycles in random regular graphs, *Journal of Combinatorial Theory, Series B*, **37** (1984), 103–112.

[39] J. Friedman, On the second eigenvalue and random walks in random d-regular graphs, *Combinatorica*, **11** (1991), 331–362.

[40] A. M. Frieze, Maximum matchings in a class of random graphs, *Journal of Combinatorial Theory, Series B*, **40** (1986), 196–212.

[41] A. M. Frieze, Finding Hamilton cycles in sparse random graphs, *Journal of Combinatorial Theory, Series B*, **44** (1988), 230–250.

[42] A. Frieze, M. Jerrum, M. Molloy, R. W. Robinson & N. C. Wormald, Generating and counting Hamilton cycles in random regular graphs, *Journal of Algorithms*, **21** (1996), 176–198.

[43] A. Frieze, M. Karoński & L. Thoma, On perfect matchings and Hamilton cycles in sums of random trees, *SIAM Journal on Discrete Mathematics*, in press.

[44] A. M. Frieze & T. Łuczak, On the independence numbers of random regular graphs, *Journal of Combinatorial Theory, Series B*, **54** (1992), 123–132.

[45] A. M. Frieze & S. Suen, On the independence number of random cubic graphs, *Random Structures & Algorithms*, **5** (1994), 649–664.

[46] Z. Gao, B. D. McKay & X. Wang, On the structure of a random tournament with a given score sequence, (submitted).

[47] Z. Gao & N. C. Wormald, The size of the largest components in random planar maps, *SIAM Journal on Discrete Mathematics*, in press.

[48] H. Garmo, Random railways modeled as random 3-regular graphs, *Random Structures & Algorithms*, **9** (1996), 113–136.

[49] H. Garmo, Asymptotic properties of the connectivity number of random railways, *Advances in Applied Probability*, **31** (1999), in press.

[50] H. Garmo, Random railways and cycles in random regular graphs, Doctoral thesis, Uppsala University, 1998.

[51] C. D. Godsil & B. D. McKay, Asymptotic enumeration of Latin rectangles, *Journal of Combinatorial Theory, Series B*, **48** (1990), 19–44.

[52] A. Goerdt, The giant component threshold for random regular graphs with edge faults, in *Mathematical Foundations of Computer Science 1997* (eds. I. Prívara & P. Ružička), *Lecture Notes in Computer Science*, 1295, Springer, Berlin (1997), pp. 279–288.

[53] G. R. Grimmett & D. R. Stirzaker, *Probability and Random Processes*, second edition, Clarendon Press, Oxford (1992).

[54] F. Harary & E. M. Palmer, *Graphical Enumeration*, Academic Press, New York (1973).

[55] S. Janson, private communication, 1997.

[56] S. Janson, Random regular graphs: Asymptotic distributions and contiguity, *Combinatorics, Probability and Computing*, **4** (1995), 369–405.

[57] S. Janson & N. C. Wormald, Rainbow Hamilton cycles in random coloured regular graphs, in preparation.

[58] M. R. Jerrum & A. J. Sinclair, Fast uniform generation of regular graphs, *Theoretical Computer Science*, **73** (1990), 91–100.

[59] M. R. Jerrum, B. D. McKay & A. J. Sinclair, When is a graphical sequence stable?, in *Random Graphs Vol. 2* (eds. A. Frieze & T. Łuczak), Wiley, New York (1992), pp. 101–115.

[60] J. H. Kim & N. C. Wormald, Random matchings which induce Hamilton cycles, and hamiltonian decompositions of random regular graphs, in preparation.

[61] J. Komlós & E. Szemerédi, Limit distributions for the existence of Hamilton circuits in a random graph, *Discrete Mathematics*, **43** (1983), 55–63.

[62] J. Lawrence, Covering the vertex set of a graph with subgraphs of smaller degree, *Discrete Mathematics*, **21** (1978), 61–68.

[63] T. Łuczak, Sparse random graphs with a given degree sequence, in *Random Graphs Vol. 2* (eds. A. Frieze & T. Łuczak), Wiley, New York (1992), pp. 165–182.

[64] T. Łuczak & N. C. Wormald, Phase transition for random graph processes, in preparation.

[65] C. McDiarmid, On the method of bounded differences, in *Surveys in Combinatorics 1989* (ed. J. Siemons), *London Mathematical Society Lecture Note Series*, 141, Cambridge University Press, Cambridge (1989), pp. 148–188.

[66] C. McDiarmid & B. Reed, Linear arboricity of random regular graphs, *Random Structures & Algorithms*, **1** (1990), 443–445.

[67] B. D. McKay, The expected eigenvalue distribution of a large regular graph, *Linear Algebra and its Applications*, **40** (1981), 203–216.

[68] B. D. McKay, Spanning trees in random regular graphs, in *Proceedings of the Third Caribbean Conference on Combinatorics and Computing (Bridgetown, 1981)* (ed. C. C. Cadogan), University of the West Indies, Cave Hill Campus, Barbados (1981), pp. 139–143.

[69] B. D. McKay, Subgraphs of random graphs with specified degrees, *Congressus Numerantium*, **33** (1981), 213–223.

[70] B. D. McKay, Spanning trees in regular graphs, *European Journal of Combinatorics*, **4** (1983), 149–160.

[71] B. D. McKay, Asymptotics for 0-1 matrices with prescribed line sums, in *Enumeration and Design* (eds. D. M. Jackson & S. A. Vanstone), Academic Press, Toronto (1984), 225–238.

[72] B. D. McKay, Asymptotics for symmetric 0-1 matrices with prescribed row sums, *Ars Combinatoria*, **19A** (1985), 15–25.

[73] B. D. McKay, Independent sets in regular graphs of high girth, in *Proceedings of the Australia-Singapore Joint Conference on Information Processing and Combinatorial Mathematics (Singapore, 1986), Ars Combinatoria*, 23A, (1987), pp. 179–185.

[74] B. D. McKay, The asymptotic numbers of regular tournaments, eulerian digraphs and eulerian oriented graphs, *Combinatorica*, **10** (1990), 367–377.

[75] B. D. McKay, E. M. Palmer, R. C. Read & R. W. Robinson, The asymptotic number of claw-free cubic graphs, preprint, Australian National University, 1998.

[76] B. D. McKay & X. Wang, Asymptotic enumeration of bipartite graphs by degree sequence, in preparation.

[77] B. D. McKay & X. Wang, Asymptotic enumeration of tournaments with a given score sequence, *Journal of Combinatorial Theory, Series A*, **73** (1996), 77–90.

[78] B. D. McKay & N. C. Wormald, Automorphisms of random graphs with specified degrees, *Combinatorica*, **4** (1984), 325–338.

[79] B. D. McKay & N. C. Wormald, Uniform generation of random regular graphs of moderate degree, *Journal of Algorithms*, **11** (1990), 52–67.

[80] B. D. McKay & N. C. Wormald, Asymptotic enumeration by degree sequence of graphs of high degree, *European Journal of Combinatorics*, **11** (1990), 565–580.

[81] B. D. McKay & N. C. Wormald, Asymptotic enumeration by degree sequence of graphs with degrees $o(n^{1/2})$, *Combinatorica*, **11** (1991), 369–382.

[82] B. D. McKay & N. C. Wormald, The degree sequence of a random graph. I. The models, *Random Structures & Algorithms*, **11** (1997), 97–117.

[83] B. D. McKay, N. C. Wormald & B. Wysocka, Asymptotic enumeration of regular graphs of high girth, in preparation.

[84] M. Molloy, The chromatic number of sparse random graphs, MMath thesis, University of Waterloo, 1992.

[85] M. Molloy & B. Reed, A critical point for random graphs with a given degree sequence, *Random Structures & Algorithms*, **6** (1995), 161–179.

[86] M. Molloy & B. Reed, The dominating number of a random cubic graph, *Random Structures & Algorithms*, **7** (1995), 209–221.

[87] M. Molloy, H. Robalewska, R. W. Robinson & N. C. Wormald, 1-factorisations of random regular graphs, *Random Structures & Algorithms*, **10** (1997), 305–321.

[88] S. E. Nikoletseas & P. G. Spirakis, Expander properties in random regular graphs with edge faults, in *STACS 95* (eds. E. W. Mayr & C. Puech), *Lecture Notes in Computer Science*, 900, Springer, Berlin (1995), pp. 421–432.

[89] P. E. O'Neil, Asymptotics and random matrices with row-sum and column sum-restrictions, *Bulletin of the American Mathematical Society*, **75** (1969), 1276–1282.

[90] L. Pósa, Hamiltonian circuits in random graphs, *Discrete Mathematics*, **14** (1976), 359–364.

[91] R. C. Read, The enumeration of locally restricted graphs (II), *Journal of the London Mathematical Society*, **35** (1960), 344–351.

[92] L. B. Richmond, R. W. Robinson & N. C. Wormald, On Hamilton cycles in 3-connected cubic maps, in *Cycles in Graphs* (eds. B. Alspach & C. D. Godsil), *Annals of Discrete Mathematics*, **27** North-Holland, Amsterdam (1985), pp. 141–150.

[93] L. B. Richmond & N. C. Wormald, Random triangulations of the plane, *European Journal of Combinatorics*, **9** (1988), 61–71.

[94] L. B. Richmond & N. C. Wormald, Almost all maps are asymmetric, *Journal of Combinatorial Theory, Series B*, **63** (1995), 1–7.

[95] H. Robalewska, 2-factors in random regular graphs, *Journal of Graph Theory*, **23** (1996), 215–224.

[96] H. Robalewska, Random graphs generated by the star 2-process, *Random Structures & Algorithms*, **11** (1997), 125–149.

[97] H. Robalewska & N. C. Wormald, Random star processes, preprint, University of Melbourne, 1997.

[98] R. W. Robinson & N. C. Wormald, Existence of long cycles in random cubic graphs, in *Enumeration and Design* (eds. D. M. Jackson & S. A. Vanstone), Academic Press, Toronto (1984), 251–270.

[99] R. W. Robinson & N. C. Wormald, Almost all cubic graphs are hamiltonian, *Random Structures & Algorithms*, **3** (1992), 117–125.

[100] R. W. Robinson & N. C. Wormald, Almost all regular graphs are hamiltonian, *Random Structures & Algorithms*, **5** (1994), 363–374.

[101] R. W. Robinson & N. C. Wormald, Hamilton cycles containing randomly selected edges in random cubic graphs, in preparation.

[102] A. Ruciński & N. C. Wormald, Random graph processes with degree restrictions, *Combinatorics, Probability and Computing*, **1** (1992), 169–180.

[103] A. Ruciński & N. C. Wormald, Random graph processes with maximum degree 2, *Annals of Applied Probability*, **7** (1997), 183–199.

[104] A. Ruciński & N. C. Wormald, The probability that a random d-process results in a connected graph, preprint, University of Melbourne, 1997.

[105] J. B. Shearer, A note on the independence number of triangle-free graphs, *Journal of Combinatorial Theory, Series B*, **53** (1991), 300–307.

[106] J. H. Spencer, Random regular tournaments, *Periodica Mathematica Hungarica*, **5** (1974), 105–120.

[107] A. Steger & N. C. Wormald, A fast algorithm for generating random regular graphs, preprint, University of Melbourne, 1998.

[108] A. Telcs & N. C. Wormald, Hamiltonicity of random 2-processes, in preparation.

[109] G. Tinhofer, On the generation of random graphs with given properties and known distribution, *Applied Computer Science, Berichte zur Praktischen Informatik*, **13** (1979), 265–297.

[110] W. T. Tutte, On the enumeration of convex polyhedra, *Journal of Combinatorial Theory, Series B*, **28** (1980), 105–126.

[111] D. W. Walkup, Matchings in random regular bipartite digraphs, *Discrete Mathematics*, **31** (1980), 59–64.

[112] H. S. Wilf, *generatingfunctionology*, Academic Press, San Diego (1990).

[113] J. B. Wilson, Methods for detecting non-randomness in species co-occurrences: a contribution, *Oecologia*, **73** (1987), 579–582.

[114] N. C. Wormald, Triangles in labelled cubic graphs, in *Combinatorial Mathematics* (eds. D. A. Holton & J. Seberry), *Springer Lecture Notes in Mathematics*, 686, Springer, Berlin (1978), pp. 337–343.

[115] N. C. Wormald, Some Problems in the Enumeration of Labelled Graphs, PhD thesis, University of Newcastle, 1978.

[116] N. C. Wormald, Enumeration of labelled graphs II: Cubic graphs with a given connectivity, *Journal of the London Mathematical Society (2)*, **20** (1979), 1–7.

[117] N. C. Wormald, The asymptotic connectivity of labelled regular graphs, *Journal of Combinatorial Theory, Series B*, **31** (1981), 156–167.

[118] N. C. Wormald, The asymptotic distribution of short cycles in random regular graphs, *Journal of Combinatorial Theory, Series B*, **31** 168–182. (1981),

[119] N. C. Wormald, Generating random regular graphs, *Journal of Algorithms*, **5** (1984), 247–280.

[120] N. C. Wormald, A simpler proof of the asymptotic formula for the number of unlabelled r-regular graphs, *Indian Journal of Mathematics*, **28** (1986), 43–47.

[121] N. C. Wormald, Generating random unlabelled graphs, *SIAM Journal on Computing*, **16** (1987), 717–727.

[122] N. C. Wormald, Differential equations for random processes and random graphs, *Annals of Applied Probability*, **5** (1995), 1217–1235.

[123] N. C. Wormald, The perturbation method and triangle-free random graphs, *Random Structures & Algorithms*, **9** (1996), 253–270.

[124] N. C. Wormald, The differential equation method for random graph processes and greedy algorithms, in *Lectures on Approximation and Randomized Algorithms* (eds. M. Karoński & H. J. Prömel), PWN, Warsaw, in press.

Department of Mathematics and Statistics
University of Melbourne
Parkville VIC 3052
Australia
nick@ms.unimelb.edu.au